TEUBNER-TEXTE zur Physik   Band 27

W. Ehrfeld, G. Wegner, W. Karthe,
H.-D. Bauer, H. O. Moser (Hrsg.)

Integrated Optics and Micro-Optics with Polymers

TEUBNER-TEXTE zur Physik

Herausgegeben von
Prof. Dr. Werner Ebeling, Berlin
Prof. Dr. Manfred Pilkuhn, Stuttgart
Prof. Dr. Bernd Wilhelmi, Jena

This regular series includes the presentation of recent research developments of strong interest as well as comprehensive treatments of important selected topics of physics. One of the aims is to make new results of research available to graduate students and younger scientists, and moreover to all people who like to widen their scope and inform themselves about new developments and trends.

A larger part of physics and applications of physics and also its application in neighbouring sciences such as chemistry, biology and technology is covered. Examples for typical topics are: Statistical physics, physics of condensed matter, interaction of light with matter, mesoscopic physics, physics of surfaces and interfaces, laser physics, nonlinear processes and selforganization, ultrafast dynamics, chemical and biological physics, quantum measuring devices with ultimately high resolution and sensitivity, and finally applications of physics in interdisciplinary fields.

Integrated Optics and Micro-Optics with Polymers

Edited by

Prof. Dr. Wolfgang Ehrfeld
IMM Institute of Microtechnology GmbH, Mainz

Prof. Dr. Gerhard Wegner
Max Planck Institute of Polymer Research, Mainz

Prof. Dr. Wolfgang Karthe
Fraunhofer Institution of Applied Optics and Precision Engineering, Jena

Dr. Hans-Dieter Bauer
IMM Institute of Microtechnology GmbH, Mainz

Dr. Herbert O. Moser
IMM Institute of Microtechnology GmbH, Mainz

 B. G. Teubner Verlagsgesellschaft
Stuttgart · Leipzig 1993

Wolfgang Ehrfeld (1938), physicist (University of Karlsruhe, Dr.-Ing. 1969), Professor of physics (University of Mainz), head of the IMM Institute of Microtechnology GmbH (Mainz). Research interests: microfabrication processes, microstructure products.

Gerhard Wegner (1940), chemist (University of Mainz, Dr. rer. nat. 1965), Professor in Physical Chemistry (University of Mainz), one of the Directors of the Max Planck Institute for Polymer Research (Mainz). Research interests: structure and properties of solid polymers, novel materials based on polymers.

Wolfgang Karthe (1938), physicist (University of Jena, Dr. rer. nat. 1967, Dr. sc. nat. 1977), Full Professor of Applied Physics (University of Jena), head of Fraunhofer Establishment of Applied Optics and Precision Mechanics (Jena). Research interests: integrated guided wave optics, micro-optics, precision micromechanics.

Hans-Dieter Bauer (1960), physicist (University of Bayreuth, Dr. rer. nat. 1990), is with IMM Institute of Microtechnology GmbH (Mainz). Working areas: micro-optics, nonlinear optics, optical sensors, organic/polymer materials.

Herbert O. Moser (1944), physicist (Karlsruhe University, Dr.-Ing. 1971, Dr.-Ing. habil. 1985, Privatdozent 1986), is with IMM Institute of Microtechnology GmbH (Mainz), as head Physical Technology Division. Main interests: microtechnology, light and particle optics, accelerator physics.

Die Deutsche Bibliothek – CIP-Einheitsaufnahme

Integrated optics and micro-optics with polymers /
ed. by Wolfgang Ehrfeld ... - Stuttgart ; Leipzig : Teubner, 1993
  (Teubner-Texte zur Physik ; Bd. 27)
    ISBN 978-3-322-93431-4   ISBN 978-3-322-93430-7 (eBook)
    DOI 10.1007/978-3-322-93430-7
NE: Ehrfeld, Wolfgang [Hrsg.]; GT

Umschlaggestaltung: E. Kretschmer, Leipzig

Preface

The present book is the account of a workshop on Integrated Optics and Micro-Optics with Polymers held in spring 1992 at Mainz and organized by IMM Institute of Microtechnology GmbH, the Max Planck Institute of Polymer Research, and the Institute of Applied Physics of Friedrich Schiller University at Jena.

The field of Integrated Optics and Micro-Optics with Polymers is receiving growing interest from multiple sides. Among the important reasons are the potential of tailoring materials for a specific application, the easy and cheap availability of those materials, and the possibilities of mass fabrication with plastics. Accordingly, materials researchers, microtechnologists, process engineers, and device builders are active in this field. Their interest is fed from prospective applications of integrated or micro-optical devices and systems in telecommunication, sensors, optical switching and routing, and, in a more distant future, optical processing.

The workshop succeeded to bring together more than 130 experimentalists and theorists, physicists and chemists, device developers and users, materials researchers and process engineers, as well as polymer scientists and those dealing with anorganic materials, coming from industry, research institutes, and universities.

The successful organization of a workshop needs a lot of volunteers and money. It is our pleasure to thank CIBA VISION, Aschaffenburg, for their financial contribution. Special thanks go to Dr. H. Freimuth, IMM, for his share in the organization. Last, but not least, we are grateful to those colleagues who accepted to serve as a referee, and to the publisher for providing the means of publishing the contributions to the workshop in this prestigious series.

We wish this book a widespread distribution.

Mainz, January 1993   The editors.

Contents

Materials for Micro-Optics and Nonlinear Optics

Characterization and Modelling

Theory

Inorganic-organic Polymers for Micro-Optic Applications

H. Krug, H. Schmidt

Institut für Neue Materialien gem. GmbH, Universitätscampus, Geb. 43, Im Stadtwald, 6600 Saarbrücken

1 Introduction

The realisation of microoptical devices is strictly correlated to the development of new materials and technologies. Simple processing techniques and multifunctionallity of the materials is mostly not available by a single but only by the combination of several materials. Materials which are used nowadays for optical applications are inorganic glasses, silicon, lithium niobate, semiconductor materials and organic polymers. For integration, these different materials have to be combined on a single substrate, which leads to very complex and cost intensive processing techniques. The industrial development of high performance materials with high development costs is strongly limited by small market volumes, if the added value of down stream systems based on these materials cannot be included.

Chemical synthesis by sol-gel techniques allows to fabricate materials with a lot of very interesting optical properties [UHL 90]. Multicomponent systems with wide variations can be synthesized and thick layers can be produced by simple techniques like spin-on and dip-coating. To achieve inorganic materials like glasses, temperatures of about the glass transition have to be employed in order to get fully dense final materials.

But the sol-gel process also allows the incorporation of organic compounds or groupings [SCH 88, SCH 89, Bri 90]. The integration of such organic groupings can reduce the densification temperature drastically [DUN 90, MAC 91, SCH 91a], and thick and dense layers can be produced at moderate temperature. Sol-gel chemistry combined with organic polymer chemistry makes it possible to synthesize materials with different compositions of organic and inorganic components. The structure of this materials can be controlled by the bonds between the organic and inorganic units, as well as phase separation phenomena to provide high optical quality. The materials can be used as matrix-materials for various types of optical active dopends. These manyfold variation possibilities in combination with their good optical properties provide an interesting application potential for this type of inorganic-organic composites (ORMOCER $\equiv$ ORganically MOdified CERamics) in integrated optics. Due to their sol-gel based synthesis they can be produced in small quantities, too, and a system and problem oriented material development are possible.

2 Structure elements

For the synthesis of inorganic materials by the sol-gel process, reactive monomers, oligomers or colloids can be used as starting materials. By polycondensation step, a polymeric network can be formed especially if silanes are used. A common route is the use of alkoxides as precursors which hydrolyse in the presence of water, condensing spontaneously to polymeric species. An other route is the hydrolysation and condensation of inorganic salts by pH-change. Precipitation can also be started by the destabilization of colloidal sols by pH-change either in organic solvents or in water. These condensation reactions lead to gels, in which solvent or air are contained after first drying. Depending from processing steps, they contain water, organic or unhydrolysed alkoxy groups. Drying and densification of the gels is made by heat treatment to convert them into glasses or ceramics.

Introducing organic groupings decrease the network connectivity of the gel and leads to dense materials at temperatures between 50 °C and 150 °C. Several chemical links and bonds between organic units and the inorganic backbone can be considered and the type of connection between organic and inorganic units is the main structure determining factor.

Covalent bonds can be used to form links to organic groups as well as to inorganic backbones (Fig. 1). Silicon for example is able to form stable bonds to oxygen as well as to carbon, but some other bonds like $=$P-C$\equiv$ or $\equiv$Sn-C$\equiv$ are possible whereas transition metal Me-C bonds, as a rule, are not stable against hydrolysis. This covalent bond can be used as a general link between organic and inorganic components. In case of non-reactive organic groups, organic network modifikation leads to so called spin-on glasses [BAG 90] which can be densified at low temperatures of 200 - 300 °C, and which can be applied in thickness of more than 20 μm.

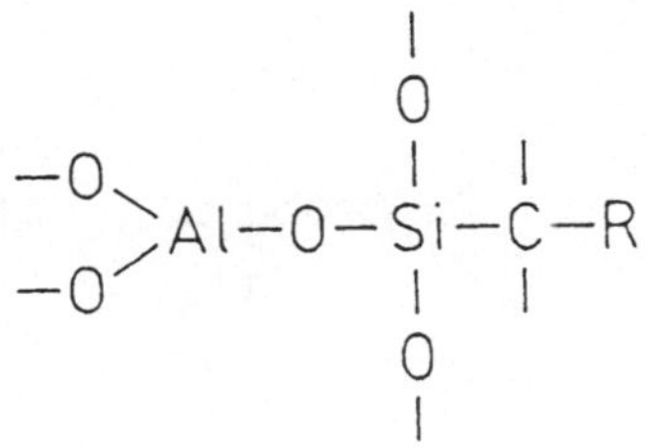

$$\begin{array}{c} | \\ O \\ | \quad | \\ -O{\diagdown} \\ \qquad {>}Al-O-Si-C-R \\ -O{\diagup} \\ \qquad\quad | \quad | \\ \qquad\quad O \\ \qquad\quad | \end{array}$$

Fig. 1:   Structure model for a covalent link between oxide network and organics

For incorporation of a wider range of oxides, complexing agents like ß-diketons, conjugated organic acids or amines can be used, forming coordinative bonds and allowing organic modification with polymerizable chelating ligands. Examples of alkoxy dicarbonates (a) or methacrylates (b) are shown in Fig. 2 a, b. These complexing ligands are acting as termination sites for condensation and, therefore are responsable for spatial extensions of the transition metal-oxo core [SAN 91]. Furthermore, inorganic ions can be incorporated by using amine complexes as chelating ligands (Fig. 2 c, d).

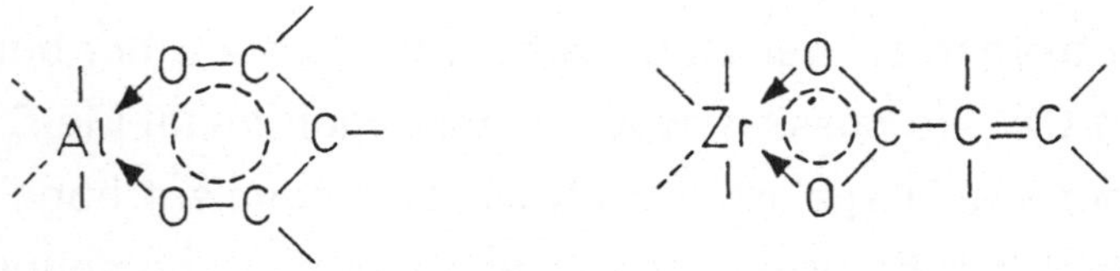

Fig. 2 a, b: Structure model of ß-dicetone and methacrylate complexes connecting transition metal oxide and organics

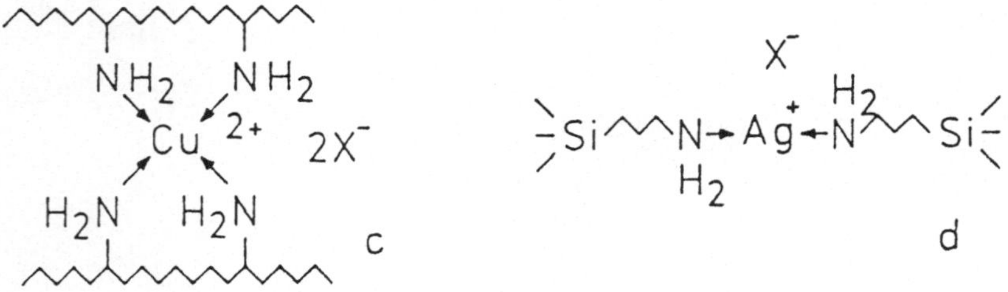

Fig. 2 c, d: Structure model of amine complexes to incorporate metal ions

The combination of the described principle with functional groups, which are able to be polymerized, polyadded or polycondensed leads to inorganic-organic polymers (hybrid polymers). These reactive groupings can be directly crosslinked, or organic chains can be built up by incorporation of organic monomers or oligomers (Fig. 3 a, b). Reactive crosslinking can be achieved by thermally or photochemically initiated polymerization and polycondensation reactions. E.g. oligomeric acrylates in combination with a photoinitiator can be used to build up crosslinked organic-inorganic polymers.

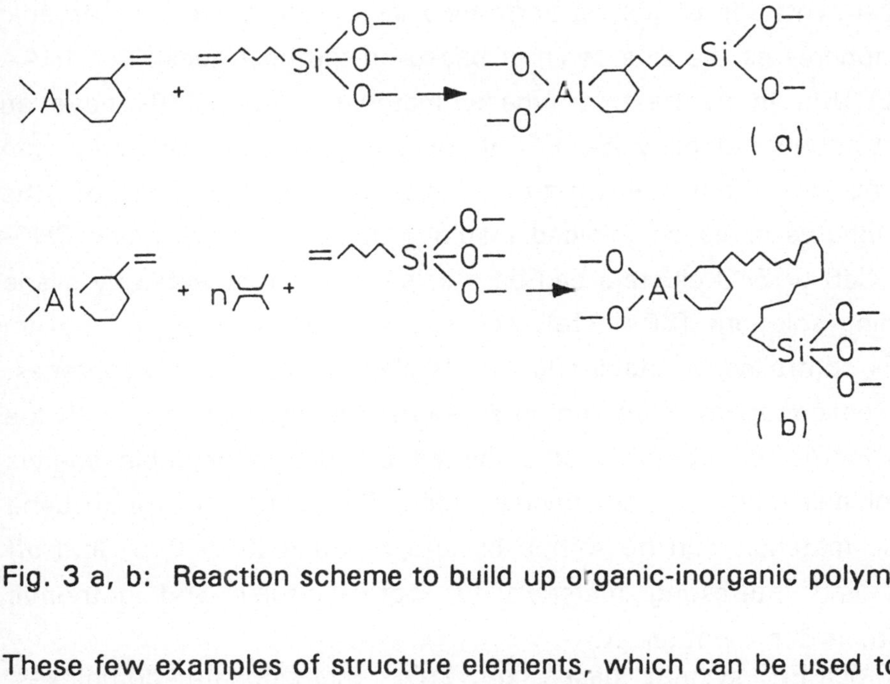

Fig. 3 a, b: Reaction scheme to build up organic-inorganic polymers

These few examples of structure elements, which can be used to build up the bulk material should demonstrate the numerous possibilities to tailor structure and thereby properties of the final product. Only small changes in synthesis conditions can result in large changes of the gel structure and organic-inorganic composites with completely different properties can be obtained. Some variations of optical properties by variation of synthesis conditions will be presented in the following examples.

3 Optical properties

As indicated above, the introduction of organic groupings into inorganic networks leads to a broad variation of structure and properties. For example, the reduced network connectivity decreases T_g and increases the thermal expansion coefficient a [SCH 90a]. T_g and a can be tailored over a wide range by synthesis parameters. Applications of such composite materials for microelectronics are for example described in [POP 90]. The low temperature at which the chemical synthesis of gels is performed allows to incorporate organic chromophores as the side chain or based on the host/guest idea [HAA 89, REI 90]. In contrast to organic materials, ORMOCERs have an exellent network stability [SCH 90b] and poling and polymerization can be done at room temperature. Relaxation phenomena of the chromophores could be avoided resulting in a long term stable SHG-signal. CdS or ZnO clusters can be precipitated in methacryloxy silane containing solutions [SPA 91a]. The size of these clusters in the nm range is controlled by stabilizing agents like amines and mercaptanes. These semiconductor quantum dots can be directly combined with the organic-inorganic network and stabilized by interconnectable ligands over polymerization and polycondensation. The sharp absorption band of these materials can be shifted by size-variation [SPA 91b] (cut-off effect), and interesting materials for optoelectronic and optooptic applications can be obtained.

An ORMOCER system based on ZrO_2 methacrylate complexes, polymerized with methacrylate derivates shows optical losses < 1 dB/cm for planar waveguides [KRU 91]. Films of 10 μm thickness prepared by spin-on have a surface roughness of less then 5 nm which is responsable for low surface scattering. The complexed ZrO_2-sols contain ZrO_2 particles of about 10 nm [NAS 89] which contribute to small Rayleigh-scattering. By variation of ZrO_2-content, the index of refraction can be change in the range $1.52 < n < 1.54$ at 633 nm.

Diffraction gratings with up to 2400 lines/mm and peak-to-tough values of 100 nm could be obtained by embossing techniques [SCH 91b]. These high amplitudes of the grating are available by the small shrinkage rate < 5 Vol-% of the liquid (non cured) material to the finally cured system. Mulitmode strip-waveguides were prepared by direct laser writing, which possess optical loss < 3 dB/cm [SCH 91c], and complex microstructures were obtained by maskaligner techniques. The higher optical loss compared to planar waveguides can be attributed to surface roughness of the written structure which is due to not yet optimized processing.

Fluorine modified ORMOCER's allow to change index of refraction to lower values $1.38 < n_D < 1.52$. Lowering the index of refraction also diminishes optical losses, as in case of density fluctuations, the loss is proportional to the eight power of the refractive index. The same is true for anisotropy fluctuations (proportional to the fourth power) which are pronounced in polymers with aromatic groups. An other interesting point of fluorine modification is the minimization of optical losses in the near infrared by suppressing C-H overtone absorption. Minimization of attenuation by fluorination of PMMA is already described in literature [GRO 89]. Fluorine modificated ORMOCER's are also already described as hydrophobic dust repellent coatings [KAS 91]: by using only 2 mole-% of fluorinated precursors, surface energies comparable with PTFE could be obtained. Optical losses of such fluorinated ORMOCER's in the near infrared are to be measured.

4 Conclusions

Sol-gel chemistry in combination with organic polymer chemistry is a promising synthesis route to get tailormaid composite materials for microoptic applications. Structure, and thereby properties of these composite materials can be varied over a wide range by synthesis parameters. These materials have high application potential for passive/active lightguidance and system and problem oriented tailoring is possible.

5 Acknowledgement

The financial support of the government of the Saarland is greatfully acknowledged.

[BAG 90]   B.G. Bagley, W. E. Quinn, P. Barboux, S. A. Khan, J. M. Tarascon
           J. Non-Cryst Solids 121 (1990), 454

[Bri 90]   C. J. Brinker, G. W. Scherer
           Sol-Gel Science, Academic Press, Inc., New York 1990

[DUN 90]   B. S. Dunn, J. D. Mackenzie, J. I. Zink, O. M. Srafsudd
           SPIE Vol. 1328 (1990), 174

[GRO 89]   W. Groh, D. Lupo, H. Sixl
           Adv. Mater. 11 (1989), 366

[HAA 89]   K. H. Haas, H. Schmidt, Roggendorf
           Top. Meeting on Glasses for optical applications 1989,
           Ceram. Soc. Japan 68.1

[KAS 91]   R. Kasemann, S. Brück, H. Schmidt
           Proceedings Eurogel 91, Saarbrücken (in print)

[KRU 91]   H. Krug, N. Merl, H. Schmidt
           Proc. 6th International Workshop on Glasses and Ceramics
           from Gels, October 6-11, 1991, Sevilla/Spain (in print)

[MAC 91]   J. D. Machenzie
           Proc. 6th International Workshop on Glasses and Ceramics
           from Gels, October 6-11, 1991, Sevilla/Spain (in print)

[NAS 89]   R. Nass, E. Arpac, H. Schmidt
             Conference on Ceramic powder processing science, Poster
             session, San Diego 1989

[POP 90]   M. Popall, H. Meyer, H. Schmidt, J. Schulz
             Mat. Res. Soc. Symp. Proc. 180 (1990), 995

[REI 90]   R. Reisfeld
             SPIE 1328 (1990), 29

[SAN 91]   C. Sanchez, M. In
             Proc. 6th International Workshop on Glasses and Ceramics
             from Gels, October 6-11, 1991, Sevilla/Spain (in print)

[Sch 88]   H. Schmidt
             J. Non-Cryst. Solids 100 (1988), 51

[Sch 89]   H. Schmidt
             J. Non-Cryst. Solids 112 (1989), 419

[SCH 90a]   H. Schmidt
             Mat. Res. Soc. Sym. Proc. Vol 171 (1990), 3

[SCH 90b]   H. Schmidt
             J. Non-Cryst. Solids 121 (1990), 428

[SCH 91a]   H. Schmidt
                 in: Chemical Processing of Advanced Materials;
                 Proceedings 1991 Ultrastructure Conference,
                 Orlando, eds.: J. Wiley & Sons (in print)

[SCH 91b]   H. Schmidt, H. Krug, N. Merl
             Proceedings Topical Meeting on Intelligent Glasses,
             Venecia, Sept. 13-14, 1991 (in print)

[SCH 91c]   H. Schmidt, H. Krug, R. Kaseman, F. Tiefensee
             SPIE 1590 (1991), 36

[SPA 91]   L. Spanhel, E. Arpac, H. Schmidt
             Proc. 6th International Workshop on Glasses and Ceramics
             from Gels, October 6-11, 1991, Sevilla/Spain (in print)

[UHL 90]   D. R. Uhlmann, J. M. Boulton, G. Teowee
             SPIE 1328 (1990), 270

Bistable Nematics - A Novel Approach towards Optical Information Processing

R. Eidenschink

NEMATEL, Galileo-Galilei-Straße 10, W-6500 Mainz, Germany

Abstract

Particles of fumed silica with primary diameter of 7 to 16 nm forming aggregates and agglomerates stabilize the orientation of nematic liquid crystals. By external influence layers of stable suspensions of 2 to 3 vol. per cent of the solid can be rendered transparent or light scattering.

1. Introduction

A number of organic compounds do not undergo a direct phase transition from the crystalline to the liquid state. Instead, they take on one or more different phases, limited to well defined temperature intervals. These phases possess anisotropic physical properties similar to crystals but at the same time remain fluids. The interest in technical applications of these so called thermotropic liquid crystals and thereby the progress in the synthesis of suitable compounds only started in the late sixties after there had been first reports on exploitable electrooptic effects.

Fig. 1 shows schematically the molecular order in some of these phases formed by rod-shaped molecules. The nematic phase in comparison with the isotropic phase is characterized by a long range molecular order without positional order. The preferred direction is given by the director **n**. In most cases the molecules can rotate freely around their longest axes. The turbid, schlierenlike appearance of a bulk of such a fluid stems from very small ordered domains of random **n** which build up and vanish permanently.

Intermolecular interactions may aggregate the rodlike molecules into layers stacked equidistantly. This feature characterizes the smectic phases[1]. The centres of mass within one single layer may be arranged statistically (S_A or S_C phase) or regularly (e.g. in a S_B phase). The rod axis can be parallel or tilted with respect to the layer normal. The thickness of one layer may vary between one and two molecule lengths. Though the layers can easily be moved against one another it is clear that the bulk viscosity of smectic phases is high compared to the nematic phase

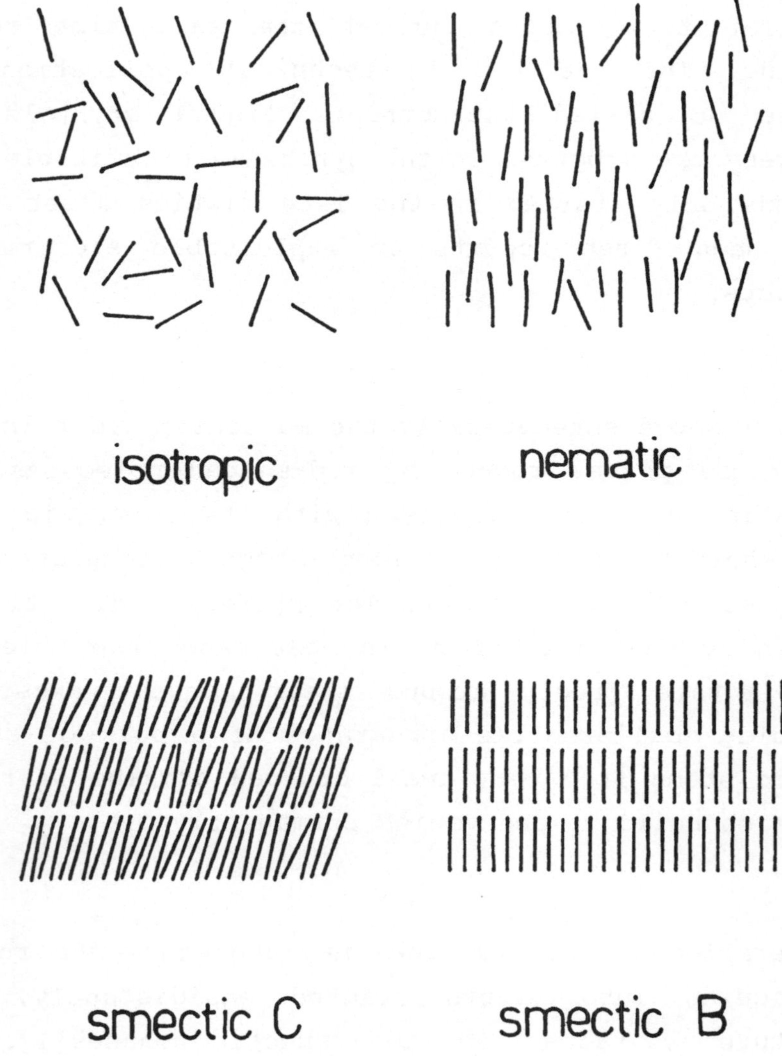

Fig. 1

Schematical order of rod-shaped molecules in the isotropic, nematic, smectic C and smectic B phase formed by the same molecular species.

Among all thermotropic liquid crystalline phases the nematic phase is of highest technical importance. Its application is limited nearly exclusively to electro-optic display systems, which usually are variations of the twisted nematic cell (TNC) [2] with two polarizer foils. The nematic phases used nowadays are mixtures often containing 10 or more components.

Fig. 2 shows the chemical structures of some components of nematic phases. The market is characterized by a hard competition in which the European and especially the German chemical industry still plays an important role. A selling price of US$ 4000 to 10000 per kilogram and an annual consumption of 15 to 20 tons demonstrate the attractivity of the market. In the production of the displays, where most of the evaluation occurs, the European and American industries do not play a significant roll any more.

Recently, a completely new principle of information display using nematics has been introduced, the so-called Polymer Dispersed Liquid Crystals (PDLC) [3]. The index of refraction of a transparent polymer is matched to the ordinary index of the encapsulated nematic material. This system is transparent in the presence of an electric field and light scattering in its absence. Here the orientation of the nematic phase is influenced by the interactions with the surrounding

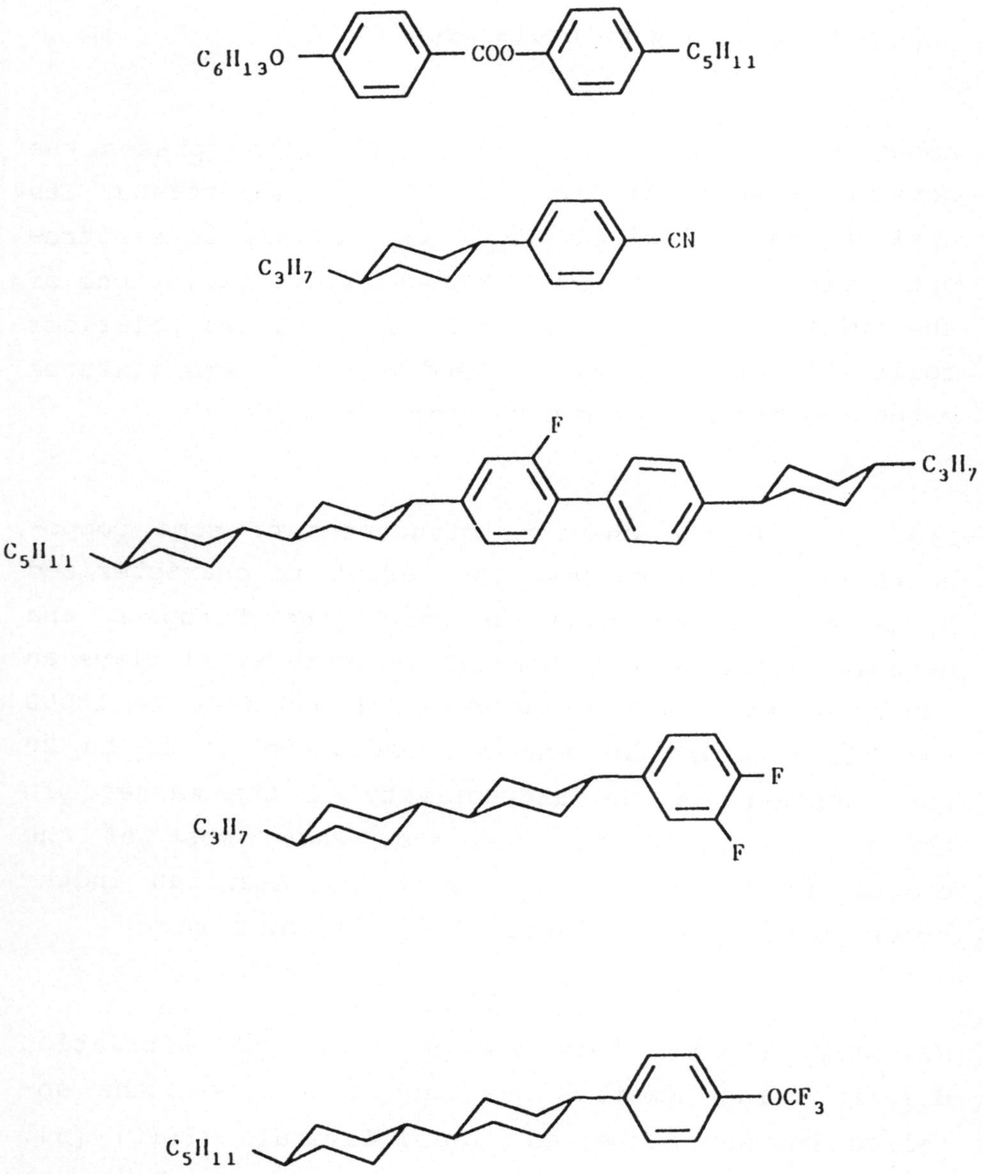

Fig. 2

Some compounds used as components for nematic phases for
TNC displays

surface, tco. This prevents the system from being bi-stable. The advantage of such scattering systems is that there is no need for polarizers avoiding major losses of light intensity. To keep the display trans-parent, a voltage has to be applied permanently, which is a clear disadvantage.

2. The bistability in filled nematics

It would be desirable to create a simple bistable system on the basis of the fast switchable nematic phase taking advantage of the high quality of light scattering with materials of high birefringence. A high quality of bistability has already been achieved by smectic A phases [4] switching between the homeo-tropic and the light scattering focal conic texture and by ferroelectric smectic C phases [5] which require polarizers.

Here an observation was helpful which can often be made during chromatographic separations of organic compounds using columns containing silica gel dis-persed in different liquids. Depending on the compo-sition of the liquid a column of, say, 3 cm in dia-meter appears fully white and in some cases almost transparent. Behind this is, of course, the matching of the refractive indices. The idea was to build a "reversed PDLC" in which the refractive indices of the nematic and the silica gel could be mismatched by an electric field. Unfortunately, there are no suitable nematic phases that permit this for silica gel as a solid, having a refractive index of 1.43. Also the handling of suspensions of such particles with diameters of 60 to 25 µm and of a volume ratio of 25% of the solid was difficult. Moreover, later on the idea proved to be not so new. Already in 1974 Hilsum was granted a patent for a display using small glass spheres [6].

The problem of index-matching should not arise with fumed silica (HDK, hochdisperse Kieselsäure). The interesting point with these extraordinary particles, which have been well-known since the forties and are used as thixotropes, is their structure. For example, a stable, not sedimentating suspension can be prepared with only 2 to 3 vol. % HDK. 10 µm thick layers of such suspensions in isotropic liquids do not scatter light to a remarkable degree. In nematics small domains of different orientation are formed causing strong light scattering. Above the clearing temperature, i.e. in the isotropic phase, the cell becomes transparent.

Transparency should also be achievable by a regular homeotropic orientation of the molecules in a nematic phase. In case of a positive dielectric anisotropy this should be feasible by an electric field applied via transparent electrodes. Unfortunately, the electric field required turned out to be higher than in usual reorientation processes. As a sort of compensation, nature surprised us with the result that, dependent on the voltage, the display stayed transparent after switching off the electric field (fig. 3). The question arose how to bring back the display into the scattering state. This can be done by shearing, which surely is not a very elegant way. In addition, repeated shearing produces a planar orientation of the liquid crystalline phase. Surprisingly, small sharply limited regions with very high light scattering can be generated within such a transparent layer by applying ultrasound. Obviously, high-frequency sound sources (800 kHz) can be well focused on the liquid crystal layer despite the relatively thick (1 mm) glass plate. However, such an "ultrasonic pen" would be technically difficult and expensive.

3. First applications

The possibility to switch the display on and off electrically seems to be given in the case of a nematic material that shows different values for the dielectric anisotropy $\Delta\epsilon$. This situation can be found in nematics that have a low-frequency relaxations of the dipole moment parallel to the long molecular axis [7] (fig. 4). When a high-frequency voltage pulse is applied to a layer of homeotropic stable orientation the molecules try to orient perpendicular to the electric field. Because there is no preferred direction in

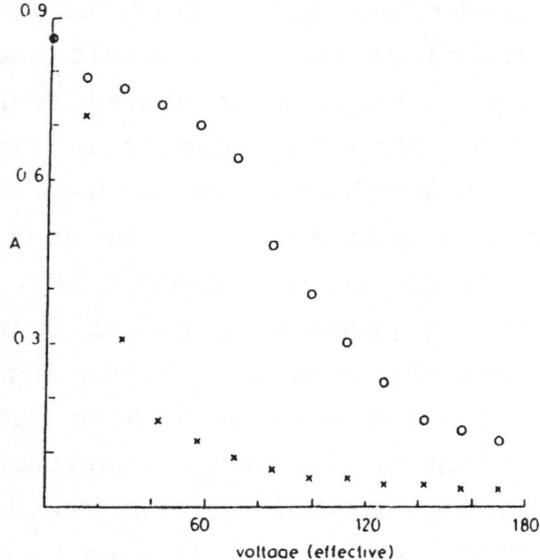

Fig. 3

Absorption A of ZLI 1132 filled with 2.8 vol.% of Aerosil R812
against effective voltage (400 Hz) measured in 14 μm layer

x   voltage on

°   after voltage switched off

Absorption defined as intensity of light beam having aperture
of 4° in front of and behind cell

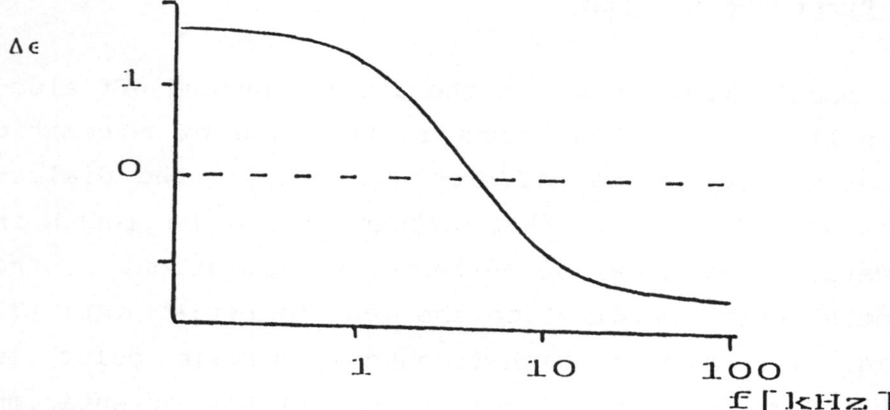

Fig. 4

Dielectric anisotropy Δε against frequency of a typical nematic
2-frequency-mixture

the plane parallel to the plates the domains produced are randomly oriented to one another and thus scatter light [8]. The contrast reached with this two-frequency procedure failed so far to be sufficient for projection displays. An improvement will only be reached by optimizing the nematic mixtures, often consisting of up to 12 different, complicated and expensive compounds.

What now is the interesting point with these bistable nematics and on which unique property could a technical application be based? To generate very small light-scattering areas in a transparent layer looks most promising. Kreuzer and Tschudi could achieve this with a laser beam. The theoretical background, which is closely related to practical application, still has to be clarified. Nevertheless, it seems clear that a minimum of electromagnetic energy has to be absorbed. Obviously a number of domains randomly oriented to one another are generated by a thermal shock, similar to the experiments using ultrasound mentioned above. The achievable contrast is very high: Kreuzer has measured an intensity ratio of 1:100 [9]. Dyes, dissolved in the liquid crystal, allow reasonably low laser energies. With an argon ion laser a picture, consisting of 500 x 500 pixels, was written into a 14 μm thick layer of a suspension of HDK in a nematic phase containing an appropriate dye. The storage area was 8 x 8 mm^2. Via intensity control 16 grey levels could be realized. This picture could be erased by a voltage pulse within some milliseconds and again written in by the aid of a scanner. Meanwhile, the storage time per pixel has been decreased to approximately 100 ns. The spatial resolution is at 2 µm. Although these first results are very motivating, a lot of work remains to be done concerning the physics, the technology and the material parameters.

For commercially relevant devices the huge and expensive argon lasers are, of course, prohibitive. The intensities necessary can also be provided by cheap and compact laser diodes emitting in the near infrared region. Thus the bothersome tinge of the liquid crystalline layer can be avoided. Fortunately, infrared-absorbing dyes which are sufficiently soluble in nematic phases have been developed in recent years.

4. The role of the solid particles

It can be stated that the quality of the contrast in the two bistable states, the switching time, the resolution and the lifetime of the stored information depend on the matching of the properties of both nematic phase and solid particles. In addition, the ration of both components cannot be varied on a broad scale.

In view of the many material parameters an optimization without intimate knowledge of the chemical-physical interrelationship would be too extensive and time-consuming. The key to the understanding of the effects lies in the shape and chemical nature of the solid particles. Particles of fumed silica [10] proved to be most advantageous ones among the highly dispersed oxides. They consist of primary particles of X-ray amorphous silica of typically 7 to 16 nm diameter. A great number of these species are fused together via $\equiv$Si-O-Si$\equiv$ bonds forming so-called aggregates that cannot be broken apart. The surface of the original hydrophilic fumed silica is covered with $\equiv$SiOH groups. By the formation of hydrogen bonds between the silanol groups of different particles so-called agglomerates are formed (fig. 5). This process which can be reversed by mechanical means makes these particles useful as thixotropes.

The hydrophilic HDK are less suited to meet our requirements, because very high voltages (often higher than the breakdown voltage of the system) are necessary to reach the transparent state. The same is the case with so-called precipitated silica which is not characterized by primary particles but by irregular skeletons of silica forming extended caves.

On the contrary, the so-called hydrophilic fumed silica is of special interest. It can be obtained e.g. by treatment of hydrophobic HDK with dichlorodimenthylsilane or hexamethyldisilazane. The density of

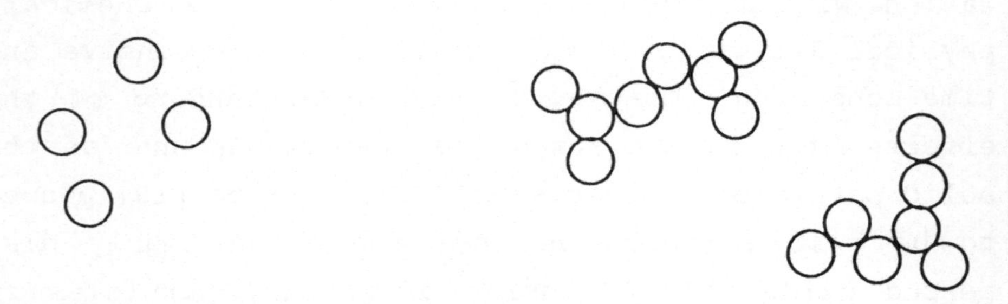

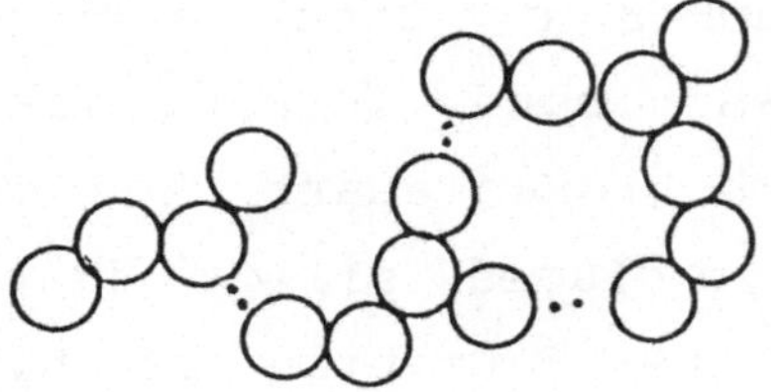

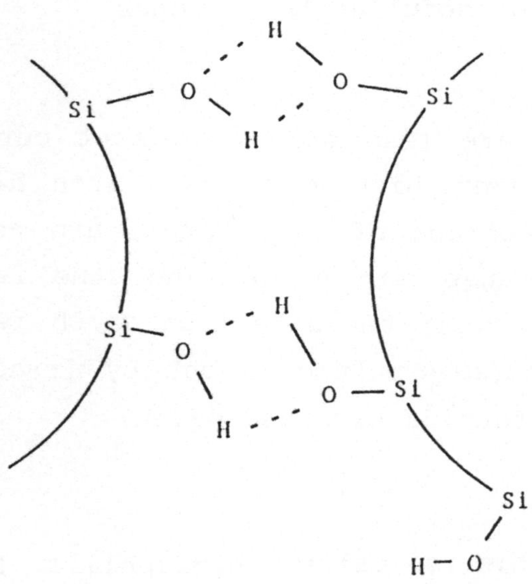

Fig. 5
Feature of the particles in fumed silica

the silanol groups on the surface is approximately $0.5/nm^2$ that is only 1/10 of the density on the hydrophilic ones. Despite the chemical modification the specific surface, measured by the BET method, is rather high: it ranges from 100 to 300 m^2/g.

What makes this system of aggregates and agglomerates with nematics bistable? The idea that the interactions between solid surfaces and organic molecules have to be very weak and therefore the solid particles act as a kind of skeleton-like support was confirmed by the finding that a stable homeotropic orientation in an electric field can be achieved by far lower voltages using the hydrophobic particles instead of the particles densely covered with silanol groups. Possibly some of the hydrogen bonds are broken during the orientation due to the solid/liquid interaction and regenerated in an energetically favoured arrangement. This skeleton stabilizes the homeotropic orientation and stays unchanged after switching off the electric field.

Although hydrogen bonds, easily formed and easily broken, provide an elegant explanation, other types of interparticle forces should not be overlooked [10]. More information can be expected from the use of silanol-free particles. Without anticipating a final assessment, it should be noted that the energy needed to orient the molecules in the field and the energy for transferring the system into the light-scattering state is sufficient to break up 3 to 6% of all conceivable hydrogen bonds [11].

It can be expected that after elucidation of the bistability mechanism the properties of both the liquid crystal and the solid particle can be matched in a more optimal way. This should lead to higher contrast ratios as well as lower threshold energies for the writing process. This might be of special interest for new projection displays as well as erasable data storages. The application as erasable masks for photolithography taking advantage of the transmittance towards UV light of some liquid crystals seems especially promising. The reproduction of photographs from stored data, e.g. a compact disk, seems also feasible because of the high optical resolution.

Acknowledgment

I want to express my thanks to the Department of Commerce of Rhineland-Palatinate for support via the Innovationsförderprogramm. I am indebted to Prof. W.H. de Jeu (FOM Institute Amsterdam) for stimulating discussions and valuable hints and also to Prof. T. Tschudi and M. Kreuzer (TH Darmstadt) for their enlightening laser works.

References

[1] G.W. Gray, J.W.G. Goodby
     Smectic Liquid Crystals, Leonard Hill, Glasgow
     1984.

[2] M. Schadt, Liquid Crystals $\underline{5}$, 57 (1989).

[3] J.L. West, Mol.Cryst.Liq.Cryst. $\underline{157}$, 427 (1988).

[4] F.J. Kahn, Physics Today, $\underline{1982}$, 66.

[5] N.H. Clark, S.T. Lagerwall, Appl.Phys.Lett. $\underline{36}$,
     899 (1980).

[6] C. Hilsum, Brit.Pat. 1 442 360 (Filed 1973).

[7] W.H. de Jeu, Solid Static Phys., Suppl. $\underline{14}$
     109-145 (1978).

[8] R. Eidenschink, W.H. de Jeu, Electronics Letters
     $\underline{27}$, 1195 (1991)

[9] M. Kreuzer, T. Tschudi, R. Eidenschink
     Mol.Cryst.Liq.Cryst. in print.

[10] Degussa AG, Frankfurt, Schriftenreihe Pigmente
     Nr. 11 (1991)

[11] R. Eidenschink, W.H. de Jeu, M. Kreuzer,
     T. Tschudi to be published.

Nonlinear Optics with Inhomogeneously Poled Polymers

Siegfried Bauer

Institut für angewandte Physik der Universität

Kaiserstr. 12, 7500 Karlsruhe, Germany

Abstract

The poling of polymers leads in general to inhomogeneous polarization distributions within the polymer film. A short review of common poling methods is given together with the resulting polarization distributions for the case of the ferroelectric polymer PVDF. It is shown that these polarization nonuniformities affect any nonlinear optical experiments based on second order nonlinearities. Second harmonic generation experiments are discussed which are able to show the polarization nonuniformities.

1 Introduction

Nonlinear optical effects have received much interest in recent years due to the many possible device applications in optical communications technology. Electro-optic modulation, multiplexers, spatial light modulators, second harmonic generation and frequency mixing are typical examples for second order nonlinearities, whereas optical bistability, optical switches, optical power limiters and regulators, tuneable filters, degenerate four wave mixing and phase conjugation schemes are based on third order nonlinearities [ULR 88]. Third order nonlinearities are common to all materials contrary to second order nonlinearities which require noncentrosymmetric materials. The commonly studied polymer systems for nonlinear optical applications include guest/host polymers containing nonlinear active chromophores (guest) dispersed in an amorphous polymer (host), linear polymers with covalently attached nonlinear active side chains, and chemically cross linked structures with covalently attached nonlinear active chromophores under electric field [EICH 90]. Usually, the polymers must be poled by the application of an electric field, to achieve a noncentrosymmetric ordering of the nonlinear active chromophores. From electret research it is well known that under certain conditions strongly nonuniform polarization distributions may result [SES 89, BIH 89], depending on a variety of parameters, as for example the electric conductivity of the polymer, the injection and the trapping of charge carriers etc. It is to be expected that these inhomogeneous distributions of polarization will influence any nonlinear optical experiment based on the second order nonlinearity. Recently an optical transfer matrix technique has been used to discuss optical harmonic generation in inhomogeneously poled polymers, showing that an inhomogeneous distribution of polarization strongly affects the intensity of the second harmonic intensity, if the polymer thickness is comparable or greater than the coherence length for second harmonic generation [BAU 92]. Thus nonlinear optical experiments with poled polymers must be accompanied by the measurement of the charge and the polarization distribution within the sample [BAU 92]. The present paper gives a short review of common poling methods for polymers together with the resulting polarization distributions for the case of the ferroelectric polymer PVDF. Then, second harmonic generation experiments are discussed which are able to show the existence of these polarization nonuniformities.

2 Polarization Profiles in Poled Polymers

Fig. 2.1 shows the common poling methods used for the preparation of electrets, Fig. 2.1.a shows the electrode poling arrangement, Fig. 2.1.b the corona poling arrangement and Fig. 2.1.c the electron beam poling method. Especially, the electron beam poling method offers the possibility to polarize small areas of the polymer by focused irradiation.

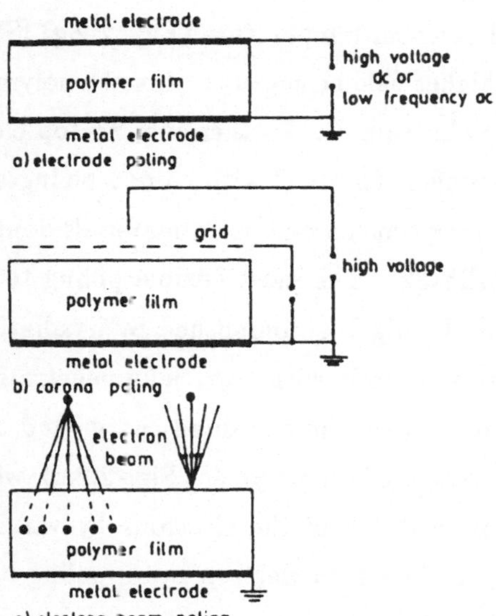

Fig. 2.1.

Common poling methods

Fig. 2.2 shows the resulting polarization distributions obtained with these poling methods for the ferroelectric polymer PVDF.

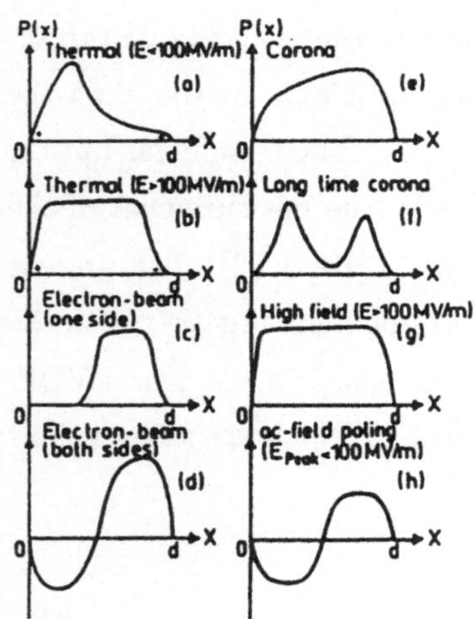

Fig. 2.2.

Resulting polarization distributions for the ferroelectric polymer PVDF

If electrode poling is performed at elevated temperatures with medium field strengths strongly nonuniform polarization profiles result (Fig. 2.2.a), which tend to become more uniform if the field strength is high (Fig. 2.2.b). Uniform polarization profiles result if very high electric fields are applied at room temperature (Fig. 2.2.g) [SES 89]. Corona poling allows the realization of higher field strengths within the polymer as it tolerances defects or shorts within the film sample. Furthermore no top electrode is needed. The resulting polarization profiles obtained with corona poling are relatively uniform (Fig. 2.2.e) [SES 89]. For long time corona poled materials double polarization zones may occur (Fig. 2.2.f) [EIS 82]. The most flexible poling technique is the electron beam method [SES 89]. Poling is accomplished by irradiating one-side metallized samples on their non-metallized side with a partially penetrating electron beam. An electric field is generated between the deposited charge and the rear electrode. This leads to a partial polarization of the polymer (Fig. 2.2.c), with the possibility to easily adjust the penetration depth of the electrons by varying the kinetic energy of the electrons [SES 89]. In fact with electron beam poling it is possible to prepare poled polymers such that the polarization reverses sign in the midplane of the samples (Fig. 2.2.d) [SES 89]. This kind of polarization is achieved by irradiating a two side metallized sample first on one and then on the other surface with an electron beam having a penetration depth which equals half the sample thickness [SES 89]. It is interesting to note that similar polarization profiles can be obtained by poling the polymer with a low frequency ac-field having a medium field strength. However, the resulting polarization is much smaller than for the electron beam method (Fig. 2.2.h) [BAU 91]. Thus electron beam poling is a flexible method which allows the preparation of different polarization profiles within polymer samples. It must be mentioned that for every poling procedure which leads to homogeneously poled films small depolarized layers near the elctrodes exist, with thicknesses ranging from $0.1\mu m$ to approximately $1 - 2\mu m$ [EMM 92].

3 Second Harmonic Generation with Poled Polymers

The calculation of the generated second harmonic light from a nonlinear slab is usually based on the boundary value problem showed in Fig. 3.1.

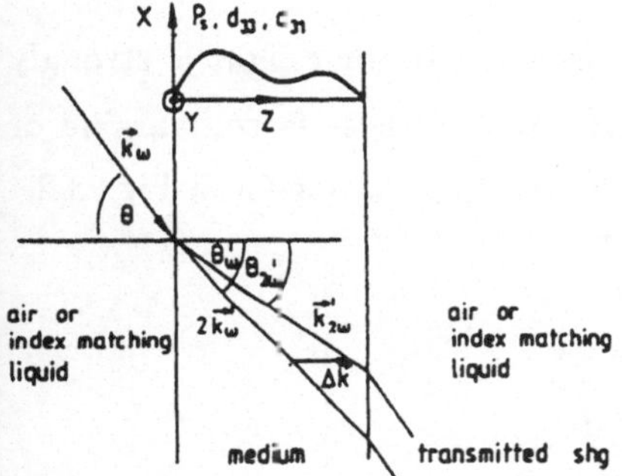

Fig. 3.1.

Schematic view of a SHG arrangement. With inhomogeneously poled polymers the polarization and the nonlinear coefficients can have any arbitrary distribution

If the polymer is represented by a $C_{v\infty}$ symmetry the nonlinear polarization is given by:

$$\vec{P}^{NL} = (2d_{31}E_1E_3, \ 2d_{31}E_2E_3, \ d_{31}E_1^2 + d_{31}E_2^2 + d_{33}E_3^2) \tag{1}$$

with d_{31} and d_{33} denoting the second order nonlinear coefficients and $\vec{E} = (E_1, E_2, E_3)$ the optical field at the fundamental frequency. For inhomogeneously poled films the spontaneous electric polarization is a function $P_s(z)$, thus the second order nonlinear coefficients are given by similar functions $d_{31}(z) = \alpha_{31}P_s(z)$ and $d_{33}(z) = \alpha_{33}P_s(z9$, with α_{ij} the proportionality constant. It can be shown, by using the matrix approach of [BAU 92] that for neglectible back reflection of the fundamental wave and neglectible pump depletion the generated second harmonic light for an inhomogeneously poled film is given by:

$$I_{SHG} = A\left| \int_0^L P_s(z)\exp(i\Delta kz)dz\right|^2 \tag{2}$$

with $\Delta k = \vec{k}'_{2\omega} - 2\vec{k}'_\omega$ denoting the wave vector mismatch, $\vec{k}'_\omega$, $\vec{k}'_{2\omega}$ the wave vectors of the fundamental and the harmonic light fields within the sample. The amplitudes of the $\vec{k}$ vectors are connected with the amplitude of the wave vector of the incident light k_0 and the refractive indices at the fundamental n_ω and the harmonic frequencies $n_{2\omega}$: $k'_\omega = k_0/n_\omega$, $k'_{2\omega} = 2k_0/n_{2\omega}$. The factor A is strongly angle dependent as it contains Fresnel transmission coefficients and an angle dependent projection factor [BER 89].

From Eq. (3.2) it is obvious that the generated second harmonic light is strongly influenced by a polarization distribution if the sample thickness is comparable or greater than $1/\Delta k$. This is confirmed by numerical examples as shown in Fig. 3.2.

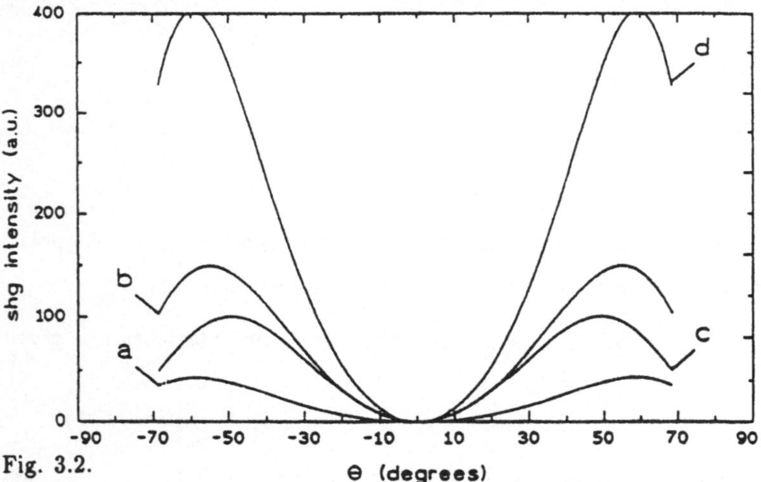

Fig. 3.2.
SHG intensity as a function of the incidence angle for a series of inhomogeneously poled films; (a) $P_s(z) = P$ for $0 \leq z \leq 0.25L$, (b) $P_s(z) = P$ for $0 \leq z \leq 0.75L$, (c) homogeneously poled film and (d) $P_s(z) = P$ for $0 \leq z \leq 0.5L$ und $P_s(z) = -P$ for $0.5L \leq z \leq L$

Fig. 3.2.c shows the calculated second harmonic light for a homogeneously poled $40\mu m$ thick PVDF sample. Fig. 3.2.a shows the calculated second harmonic signal for a sample poled only from $0 \leq z \leq 0.25L$, Fig. 3.2.b for a sample poled from $0 \leq z \leq 0.75L$. It can be seen from the figures that not necessarily the homogeneously poled film gives the highest second harmonic output. Fig. 3.2.d shows the calculated second harmonic intensity for a film with a polarization profile which reverses sign in the midplane of the film. Obviously this film gives the highest second harmonic output as approximately quasi phase matching conditions are realized [BAU 92]. For all calculations p-polarized fundamental beams are assumed. The parameters for the calculation are given by $n_\omega = 1.42$, $n_{2\omega} = 1.4277$, $d_{33} = 0.47 d_{11}(\text{quartz})$ and $d_{31} = 0.11 d_{11}(\text{quartz})$. As Eq. (3.2) contains no information on the phase of the second harmonic light, it is not possible to decide as an example at which electrode a thermally poled PVDF film (Fig. 2.2.a) is poled. As a further example, samples poled at $0 \leq z \leq 0.25L$, at $0.35L \leq z \leq 0.60L$ or at $0.75L \leq z \leq L$ give the same second harmonic signals.

It is now interesting to ask, if it is possible to obtain more information about the polarization distribution from nonlinear optical experiments. Fig. 3.3 shows a series of arrangements with inhomogeneously poled film samples and homogeneously poled film samples.

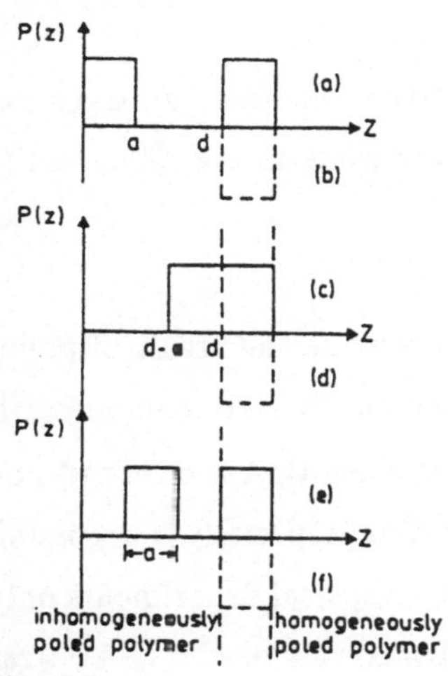

Fig. 3.3.

Possible series arrangements by combining a inhomgeneously and a homogeneously poled film

For asymmetric polarization profiles four different polarization distributions result (Fig. 3.3.a-d), thus four different angle dependent second harmonic signals must be seen. For a symmetrically poled polymer only two different arrangements result (Fig. 3.3.e-f), thus only two different angle dependent signals result. Fig. 3.4 shows the result of a calculation for a partially poled $40\mu m$ thick PVDF film ($10\mu m$ thick polarized region) and a homogeneously poled $10\mu m$ thick PVDF film. The calculations show four different angle dependent second harmonic signals. Thus it should be possible to show asymmetric polarization profiles by nonlinear optical measurements only.

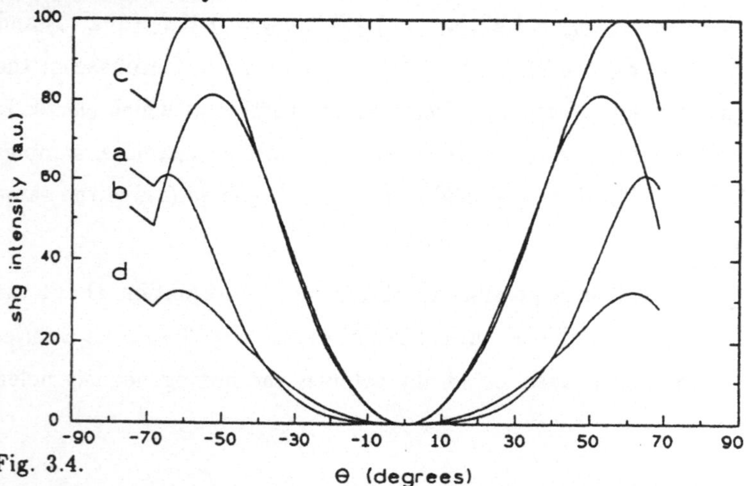

Fig. 3.4.
SHG intensity as a function of the incidence angle for the series connection arrangements; (a) according to Fig. 3.3.a (b) according to Fig. 3.3.b (c) according to Fig. 3.3.c and (d) according to Fig. 3.3.d

4 Conclusion

It has been shown, that polarization distributions can occur as the result of poling procedures, especially with the electron beam poling method strongly nonuniformly poled samples can be prepared. The numerical examples show that the second harmonic signals are sensitive to this distributions of polarization, especially it is possible to show asymmetries in the polarization profile by nonlinear optical experiments only. Experiments with partially poled PVDF films have confirmed this results [BAU 92a].

Acknowledgements

The author gratefully acknowledges financial support from the Deutsche Forschungsgemeinschaft (DFG). Thanks are due to Dr. R. Gerhard-Multhaupt, Heinrich-Hertz Institut, Berlin for his stimulating interest and fruitful discussions.

References

[BAU 91] Bauer S., Ploss B.: Polarization Distribution of Thermally Poled PVDF Films Measured with a Heat Wave Method (LIMM), Ferroelectrics 118 (1991) 435-450

[BAU 92] Bauer S.: Second Harmonic Generation of Light in Ferroelectric Polymer Films with a Spatially Nonuniform Distribution of Polarization, IEEE Trans. Electr. Insul., accepted

[BAU 92a] Bauer S., Schlaich H.: Second Harmonic Generation with Partially Poled Polymers, in preparation [BER 89] Berge B., Wicker A., Lajzerowicz J., Legrand J. F.: Second Harmonic Generation of Light and Evidence of Phase Matching in Thin Films of P(VDF-TrFE) Copolymers, Europhys. Lett. 9 (1989) 657-662

[BIH 89] Bihler E., Holdik K., Eisenmenger W.: Polarization Distributions in Isotropic, Stretched or Annealed PVDF Films IEEE Trans. Electr. Insul., 24 (1989) 541-545

[EICH 90] Eich M., Bjorklund G. C., Yoon D. Y.: Poled Amorphous Polymers for Second-Order Nonlinear Optics Polymers for Advanced Technol., 1, (1990) 189-198

[EIS 82] Eisenmenger W., Haardt M.: Observation of Charge Compensated Polarization Zones in Polyvinylidenfluoride (PVDF) Films by Piezoelectric Acoustic Step-Wave Response, Sol. State Commun. 41 (1982) 917-920

[EMM 92] Emmerich R., Bauer S., Ploss B., in preparation [SES 89] Sessler G. M.: Charge Storage in Dielectrics IEEE Trans. Electr. Insul., 24, (1989) 395-402

[ULR 88] Ulrich D. R.: Nonlinear Optical Polymer Systems and Devices Mol. Cryst. Liq. Cryst. 160, (1988) 1-31

Esterification of Polymers with Acid Chloride Groups - A New Route to Polymethacrylates with Nonlinear Optically Active Side Groups

Peter Strohriegl, Harry Müller, Irene Müller, Oskar Nuyken

Lehrstuhl Makromolekulare Chemie I und Bayreuther Institut für

Makromolekülforschung (BIMF)

Universität Bayreuth, Postfach 101251, D – 8580 Bayreuth

ABSTRACT

Polymethacrylates with pendant donor–acceptor substituted azobenzene groups have been investigated by several research groups and show excellent performance in nonlinear optical (NLO) applications. The usual method for the synthesis of these polymers is the free radical copolymerization of the corresponding methacrylates with NLO–active side groups and methyl methacrylate (MMA). However, the NLO–chromophores often contain a number of functional groups, e.g. nitro– or azo groups, which may act as inhibitors or retarders in a free radical polymerization. So in some cases the yields are not quantitative and the molecular weights are quite low.

The polymeranalogous esterification of poly(methacryloyl chloride) or copolymers of methacryloyl chloride (MACl) and MMA is an alternative route to polymethacrylates with NLO–active side groups. In a first step, reactive prepolymers are prepared by free radical copolymerization of MACl and MMA. These polymers are subsequently esterified by NLO–active side groups with a hydroxy–terminated spacer group.

Well defined, high molecular weight polymethacrylates with high dye contents can be prepared by that procedure. The polymers have electro-optical coefficients up to 19 pm/V (1500 nm) after poling. The novel method also provides easy access to copolymers with both NLO–active azobenzene and photocrosslinkable cinnamoyl groups.

1. Introduction

In recent years, advances in optical communications have created large interest in nonlinear optical (NLO) materials [1,2]. The first class of materials from which optical devices became available were inorganic single crystals like $LiNbO_3$. Since it is known that organic NLO materials possess several advantages like larger second order susceptibilities, faster switching times and simple processing methods, a lot of work has been focussed on organic materials in recent years [3]. Among these, polymers doped with NLO-active molecules (guest-host systems) [4] and polymers in which the NLO-active group is covalently attached to the polymer either as a side group [5-7] or as part of the main chain [8] have achieved much interest because of their relatively high second order susceptibilities combined with simple processing characteristics. The polymers doped with NLO-active molecules suffer from the limited solubility of the chromophores in the polymer matrix. By covalent attachment of the NLO-moieties polymers with high loadings of the NLO-active groups can be prepared. Because of their excellent optical properties, a variety of polymethacrylates with pendant NLO-chromophores has been prepared by different research groups [5-7,9].

After orientation of the NLO-active side groups in an electrical field the second order nonlinear susceptibilities of these materials are quite large. So a frequency doubling coefficient of 89 pm/V which is three times larger than in lithium niobate has been reported for a polymer with 4-amino-4´-nitrotolane side groups [10].

One of the major problems of polymers with NLO-active side groups is the slow relaxation of the field induced noncentrosymmetric alignment and hence the nonlinearity[11-16]. A slow relaxation of the chromophores is observed even at temperatures well below the glass transition of the polymer. Two strategies have been developed to overcome the problem of chromophore relaxation. Preparing polymers with high glass transition temperatures is one way to increase the stability of the poled system, crosslinking after poling is another way. If crosslinking is performed after the poling process has been completed, the polymer matrix becomes less flexible and the relaxation of the chromophores is expected to slow down. This has been demonstrated in a thermally crosslinkable epoxy resin[17,18]. In this material no significant decay of the second harmonic coefficient is observed at 80°C over a period of two weeks[19]. However, an epoxy

44

system has the disadvantage that both poling and crosslinking are thermal processes and occur simultaneously. The viscosity of the polymer increases drastically while crosslinking takes place and this makes the whole process difficult to control.

Recently, a photocrosslinkable guest/host system has been described[20-22]. Commercially available poly(vinylcinnamate) forms the host in which NLO-active azo dyes bearing two cinnamoyl groups are incorporated. Crosslinking occurs by [2+2] cycloaddition of the cinnamoyl groups upon irradiation with UV-light. No detectable decay of the second harmonic generation (SHG) signal was observed at ambient temperature over a period of several hours[21]. However, one of the major problems of guest/host systems, the limited solubility of the NLO-chromophore in the host polymer, cannot be overcome with this system.

The method normally used for the synthesis of polymers with covalently attached side groups is the free radical polymerization of the corresponding methacrylates. However the NLO-chromophores, usually conjugated aromatic molecules with an electron donor and an electron acceptor substituent often contain a number of functional groups, e.g. nitro- or azo groups. These may act as retarders or inhibitors in free radical polymerization. So in many cases the yields are not quantitative and the molecular weights are quite low.

We present an alternative method for the preparation of polymethacrylates with pendant NLO-chromophores, the polymeranalogous esterification of poly(methacryloyl chloride) (PMACl) and of copolymers of methacryloyl chloride (MACl) with methyl methacrylate (MMA). Using this method, we have also prepared terpolymers which contain both NLO-active chromophores and photocrosslinkable cinnamoyl groups. Fig. 1.1 schematically shows the processing of such a polymer. At first, the NLO-chromophores are randomly distributed in a thin film of the polymer. If this film is heated to the glass transition temperature under the influence of an electrical field the chromophores are more or less oriented. After poling, the oriented film can be photochemically crosslinked by UV-light. The crosslinking of the polymer matrix should suppress the reorientation of the chromophores[17-22]. Furthermore it is noteworthy, that small structures which are essential in integrated optical devices can be produced from these materials using well established photolithographic techniques.

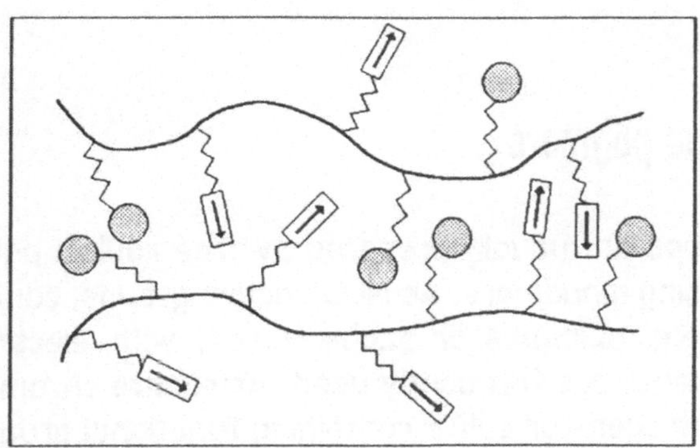

a) Unoriented sample

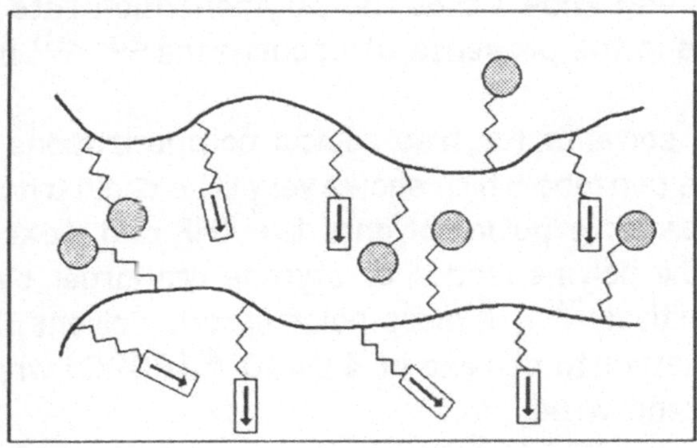

b) After orientation of the NLO-chromophores in an electrical field

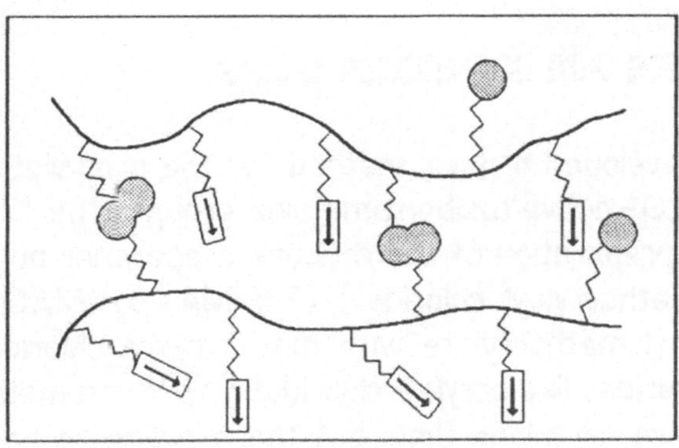

c) After irradiation of the photocrosslinkable groups

Fig. 1.1) Schematic representation of the processing of a photocross-linkable NLO - material

2. Results and discussion

2.1. Synthesis of NLO-active polymers

Polymers with NLO side groups are usually prepared by free radical polymerization of the corresponding monomers. As NLO-active groups, conjugated aromatic molecules, e.g. stilbenes or azobenzenes, with electron donor and acceptor substituents are frequently used. Attractive chromophores often include some nitrogen- or sulfur containing functional groups that may act as retarder, inhibitor or chain-transfer reagent in free radical polymerization. So it is well known that the polymerization rate of styrene drastically decreases in the presence of azobenzene[23-25] and nitrobenzene.

Furthermore the number of solvents for free radical polymerizations is limited. The classical solvent is benzene which shows very little chain transfer to various monomers. For more polar solvents like THF and dioxane the transfer constants for the polymerization of styrene are larger by a factor of 25 and 140, respectively[26]. A more polar aprotic solvent like DMF has a chain transfer constant to styrene of $4.0*10^{-4}$ (60ºC) which is the 200-fold of the benzene value.

2.1.1. Synthesis of prepolymers with acid chloride groups

With this in mind, we have developed a novel method for the preparation of polymethacrylates with NLO-active azobenzene side groups. The first step of our synthesis is the preparation of the reactive prepolymer poly-[(methylmethacrylate-co-methacryloyl chloride)] (P(MMA-co-MACl)) by copolymerization of methyl methacrylate with methacryloyl chloride. The polymerization of acid chlorides like acryloyl chloride[27-30] and methacryloyl chloride[31,32] is known for a long time, but these polymers have only very seldom been used in polymeranalogous reactions[33-37].

The poly(methacryloyl chloride) and the copolymers with MMA were prepared by free radical polymerization in dioxane with AIBN as initiator and had molecular weights $\bar{M}_w$ in the range from 66000 to 100000 and $\bar{M}_n$ between 33 000 and 44 000. The polydispersities $\bar{M}_w / \bar{M}_n$ of PMACl and the various copolymers are between 2,1 and 2,4.

2.1.2. Polymeranalogous esterification of the prepolymers

Recently, we have investigated the preparation of polymethacrylates with pendant carbazolyl groups by esterification of PMACl with a number of (ω-hydroxyalkyl)carbazoles at our institute. Due to the low reactivity of the polymer-bound acid chloride groups, the hydroxy groups were converted to the much more reactive alcoholates with butyl lithium prior to the reaction with PMACl. By this method polymethacrylates with a degree of substitution of more than 95% were obtained under mild conditions (3 hours, room temperature)[38].

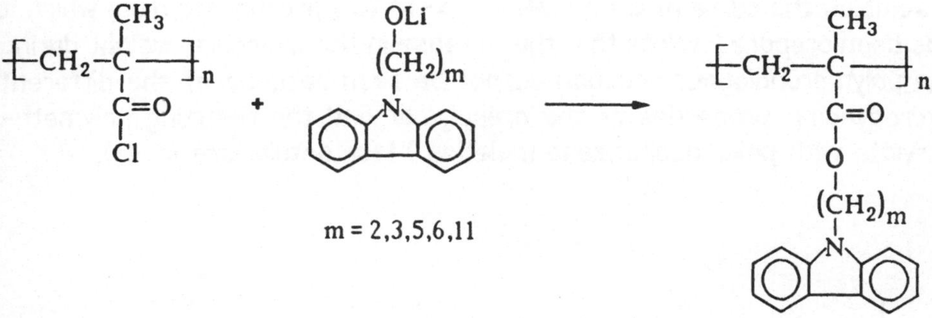

$$m = 2,3,5,6,11$$

Attempts to prepare the alcoholates of the NLO-active 4-[(2-hydroxy-ethyl)methylamino]-4´-nitroazobenzene **4a** with butyl lithium failed because of side reactions with the polar NLO-chromophore. If potassium *tert*-butoxide which is strongly basic but much less nucleophilic compared to butyl lithium is used, side reactions with the azobenzene moiety are minimized. So a copolymer of MMA and MACl with 10 mol% of acid chloride groups P(MMA-co-MACl) (mole ratio 90/10) **3a** was reacted with the potassium salt of the azochromophore **4b** to yield the copolymer **5a** with 9 mol% of NLO-active groups (see Scheme 1). The degree of substitution was determined by ^{1}H NMR spectroscopy and by elemental analysis. The results of the two methods were almost identical. Nevertheless we were not able to prepare soluble copolymers **5** with higher loadings of NLO-chromophores because of crosslinking reactions when we used alcoholates. Even in the GPC curve of polymer **5a** (Fig. 2.1) a tail at molecular weights of 10^6 to 10^7 can be seen which is not present in the prepolymer and which probably stems from a branched polymer fraction.

To circumvent the problem of crosslinking we have investigated the reaction of P(MMA-co-MACl) (mole ratio 80:20) **3b** with the alcohol **4c**. Due to the low reactivity of the polymer bound acid chloride groups [39], more drastic conditions are necessary. The reaction was carried out in boiling dioxane with pyridine as HCl-acceptor. The reaction was monitored by IR-spectroscopy. After two days at 100°C, the absorption of the acid chloride group at 1788 cm^{-1} has totally disappeared. In the carbonyl region of the IR-spectrum two small peaks at 1805 cm^{-1} and 1759 cm^{-1} are present in addition to the large absorption of the ester group at 1730 cm^{-1}. These two absorptions can be assigned to a small amount of anhydride groups. The GPC curve of polymer **5b** (Fig. 2.2) which contains 19 mol% of chromophore shows no high-molecular weight tail and closely resembles the curve of the P(MMA-co-MACl) prepolymer from which it has been prepared (Note that the increase in the molecular weight during the polymeranalogous reaction cannot be seen because of the different hydrodynamic properties of the prepolymer and the resulting polymethacrylate with polar azobenzene moieties.) (for details see lit. 41).

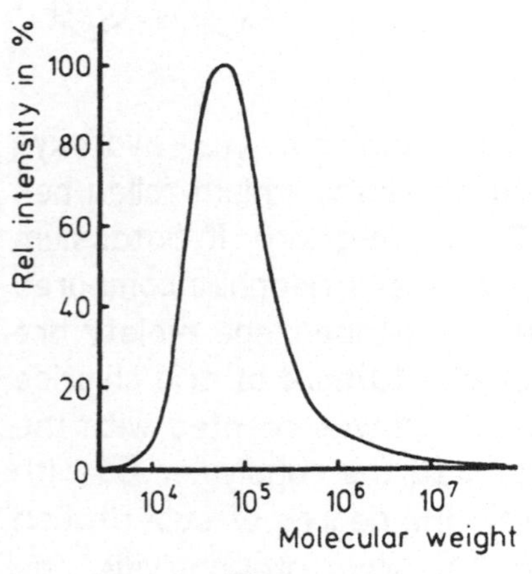

Fig. 1.

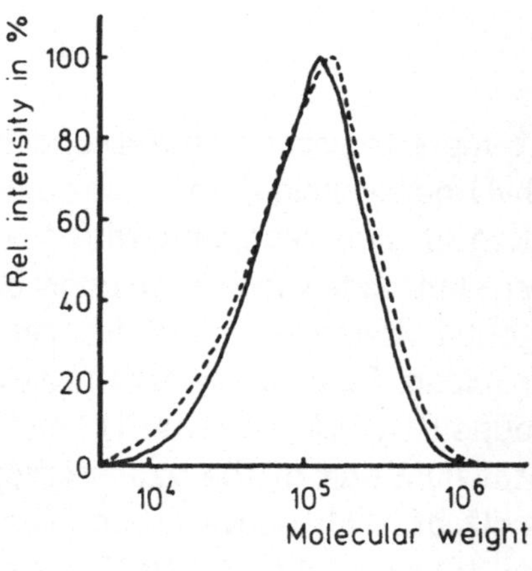

Fig. 2.

Fig. 2.1) GPC-diagram of polymethacrylate **5a** (PMMA calibration)

Fig. 2.2) GPC-diagram of poly[(methylmethacrylate)-co-(methacryloyl chloride)] **3b** (----) and the polymethacrylate **5b** (———) (PMMA calibration)

A polymer with a large amount of NLO-chromophores was prepared by the esterification of pure PMACl with the alcohol **4c**. The reaction in boiling dioxane with pyridine as HCl-acceptor yields the copolymer **6** with about 90 mol% of substituted azobenzene side groups.

The spectroscopic investigations show that **6** is a copolymer with ester and anhydride structures. The polymer is soluble in cyclohexanone. So the anhydride formation is only an intramolecular reaction that leads to six membered rings in the polymer backbone. No intermolecular crosslinking via anhydride formation takes place under the reaction conditions.

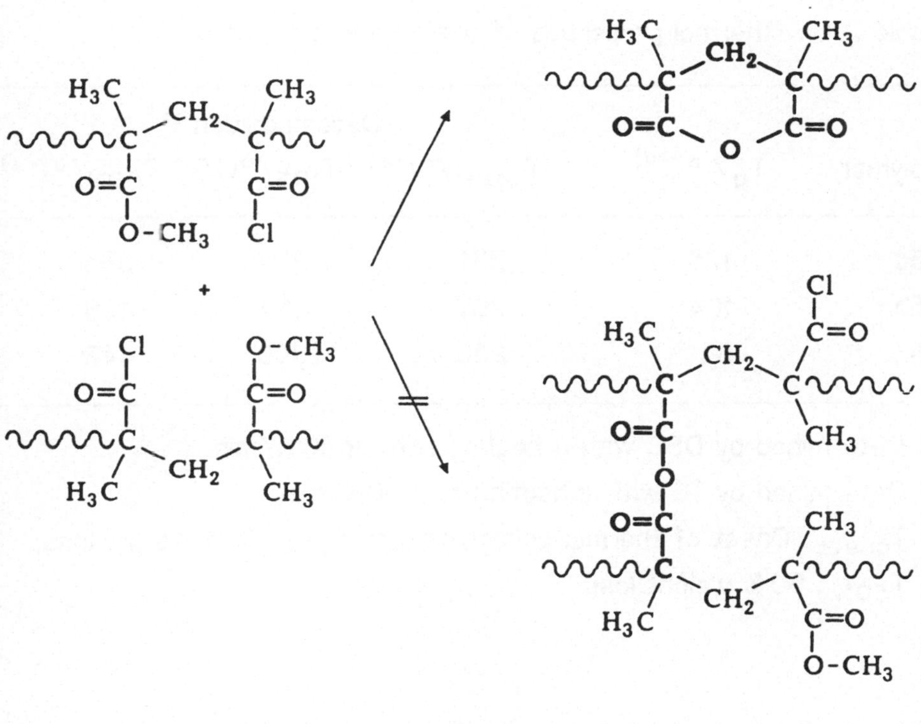

The thermal properties of the polymers **5** and **6** were investigated by differential scanning calorimetry (DSC) and thermogravimetric (TG) measurements. The data are summarized in Tab 2.1. Polymer **5a** with 9 mol% of the azochromophore has a glass transition temperature of 125°C which is considerably higher than the 105°C found for PMMA. This increase of the transition temperature can be explained by some side reactions that lead to a high molecular weight, probably branched or slightly crosslinked polymer fraction which can be seen in the GPC diagram

(Fig. 2.1). The polymethacrylate **5b** with 19 mol% of chromophores has a glass transition temperature of 104°C which is almost the same as in PMMA. The incorporation of larger amounts of NLO-moieties with a spacer of six methylene units leads to a decrease of the glass transition temperature. So polymer **6** with 90 mol% of chromophores exhibits a glass transition at 90°C. Thermogravimetric investigations show that all the polymers start to decompose well above 200°C and therewith show a sufficiently high thermostability for the poling process.

Table 2.1 : Thermal properties of the polymers **5** and **6**

| Polymer | T_g / °C [a] | Decomposition [b] | | |
		T_{onset}/°C [c]	$T_{10\%}$/°C [c]	$T_{50\%}$/°C [c]
5a	125	231	284	384
5b	104	267	310	428
6	84	230	300	442

[a] Determined by DSC with a heating rate of 10 K/min.

[b] Determined by TG with a heating rate 10 K/min.

[c] T_{onset}: Onset of thermal decomposition, $T_{10\%}$: 10 % weight loss, $T_{50\%}$: 50 % weight loss

2.1.3. NLO-measurements

The NLO-properties of the polymers **5b** and **6** have already been tested. After poling in an electrical field of 110 V/μm an electrooptical coefficient ($r_{eff} = r_{33} - r_{31}$) of 9 pm/V at a wavelength of 1300 nm was measured. Polymer **6** with a high content of NLO-active groups showed an electro-optical coefficient of 18 pm/V at 1500 nm after poling. Detailed investigations of the poling process are carried out at the moment.

Scheme 1

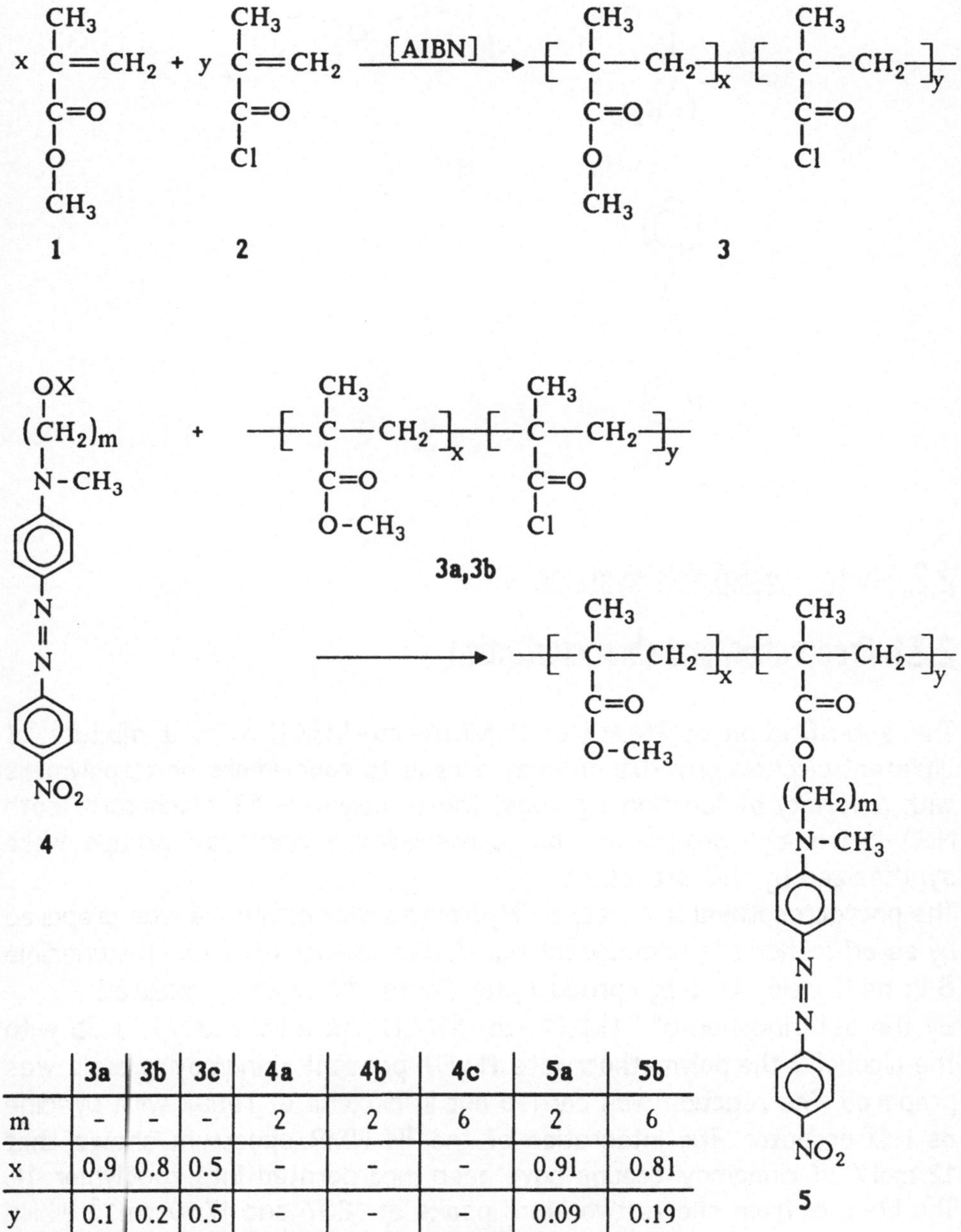

	3a	3b	3c	4a	4b	4c	5a	5b
m	–	–	–	2	2	6	2	6
x	0.9	0.8	0.5	–	–	–	0.91	0.81
y	0.1	0.2	0.5	–	–	–	0.09	0.19
X	–	–	–	H	K	H		

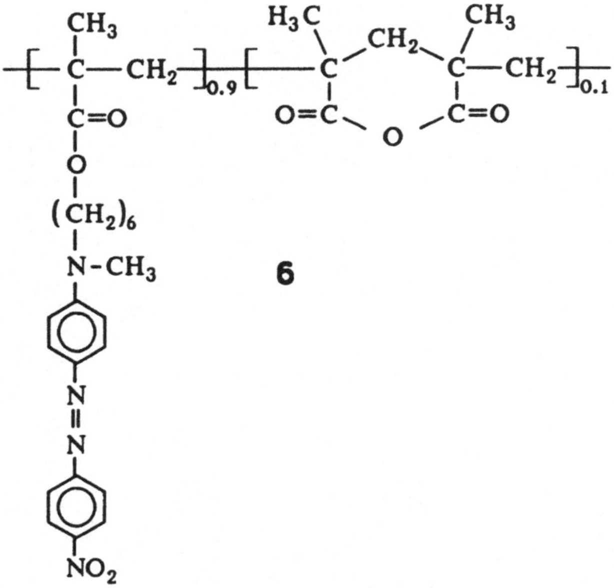

6

2.2. Photocrosslinkable systems

2.2.1. Preparation and characterization

The esterification of PMACl or P(MMA-co-MACl) with a mixture of different alcohols provides an easy access to copolymers or terpolymers with a variety of functional groups. The terpolymers **12** which carry both NLO-active dye groups and photocrosslinkable cinnamoyl groups were synthesized by this procedure.

The photocrosslinkable moiety 6-Hydroxyhexylcinnamate **9** was prepared by esterification of cinnamoyl chloride **7** with an excess of 1,6-hexanediole **8** in 64% yield. As a by-product the diester **10** could be isolated.

By the esterification of P(MMA-co-MACl) (mole ratio 80/20) **3b** with the alcohol **9** the polymethacrylate **11** with pendant cinnamoyl groups was prepared. The reaction was carried out in dioxane at reflux with pyridine as HCl acceptor. The integration of the ^{1}H NMR spectrum shows that 12 mol% of cinnamoyl groups have been incorporated into copolymer **11**. The IR-spectrum shows two small peaks at 1807 and 1760 cm^{-1} which indicate the presence of a small amount of cyclic anhydride structures in the polymer chain as shown in Scheme 2. Polymer **11** is fully soluble and shows no high molecular weight fraction in the GPC diagram. This again shows that only intramolecular anhydride formation and no crosslinking by intermolecular anhydride formation has occured.

The terpolymer **13a** was prepared by simultaneous addition of the azo-chromophore **12a** and the cinnamate **9** to a copolymer of MMA and MACl with 50 mol% of methacryloyl units **3c**. The reaction was again carried out in dioxane/pyridine at reflux temperature.

Scheme 2

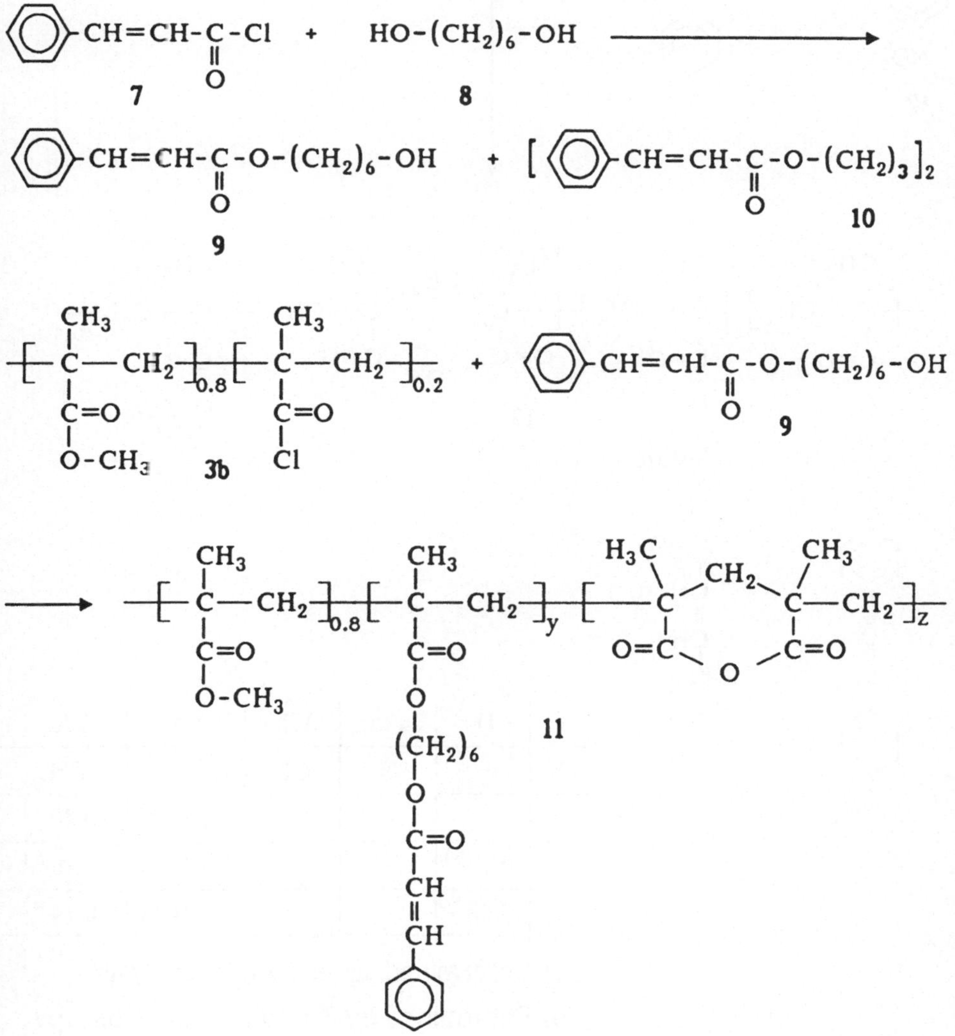

$$C_6H_5-CH=CH-\underset{\underset{O}{\|}}{C}-Cl \quad (7) \quad + \quad HO-(CH_2)_6-OH \quad (8) \quad \longrightarrow$$

$$C_6H_5-CH=CH-\underset{\underset{O}{\|}}{C}-O-(CH_2)_6-OH \quad (9) \quad + \quad \left[C_6H_5-CH=CH-\underset{\underset{O}{\|}}{C}-O-(CH_2)_3 \right]_2 \quad (10)$$

$$\left[\underset{\underset{O-CH_3}{\underset{|}{\underset{C=O}{\underset{|}{\overset{CH_3}{\underset{|}{C}}}}}}}{}-CH_2 \right]_{0.8} \left[\underset{\underset{Cl}{\underset{|}{\underset{C=O}{\underset{|}{\overset{CH_3}{\underset{|}{C}}}}}}}{}-CH_2 \right]_{0.2} \quad (3b) \quad + \quad C_6H_5-CH=CH-\underset{\underset{O}{\|}}{C}-O-(CH_2)_6-OH \quad (9)$$

$$\longrightarrow \left[\underset{\underset{O-CH_3}{\underset{|}{\underset{C=O}{\underset{|}{\overset{CH_3}{\underset{|}{C}}}}}}}{}-CH_2 \right]_{0.8} \left[\underset{\substack{C=O\\|\\O\\|\\(CH_2)_6\\|\\O\\|\\C=O\\|\\CH\\\|\\CH-C_6H_5}}{\overset{CH_3}{\underset{|}{C}}}-CH_2 \right]_{y} \left[\text{anhydride ring} \right]_{z} \quad (11)$$

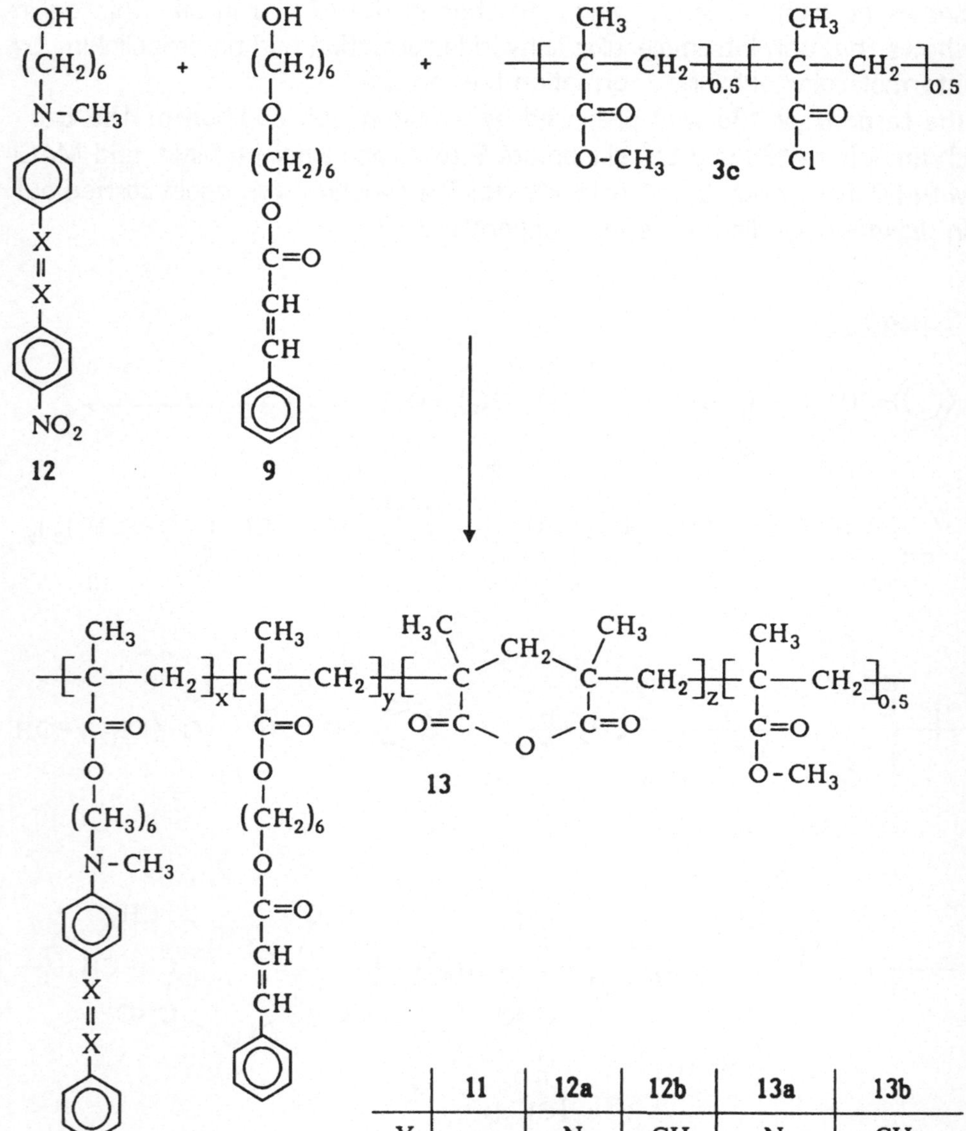

12   **9**   **3c**

13

	11	12a	12b	13a	13b
X	–	N	CH	N	CH
x	–	–	–	0,19 [a]	0,20 [a]
y	0,12 [b]	–	–	0,13 [a]	0,16 [a]
z	0,08 [b]	–	–	0,18 [a]	0,14 [a]

a) Determined by elemental analysis

b) Determined by ^{1}H NMR spectroscopy

From the ^{1}H NMR spectrum of polymer **13a** an amount of 19 mol% of azo-chromophore and 14 mol% of cinnamoyl groups was calculated. Again some anhydride groups are formed during the reaction. The elemental analysis of polymer **13a** fits well with the calculated values for a composition of 50 mol% of MMA, 19 mol% of **12a**, 13 mol% of **9** and 18 mol% of anhydride groups. The GPC analysis shows that polymer **13a** has a $\bar{M}_w$ of 55 000 and a $\bar{M}_n$ of 21 000. These values are no absolute molecular weights and are based on a PMMA calibration (for details see lit. 42).

2.2.2. Irradiation of the photocrosslinkable polymers

The photochemical crosslinking of polymers with pendant cinnamoyl groups is a well known reaction[40]. By [2+2] cycloaddition of two neighbouring cinnamoyl groups truxilic and truxinic acid derivatives are formed. Polymers like poly(vinyl cinnamate) have been frequently used as photoresists in the fabrication of integrated circuits some years ago.

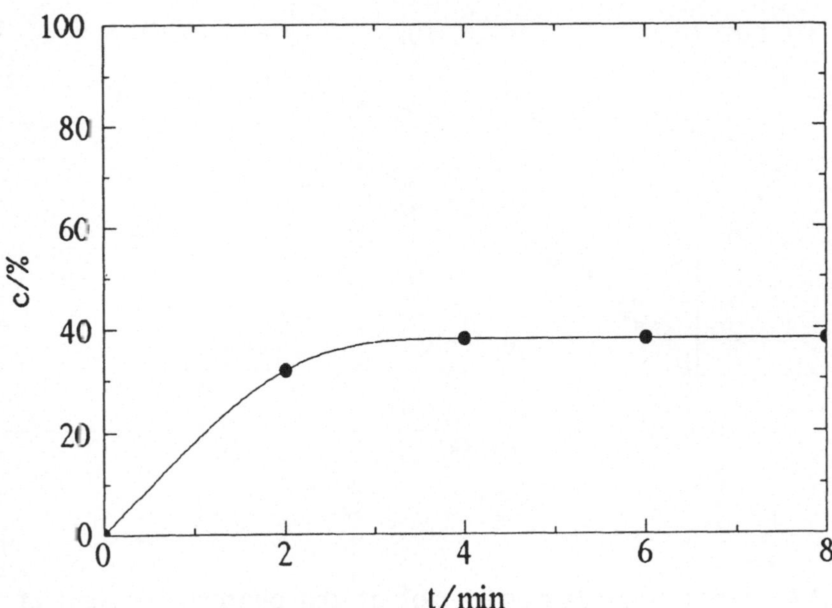

Fig. 2.3)   Time (t) – conversion (c) plot of the photocrosslinking of polymer **11**

We have first studied the photoreaction of the polymethacrylate **11** with 12 mol% of cinnamoyl groups. The progress of the reaction was monitored by the disappearence of the $-C=C-$ IR absorption at 1638 cm^{-1} The results of these measurements are shown in Fig. 2.3.

After three minutes of UV irradiation 38% of the cinnamoyl groups have reacted. Further irradiation does not increase the conversion any more, because the dimerization in the solid state is topochemically controlled and only the fraction of cinnamoyl groups with a second group in the nearest neighbourhood is able to undergo a [2+2] cycloaddition.

The situation in the terpolymers **13** with NLO-active groups is different. Fig. 2.4. shows the results obtained with copolymer **13a** that contains 19 mol% of the azobenzene chromophore **12a** and 13 mol% of cinnamoyl groups. In the terpolymer crosslinking occurs much slower compared to the copolymer **11** without azobenzene moieties. A conversion of 40% is obtained after one hour compared to three minutes in the case of poly-

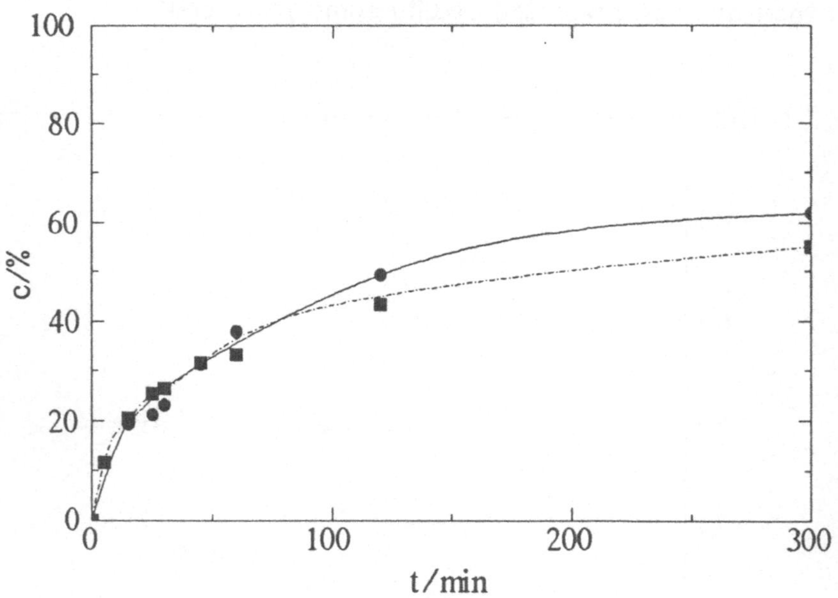

Fig. 2.4) Time(t)-conversion(c) plot of the photocrosslinking of polymer **13a** and the bleaching of the azo chromophore
(————) conversion of cinnamoyl groups
(–·–·–·) decomposition of the azo benzene groups

mer **11**. Nevertheless about 20 % of the cinnamoyl groups have reacted after 15 minutes. Already after one minute of irradiation a polymer film becomes completely insoluble as a result of crosslinking and images can be produced by photolithographic techniques.

The azobenzene derivative **12a** is not stable when irradiated with UV-light. The decomposition of the azochromophore as a function of time is also included in Fig. 2.4. It can be seen, that at longer exposure times large amounts of the chromophore decompose. However, if the polymer is irradiated only for a few minutes, which is enough to make it insoluble, the decomposition of azobenzene groups can be kept very low.

Normally the reaction rate of the [2+2] cycloaddition can be enhanced by the addition of triplet sensitizers like Michlers´ ketone[40]. Unfortunately, no effect on the reaction rate was found when Michlers´ ketone was added to polymer **13a**. Attempts to crosslink polymer **13a** at wavelength > 320 nm also failed. The azo groups are stable under these conditions, but almost no crosslinking occured.

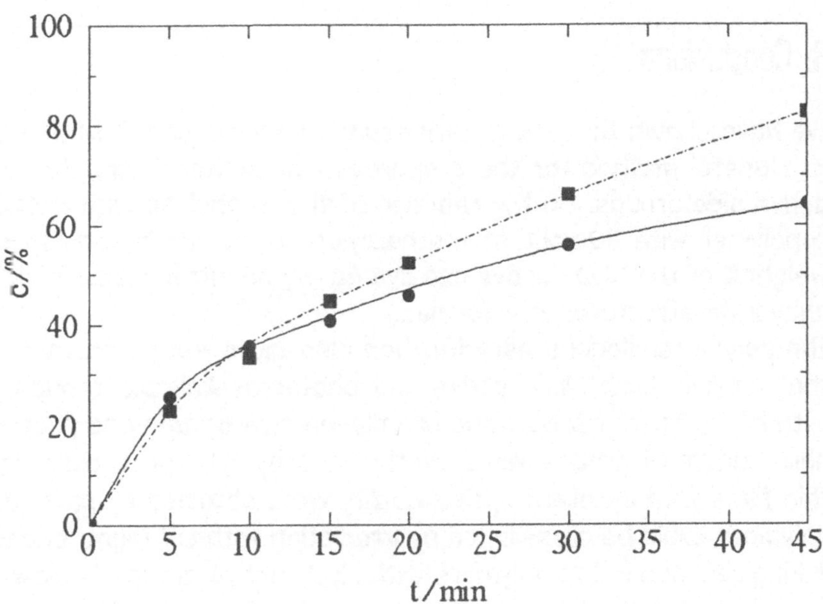

Fig. 2.5) Time (t)-conversion (c) plot of the photocrosslinking of polymer **13b** and the bleaching of the stilbene chromophore
(———) conversion of cinnamoyl groups
(–·–·–·) decomposition of the stilbene groups

The crosslinking reaction of the terpolymer **13b** with stilbene chromophores is similar to the polymer with azobenzene groups. Fig. 2.5. shows that the crosslinking reaction is much slower than in polymer **11** without NLO chromophore, but slightly faster than in polymer **13a**. Unfortunately the stilbene chromophore **12b** is even less stable towards UV light than the azobenzene moiety in polymer **13a**. After ten minutes, 33% of the stilbene groups have decomposed.

2.2.3. NLO-measurements

Preliminary NLO measurements have been carried out with polymer **13a**. After poling at the glass transition temperature of 125°C in an electrical field of 95 V/μm an electrooptical coefficient ($r_{eff.} = r_{33} - r_{13}$) of 7 pm/V at a wavelength of 1150 nm was measured. Investigations of the long-time stability of the chromophore orientation in the crosslinked matrix are in progress.

3. Conclusions

We have shown that the esterification of PMACl and P(MMA-co-MACl) is a useful method for the preparation of polymethacrylates with NLO-active side groups. So the reaction of the alcohol **4c** with PMACl yields a copolymer with 90mol% of methacrylate units which corresponds to 85 weight% of the NLO-active azo dye **4c**. By an intramolecular side reaction anhydride structures are formed.

The polymeranalogous esterification also gives easy access to polymers that contain both NLO-active and photocrosslinkable groups. Polymers with NLO-active azobenzene or stilbene side groups and photocrosslinkable cinnamoyl groups were synthesized by that procedure, from which thin films with excellent optical quality were obtained by spin coating. The polymers could be crosslinked by irradiation with UV-light, but the crosslinking reaction in the polymers with NLO-active groups is slow compared to polymers which carry only cinnamoyl side groups. Therefore photochemical decomposition of a small amount of NLO chromophores during crosslinking could not be avoided in these systems. Further studies with new NLO-chromophores and photocrosslinkable groups absorbing at longer wavelength where the NLO moieties are stable are in progress.

Acknowledgement:

We thank Heike Kilburg and Karl - Heinz Etzbach (BASF AG, Ludwigshafen) who prepared the NLO - active dyes and Peter Kersten (SEL, Stuttgart) for measuring the electrooptical coefficient.
This work was supported by the "Bundesministerium für Forschung und Technologie" (BMFT) as part of the materials research program.

4. Literatur

1) P.N. Prasad, D.J. Williams, *Introduction to nonlinear optical effects in molecules and polymers*, J. Wiley and Sons, New York 1991

2) S.R. Marder, J.E. Sohn, G.D. Stucky (Eds.), ´*Materials for nonlinear optics: Chemical perspectives*´ ACS Symposium Series, Washington 1991

3) D.S. Chemla, J. Zyss (Eds.), ´*Nonlinear optical properties of organic molecules and crystals*´, Academic Press, New York 1987

4) G.R. Meredith, J.G. van Dusen, D.J. Williams, *Macromolecules* **15**, 1385 (1982)

5) D.R. Robello, *J. Polym. Sci., Polym. Chem. Ed.* **28**, 1 (1990)

6) R.N. De Martino, E.W. Choe, G. Khanarian, D. Haas, T. Leslie, G. Nelson, J. Stamatoff, D. Stuetz, C.C. Teng, H.N. Yoon in P.R. Prasad, D.R. Ulrich (Eds.) ´*Nonlinear optical and electroactive polymers*´, p. 169, Plenum Press 1988

7) G.R. Möhlmann, *Synthetic Metals* **37**, 207 (1990)

8) W. Köhler, D.R. Robello, C.S. Willand, D.J. Williams, *Macromolecules* **24**, 4589 (1991)

9) M. Amano, T. Kaino, *Electron. Lett.* **26**, 981 (1990)

10) D. Jungbauer, I. Teraoka, D.Y. Yoon, B. Reck, J.D. Swalen, R. Twieg, C.G. Willson, *J. Appl. Phys.* **69**, 8011 (1991)

11) M.A. Mortazavi, A. Knoesen, S.T. Kowel, B.G. Higgins, A. Dienes, *J. Opt. Soc. Am. B*, **6**, 733 (1989)

12) H.L. Hampsh, J. Yang, G.K. Wong, J.M. Torkelson, *Macromolecules* **21**, 526 (1988)

13) H.L. Hampsh, J. Yang, G.K. Wong, J.M. Torkelson, *Polym. Commun.* **30**, 40 (1989)

14) G.R. Möhlmann, W.H.G. Horsthuis, A.Mc. Donach, M.J. Copeland, C. Duchet, P. Fabre, M.B.J. Diemeer, E.S. Trommel, F.M.M. Suyten, E. van Tomme, B. Baquers, P. van Daele, *Proc. SPIE 1337, ´Nonlinear optical properties of organic materials III´*, p. 215 (1990)

15) K.D. Singer, M.G. Kuzyk, W.R. Holland, J.E. Sohn, S.J. Lalama, R.B. Comizzoli, H.E. Katz, M.L. Schilling, *Appl. Phys. Letters* **53**, 1800 (1988)

16) M. Eich, A. Sen, H. Looser, G. Bjorklund, J.D. Swalen, R. Twieg, D.Y. Yoon, *J. Appl. Phys.* **66**, 2559 (1989)

17) M. Eich, B. Reck, D.Y. Yoon, C.G. Willson, G.C. Bjorklund, *J. Appl. Phys.* **66**, 3241 (1989)

18) D. Jungbauer, B. Reck, R. Twieg, D.Y. Yoon, C.G. Willson, J.D. Swalen, *Appl. Phys. Letters* **56**, 2610 (1990)

19) G.C. Bjorklund, S. Ducharme, W. Fleming, D. Jungbauer, W.E. Moerner, J.D. Swalen, R.J. Twieg, C.G. Willson, D.Y. Yoon in *´Materials for Nonlinear Optics´, ACS Symposium Series* 455 (1991)

20) B.K. Mandal, J. Kumar, J.C. Huang, S. Tripathy, *Makromol. Chem., Rapid Commun.* **12**, 63 (1991)

21) B.K. Mandal, Y.M. Chen, J.Y. Lee, J. Kumar, S. Tripathy, *Appl. Phys. Letters* **58**, 2459 (1991)

22) B.K. Mandal, J.Y. Lee, X.F. Zhu, F. Xiao, Y.M. Chen, E. Prakeenavincha, J. Kumar, S. Tripathy, *Synth. Metals* **43**, 3143 (1991)

23) D. Braun, G. Arcache, R.J. Faust, W. Neumann, *Makromol. Chem.* **114**, 51 (1968)

24) D. Braun, G. Arcache, *Makromol. Chem.* **148**, 119 (1971)

25) O.F. Olaj, J.W. Breitenbach, I. Hofreiter, *Makromol. Chem.* **91**, 264 (1966)

26) J. Brandrup, E.H. Immergut (Eds.), *´Polymer Handbook´*, 3rd Ed., J. Wiley and Sons 1989

27) C.S. Marvel, C.L. Levesque, *J. Am. Chem. Soc.* **61**, 3244 (1939)

28) M. Vrancken, G. Smets, *J. Polym. Sci.* **14**, 521 (1954)

29) S. Boyer, A. Rondeau, *Bull. Soc. Chim. Fr.* **25**, 240 (1958)

30) R.C. Schulz, P. Elzer, W. Kern, *Makromol. Chem.* **42**, 189 (1960)

31) S. Rondeau, G. Smets, M.C. De Wilde-Delvaux, *J. Polym. Sci.* **24**, 261 (1957)

32) P.E. Blatz, *J. Polym. Sci.* **58**, 755 (1962)

33) C.M. Paleos, S.E. Filippakis, G. Margomenou-Leonidopoulou, *J. Polym. Sci., Polym. Chem. Ed.* **19**, 1427 (1981)

34) C.M. Paleos, G. Margomenou-Leonidopoulou, S.E. Filippakis, A. Malliaris, P. Dais, *J. Polym. Sci., Polym. Chem. Ed.* **20**, 2267 (1982)
35) H. Kamogawa, *J. Polym. Sci. A1* **7**, 2458 (1969)
36) I. Yahagi, H. Watanabe, K. Sanui, N. Ogata, *J. Polym. Sci., Polym. Chem. Ed.* **25**, 727 (1987)
37) N. Ogata, K. Sanui, H. Watanabe, I. Yahagi, *J. Polym. Sci., Letters Ed.* **23**, 349 (1985)
38) P. Strohriegl, *Mol. Cryst. Liq. Cryst.* **183**, 261 (1989)
39) St. Polowinski, *Acta Polymerica* **35**, 193 (1984)
40) A. Reiser, ´Photoactive Polymers´, J. Wiley and Sons, New York 1989
41) H. Müller, O. Nuyken, P. Strohriegl, *Makromol. Chem., Rapid Commun.* **13**, 125 (1992)
42) H. Müller, I. Müller, O. Nuyken, P. Strohriegl, *Makromol. Chem., Rapid Commun.*, 1992, in press

Polymer Filled Porous Glass Plates - A Composite Material for Nonlinear Optics

R. Schubert, G. Franke, Ch. Kaps
Institute of Inorganic and Analytical Chemistry
Friedrich Schiller University of Jena, F.R.G.

Abstract

Porous silica glasses with a branched system of connected nanometer pores will be useful for the fabrication of optical composite materials, if a new technology is applied: the reactive deposition of small particles (<6nm) of inorganic semiconductors or the physical deposition of organic dyes and the filling of the rest volume of the pores with organic polymers. The preparation and optical properties of such composite materials like the cut-off behaviour and the transparency in the UV-VIS range of the light spectrum are described.

1 Introduction

The development of efficient devices for electro-optical and opto-optical processing requires new materials with large nonlinear optical effects [STE 88]. Inorganic semiconductors, π-conjugated organic polymers and special organic dyes have been investigated as basic materials for manufacture of materials with third order nonlinear optical properties (e.g. intensity-dependent refractive index). Microcrystalline inorganic semiconductors seem to be especially suitable, because of their large optical nonlinearities and ultrafast nonlinear response times [STE 88, WIL 88]. These properties can be applicated, if materials would be available, where the semiconductor microcrystallites are stabilised in a solid and transparent matrix.

A new manufacturing method - the reactive deposition of semiconductors particles under soft chemistry conditions (e.g. at temperatures up to 250°C), but also the physical deposition of organic dyes in mesoporous silica glasses (pore diameter 3-10 nm) - offers the possibility to fabricate new optical composite materials with tailored properties.

For materials with optical quality three technological steps are necessary:

1) the preparation of a mechanical stable porous glass with defined parameters of the pore system
2) the deposition of the optical active components (e.g. inorganic semiconductors, organic dyes) in the pores
3) complete filling of the rest volume of pores with organic polymers

2 Preparation of optical composite materials

The first technological step, the manufacture of porous silica glass plates is based on the well known Vycor-process [NOR 44]:
A sodium boro-silicate glass with a composition in the range of 70 mol% SiO_2, 20-25 mol% B_2O_3 and 5-10 mol% Na_2O is phase separated by a annealing process at temperatures in the range of glass transformation and the water soluble sodiumborate phase is leached with an 3 M aqueous solution of HCl and NH_4Cl at a temperature of 90°C. Systematical experiments confirm a very sensitive dependence of the parameters of the pore system (pore size, pore volume and pore surface area) and of the mechanical and optical properties of the porous glass substrates on glass composition and annealing conditions. Preparation conditions for homogenous, mechanically stable samples with defined porosity are deduced.
Concerning the technological step 2, we used two methods for the deposition of metal chalcogenides in the pores of porous silica glass plates:

- the precipitation of heavy metal salts in the pores from aqueous solutions (e.g. 0.1 M aquous solutions of $CdCl_2$, $HgCl_2$, or $Pb(CH_3COO)_2$) or 0.1 M solutions $SbCl_3$ or $Bi(NO_3)_3$ in 3 M HCl and the following reaction with chalcogen-hydrogen gas (H_2S, H_2Se) to the chalcogenides at room temperature (equ. 2.1-2.3). Other reactions of this type are described by BORRELLI [BOR 87].

$$CdCl_2 + H_2Se \longrightarrow CdSe + 2\ HCl \tag{2.1}$$

$$HgCl_2 + H_2Se \longrightarrow HgSe + 2\ HCl \tag{2.2}$$

$$Pb(CH_3COO)_2 + H_2S \longrightarrow PbS + 2\ CH_3COOH \tag{2.3}$$

- thermal decomposition of aqueous solutions of metal-chalcogen complexes in the pores (equ. 2.4, 2.5)

$$[Cd(NH_3)_6]SSeO_3 + H_2O \xrightarrow{70°C} CdSe + (NH_4)_2SO_4 + 4NH_3 \tag{2.4}$$

$$[Hg(SC(NH_2)_2)_2]Cl_2 + H_2O \xrightarrow{100°C} HgS + OC(NH_2)_2 + SC(NH_2)_2 + 2HCl \tag{2.5}$$

The seleno-sulphate complex in equ. 2.4 was prepared, solving 0.025 mol selen powder in 100 ml of an 1 M aqueous Na_2SO_3 solution under heating and then adding of ammonia and of an 0.01 M aqueous solution of $CdCl_2$. The thio-urea complex of mercury is obtained when adding a weak acidic 0.1 M aqeous solutions of thio-urea and an 0.01 M aqueous solution of $HgCl_2$.

Both methods allow a wide variation of applicable semiconductors, especially thermical sensitive compounds like HgSe (equ. 2.2) and HgS (equ. 2.5) could be precipitaded in the porous substrate glass.

Organic dyes like rhodamine 6G or dispersion red have been deposited in the pores, soaking the porous glass plates with saturated solutions of the dyes in

66

ethylacetate and then drying the glass plates at 80°C.

For the third technological step - the stuffing of the porous glass plates by organic polymers, the porous glasses with the deposited semiconductors particles are filled with triethylene-glycol-dimethacrylate (TEGMA) and then polymerized thermally at 100°C. A pretreatment of the inner silica surface of the dried porous glass with a vapour of 3-methacryl-oxypropyletrimethoxysilane (MAOPTS) at 100°C under a reduced pressure of 1 kPa for 2 hours has been carried out before filling the pores with TEGMA (equ. 2.6). This procedure causes a hydrophobation of the silica surface and an introduction of functional groups ($CH_2=CH-$) on the glass surface, wich are able to polymerize. This is necessary to improve the contact between the hydrophilic inner surface of the porous silica and the hydrophobic polymer and to achieve a fixing of the polymer in the pores by forming of chemical bonds to the glass surface [KAP 92].

$$O_{3/2}\underset{glass}{Si}-OH \ + \ CH_2=\overset{CH_3}{C}-COO-(CH_2)_3-\underset{OCH_3}{\overset{OCH_3}{Si}}-OCH_3 \ \xrightarrow{100°C} \ (2.6)$$

$$O_{3/2}\underset{glass}{Si}-O-\underset{OCH_3}{\overset{OCH_3}{Si}}-(CH_2)_3-OOC-\overset{CH_3}{C}=CH_2 \ + \ CH_3OH$$

3 Characterization of the optical composite materials

The preparation of the porous glass from the base glass, containing 7 mol% Na_2O, 23 mol% B_2O_3 and 70 mol% SiO_2, annealed at 510°C for 4 hours and leached as descripted before lead to mechanical stable glass plates up to 12 mm thickness with a pore diameter of (6±1) nm, a narrow pore size distribution (fig. 3.1), a pore volume

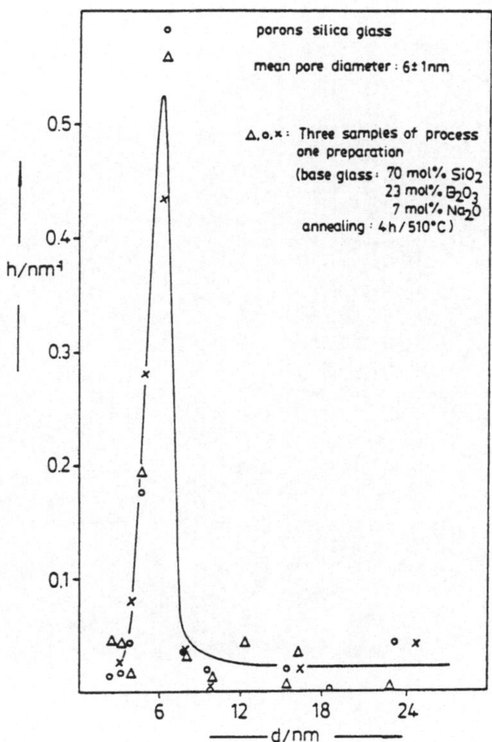

Fig. 3.1 Distribution probability of the pore diameter of three samples of one preparation process correspondending to N_2-sorption measurement

of 0.15 cm^3/g and a BET-surface area of 100 m^2/g. These parameters of the pore system were measured with N$_2$-sorption method.

The transmission electron micrograph of such a porous glass shows a regular branched system of connected cylindrical pores in the bulk glass (fig. 3.2) .

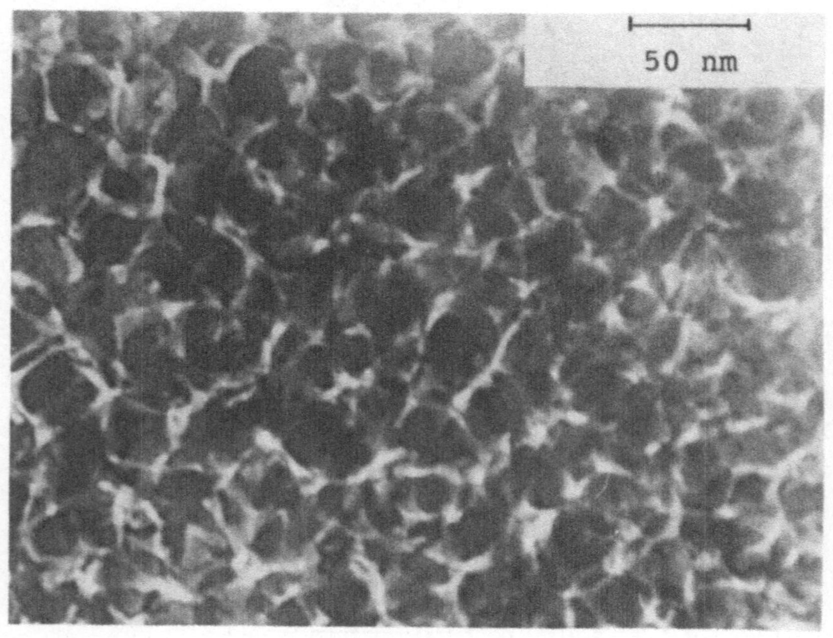

Fig. 3.2 Image of the pore system, estimated by direct transmission electron microscopy (microscope: Jeol CX 100, glass preparation see text)

In the UV-VIS spectra of composite materials, containing various sulphides (fig. 3.3) and selenides (fig. 3.4), precipitated by reaction with hydrogenchalcogenides, different transmission edges occure, which are caused by the band gaps of the semiconductors. The porous glasses with CdS, Sb_2S_3 and Bi_2S_3 show a transmission behaviour corresponding to the band gap of the bulk materials.

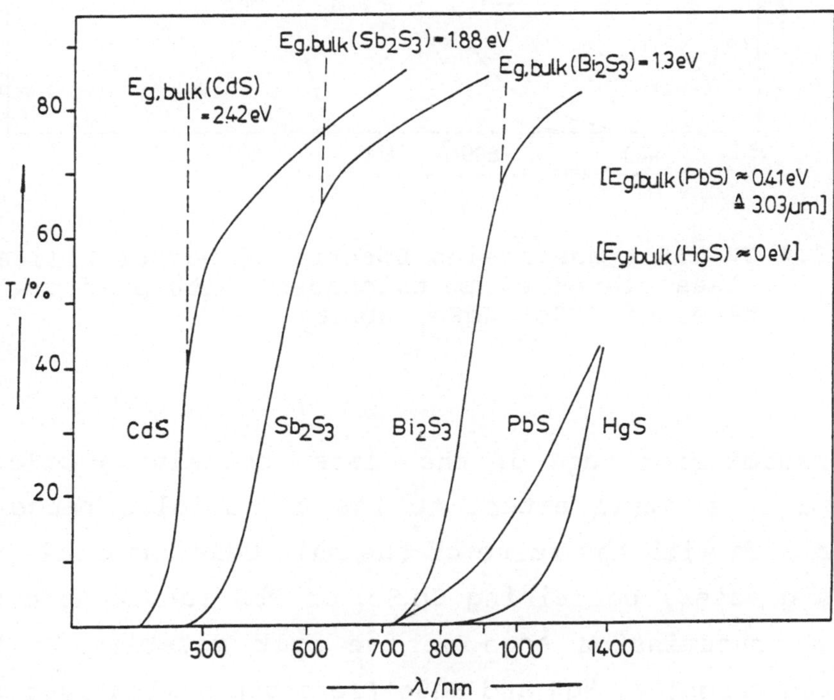

Fig. 3.3 UV-VIS transmission spectra of porous silica glass plates (3 mm thickness) with precipitates of CdS, Sb_2S_3, Bi_2S_3, PbS and HgS

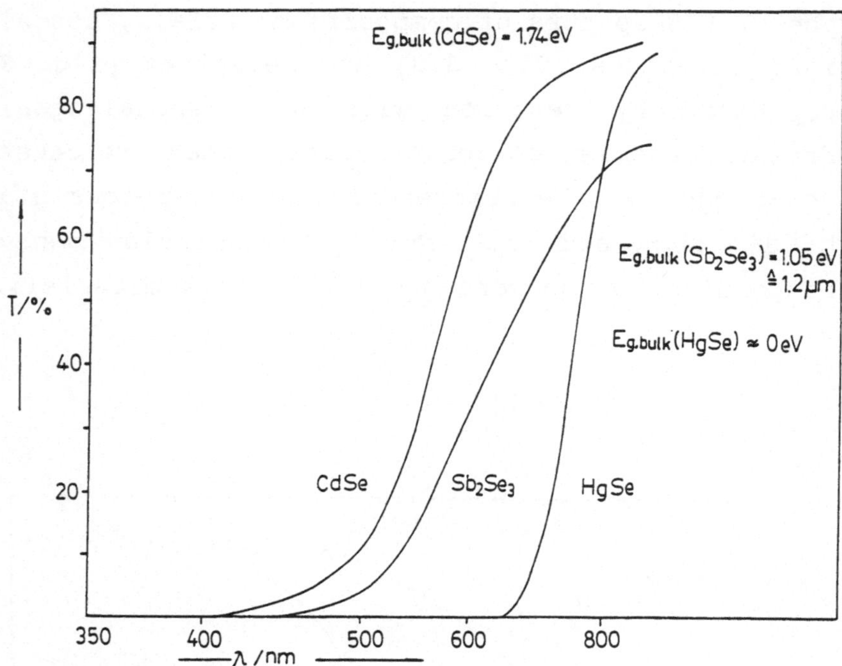

Fig. 3.4 UV-VIS transmission spectra of porous silica
glass plates (3 mm thickness) with precipitates of CdSe, HgSe, Sb_2Se_3

The transmission edge of the glass, containing CdSe is
shifted in a small extent to the ultraviolet range in
comparision with the value of the bulk CdSe material. The
porous glasses, containing Sb_2Se_3 or PbS (semiconductors
with a transmission edge of the bulk material in the
infrared range) or HgS and HgSe (zero gap semiconductors)
exhibit a large extent of blue shift of the transmission
edge of the precipitated particles in comparision with
the bulk material. This behaviour is also eventually
influenced by additional presence of a small amount of
elementary selen.

First hints of microcrystallinity of precipitated CdS in the pores were found, when using a 10^{-3} M aqueous solution of $CdCl_2$, to introduce the heavy metal ions in the pores and then precipitating CdS by the reaction with H_2S. The UV-VIS spectrum of such a glass plate shows an exciton-like structure in the transmission edge probably due quantum confinement.

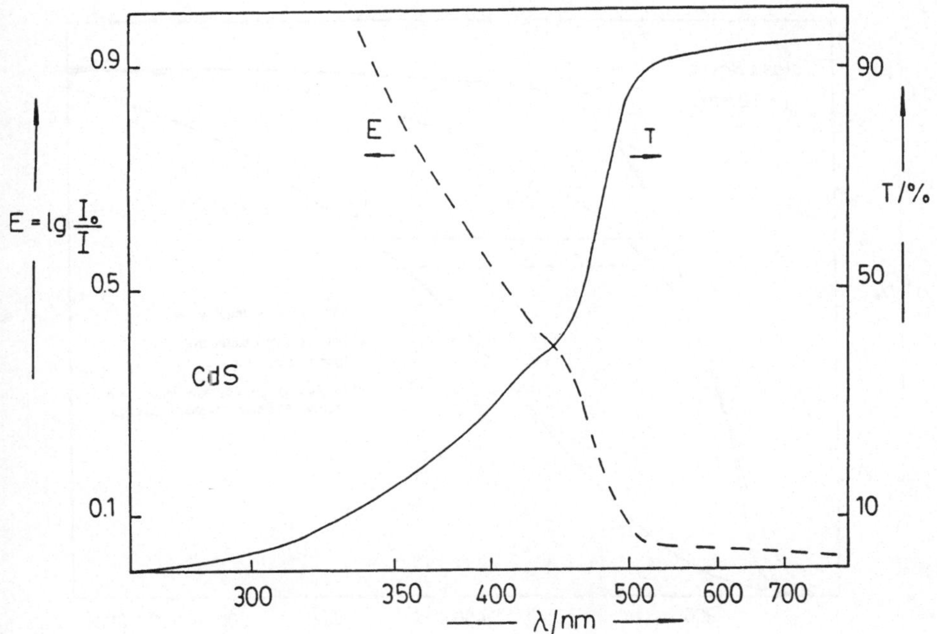

fig. 3.5 UV-VIS spectrum of a porous glass plate (0.3mm thickness) with deposited CdS particles, proceeding from the penetration of an 10^{-3}M aqueous solution of $CdCl_2$ into the pores

Especially in the spectra of the porous glasses, filled with CdS, Sb_2S_3 and Bi_2S_3 a bending of the transmission edge at transmissions higher than 60% appears, demonstrating the influence of the light scattering effects of the open pore system of the substrate silica glass and the necessity of the third technological step of manufacture of an optical material - the polymer filling of the rest pore volume.

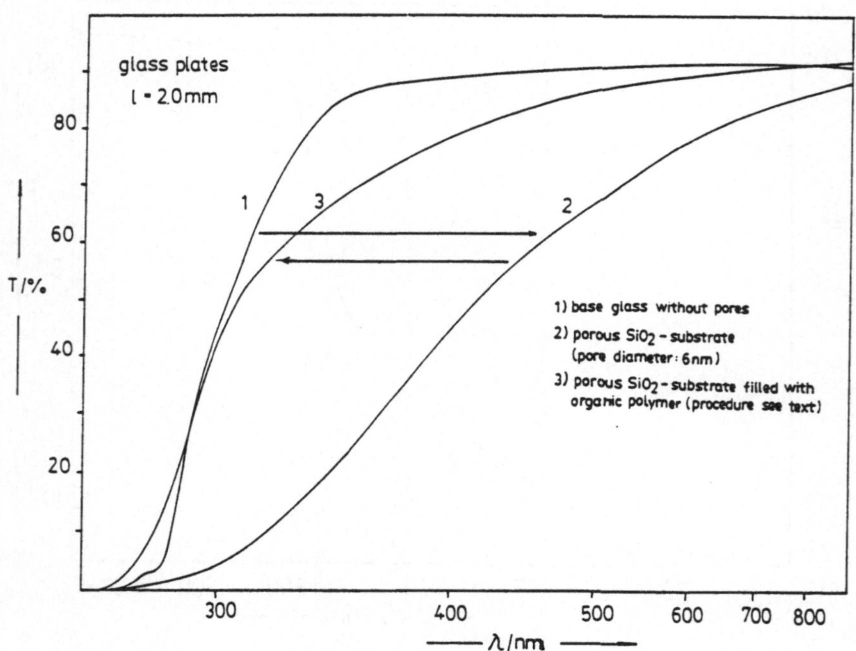

Fig. 3.6 UV-VIS transmission spectra of glass plates in different states of the preparation technology for composite glasses

The comparation of UV-VIS spectra of the unleached base glass, the porous glass and the glass with polymer filled pores (fig. 3.6) demonstrates that the polymer filling of the empty rest volume of the pores in the glass plates leads to a remarkable advance of the optical transparency near to the unporous base glass in a wide range of visible spectrum.

The intensity dependence of transmission at a wavelength in the range of the transmission edge as an resonant third order nonlinear optical effect has been observed at porous glasses with reactive deposited CdSe or physically

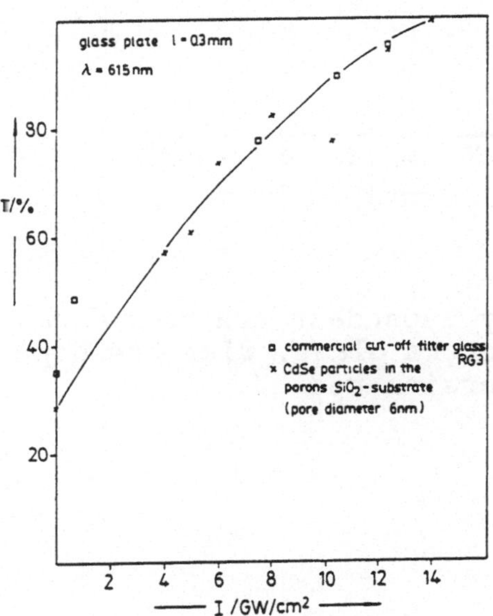

Fig. 3.7 Intensity dependent transmission of a polymer filled porous glass, with deposited CdSe particles

deposited dispersion red (fig. 3.7, 3.8). Fig 3.7 shows, that the magnitude of the intensity dependence of the transmission of the CdSe doped porous glass is comparable with that of a commercial cut-off filter glass with similar linear optical properties.

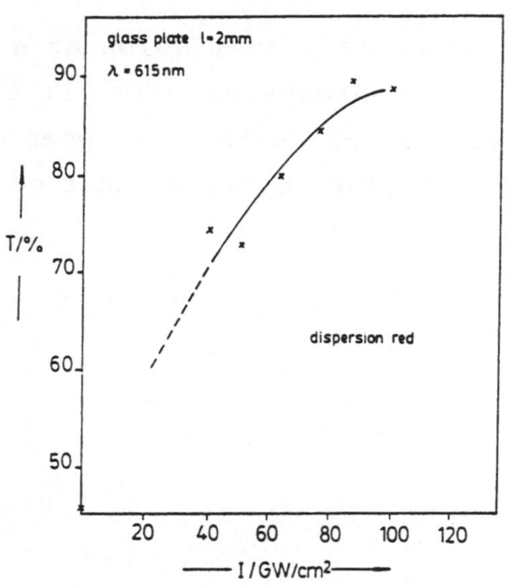

Fig. 3.8 Intensity dependent transmission of a polymer filled porous glass, with the deposited organic dye dispersion red

4 Conclusions

Resuming, it can be stated that optical composite glasses consisting of semiconductor particles or organic dye molecules, dispersed though a mesoporous silica matrix and embedded in a convenient organic polymer, represent an attractive type of new optical materials. In contrast to the limited melting technique of the commercial cut-off filter glasses, in wich only CdS and CdSe can be precipitated, the soft chemistry of reactive deposition of semiconductors or the physical deposition of organic dyes in porous glasses reveals a pronounced multiplicity of optical active compounds with a useful variety of different optical parameters.

Acknowledgements

This work was supported by the Bundesministerium für Forschung und Technologie under Project No. 03M2729 3.

Literature

[BOR 87]   BORELLI N.F., LUONG J.C., SPIE <u>866</u> (1987) 104

[KAP 92]   KAPS CH., FRANKE G., SCHUBERT R., P 4225952.5, 1992

[NOR 44]   NORDBERG E.M., J.Amer.Ceram.Soc. <u>27</u> (1944) 299

[STE 88]   STEGEMAN G.I. etal, J.Light Wave Techn. <u>6</u>, 6 (1988) 953

[WIL 88]   WILLIAMS V.S. etal, J.Mod.Opt. <u>35</u>, 12 (1988) 1979

Crystalline Organic Semiconductors: A New Class of Materials for Photonic Devices

W. Kowalsky and C. Rompf
Abteilung Optoelektronik, University of Ulm,
D-7900 Ulm, Federal Republic of Germany

1. Introduction

Today's electronic and photonic devices are mainly based on inorganic materials. In the field of optoelectronic devices III-V compound semiconductors are predominant because of the direct band structure of these alloys. However, organic compounds may emerge to a remarkable enrichment. In the near future they will support or partly replace inorganic materials. Especially polymers and cristalline organic semiconductors are promising candidates for application in photonic devices. The experimental characterization of two well established aromatic compounds, 3,4,9,10-perylenetetracarboxylic dianhydride (PTCDA) and 1,4,5,8-naphthalenetetracarboxylic dianhydride (NTCDA), and optoelectronic device fabrication are subjects of this contribution. We first discuss the deposition of crystalline organic layers by UHV sublimation. Optimized growth conditions even facilitate the growth of homogeneous quantum well structures. In view of applications in optoelectronic devices ohmic and Schottky contacts and organic-inorganic heterostructures are investigated. A planar photodetector is presented as a first example.

2. Sample Preparation

The crystalline layers are deposited by UHV sublimation of the purified materials. Because of strong covalent bonding of the atoms of inorganic semiconductors epitaxial growth of these materials requires lattice matching to avoid dislocations. In contrast, the molecules of organic crystals are bonded by relatively weak van der Waals forces. This advantage facilitates the multilayer deposition of organic solids with different lattice properties to heterostructures and the use of any substrate.

2.1 Materials Survey

The planar molecular structures of PTCDA and NTCDA are shown in Fig. 2.1. Optimized growth conditions allow the deposition of very homogeneous films. PTCDA (NTCDA) forms molecular stacks with an intermolecular separa-

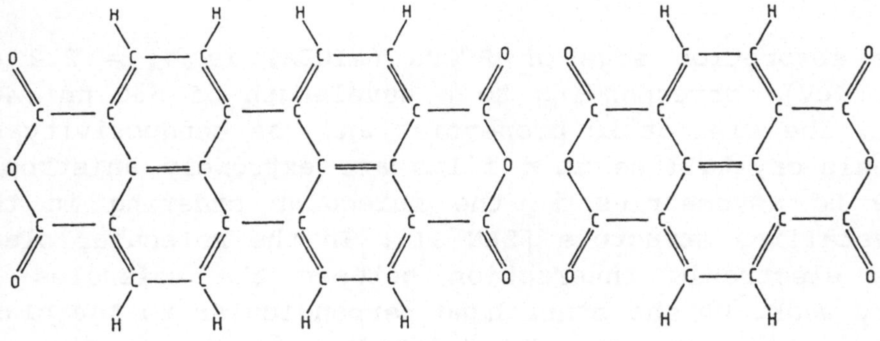

Fig. 2.1: Molecular structures of PTCDA and NTCDA.

tion of 0.321 nm (0.3506 nm). The unit cell sketched in Fig. 2.2 is monoclinic, with the crystal axis a = 0.372 nm (0.7867 nm), b = 1.196 nm (0.5305 nm), c = 1.734 nm (1.2574 nm), and ß = 98.8° (72.73°). Each cell contains two molecules which are oriented nearly perpendicular. In vacuum deposited films the (102) planes are 11° tilted (parallel) to the substrate surface [BOR 90, FOR 84, MÖB 89].

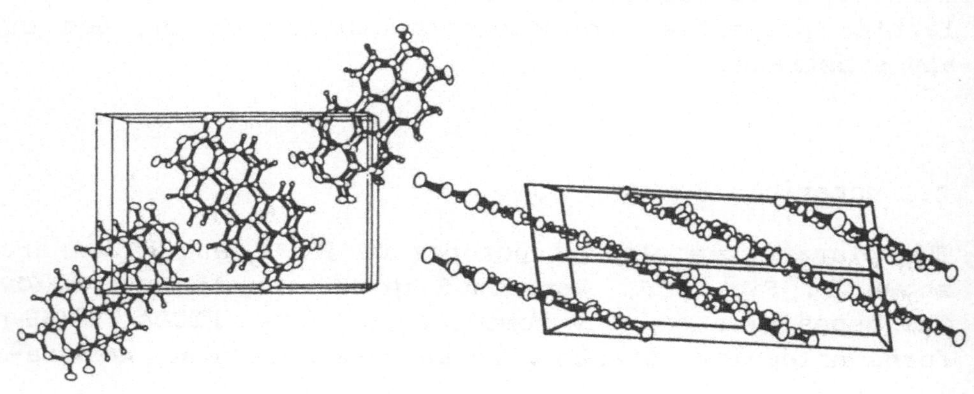

Fig. 2.2: Unit cell of PTCDA.

The absorption edge of PTCDA (NTCDA) is W_a = 2.2 eV (3.1 eV) corresponding to a wavelength of 560 nm (400 nm). The dielectric properties and the conductivity of single-crystalline thin films are extremely anistropic due to asymmetries in the molecular ordering in the crystalline structure [ZAN 91]. In the molecular plane the electronic interaction between the molecules is very weak. On the other hand perpendicular to the plane the coupling of the delocalized electrons in the extended n system facilitates carrier transport along the stacks. As a result, the conductivity in the molecular plane is about three to six orders of magnitude lower than the conductivity along the molecular stacks. This molecular stacks. This characteristic is promising for current injection into optoelectronic device because current spreading is suppressed without additional isolation or lateral structuring.

In the long wavelength region beyond the gap wavelength the absorption of PTCDA and NTCDA is negligible. This is encouraging for waveguide applications in OEICs at 800 nm or 1.3 μm to 1.55 μm wavelength for optical communication systems.

2.2 Layer Deposition

The UHV deposition of crystalline organic films at growth rates of about 0.1 nm/s is reproducible and allows the growth of complex multilayer structures. First experiments were carried out in a conventional evapo-

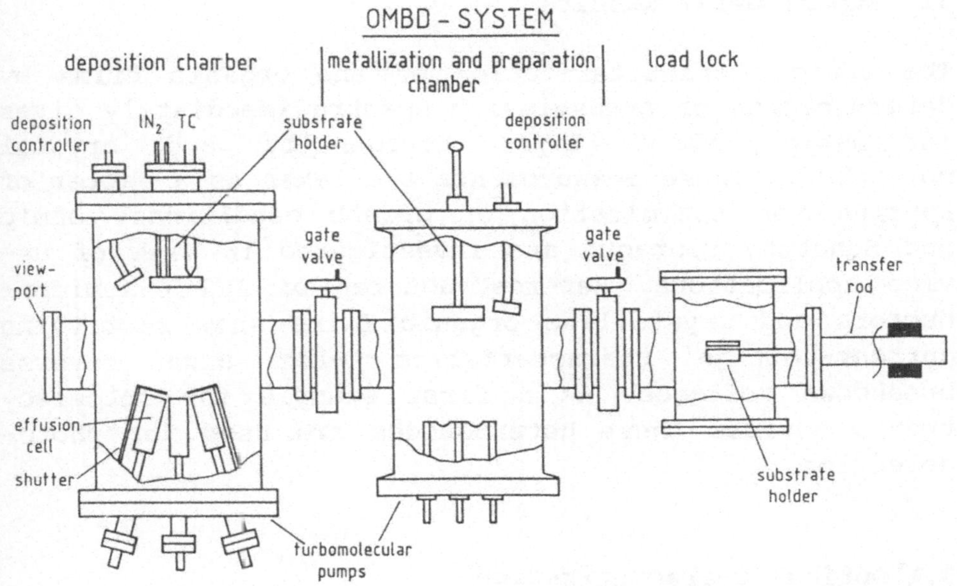

Fig. 2.3: OMBD system.

ration system. To attain improved vacuum conditions, more stable evaporation sources, and to avoid ventilation of the system for sample loading a three chamber OMBD (organic molecular beam deposition [SO 90/1]) system is developed. This concept essentially corresponds to the MBE (molecular beam epitaxy) of inorganic III-V semiconductors. As shown in Fig. 2.3 it consists of a load lock module with magnetically coupled transfer rod, a preparation and metallization chamber, and the growth chamber. The prepurified PTCDA and NTCDA powder are sublimed from pyrolithic boron nitride crucibles supplied with mechanical shutters. The substrate temperature can be varied from 300 K to 77 K by liquid nitrogen cooling. At a PTCDA (NTCDA) source temperature of 375 °C to 425 °C (215 °C to 245 °C) a deposition rate of 0.05 nm/s to 0.3 nm/s (0.1 nm/s to 0.8 nm/s) is obtained.

3. Experimental Results

The optical characterization of the organic films by determination of transmission spectra immediately gives information about layer homogeneity and crystal morphology. These measurements are taken as a matter of routine for optimization of growth conditions. Ohmic and Schottky contacts are investigated in view of device applications. Heterostructures of III-V semiconductors and crystalline organic films show rectifying current-voltage characteristics with high reverse breakdown voltages. As a first example for optoelectronic devices these heterodiodes are used for photodetection.

3.1 Optical Characterization

To obtain transmission spectra organic layers of about 500 nm thickness were grown on quartz substrates. As a tunable light source a monochromator set up with halogen lamp is used. The output beam is focused onto the organic film. The transmitted power is detected by a silicon photodetector using lock-in technique.

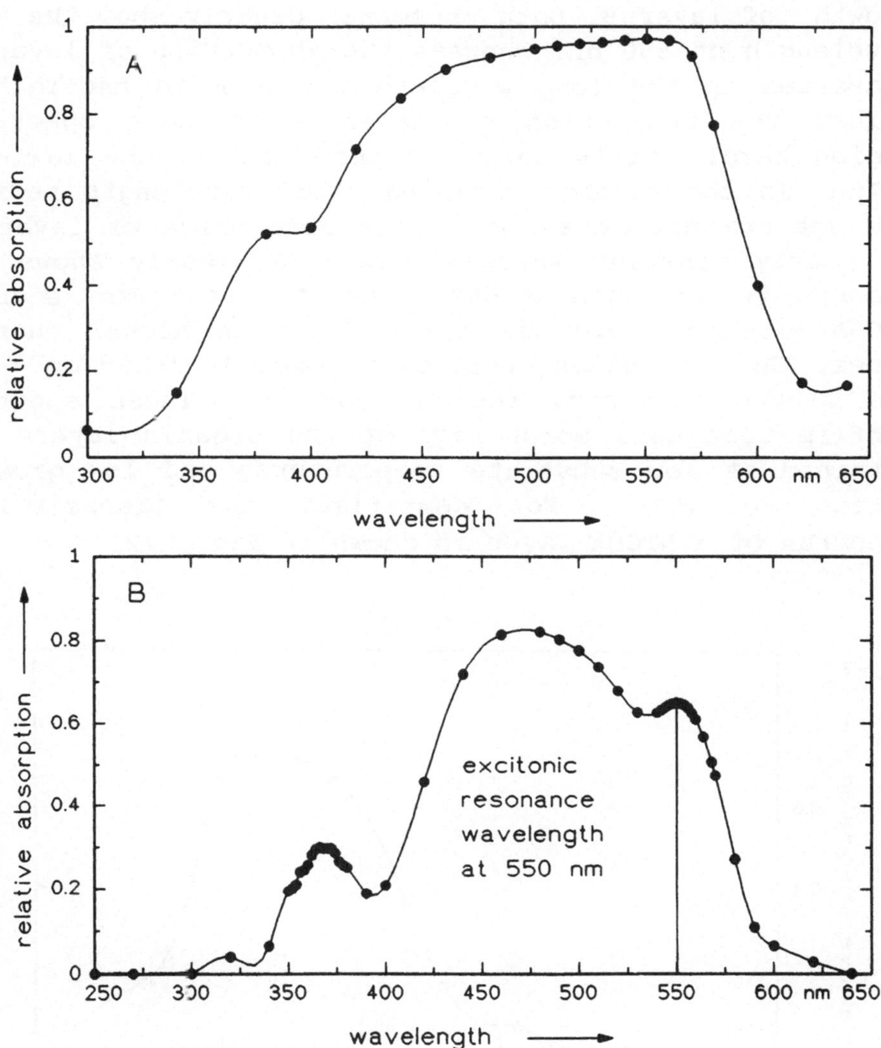

Fig. 3.1: Absorption spectra of PTCDA layers (Layer A: T_S = 300 K, T_{PTCDA} = 415 °C, layer B: T_S = 77 K , T_{PTCDA} = 376 °C).

Two absorption spectra of PTCDA layers are shown in Fig. 3.1. Layer A has been grown at a PTCDA source temperature of T_{PTCDA} = 415 $^{\circ}$C and at a substrate temperature of T_S = 300 K whereas the temperatures have been reduced to T_{PTCDA} = 376 $^{\circ}$C and T_S = 77 K for growth of layer B. Both spectra clearly show the gap wavelength of 560 nm. Whereas the absorption of layer B decreases in the long wavelength region to negligible values the absorption of layer A in this spectral region hardly falls below 20 per cent of the maximum value. In the strong absorbing short wavelength region the spectra differ as well. The absorption of layer A is nearly constant whereas sample B clearly shows an absorption resonance at 550 nm due to the lowest energy PTCDA exciton, which is split from the higher energy lines. The absorption coefficient exceeds $2 \cdot 10^5$ cm^{-1} in the strong absorbing region [FOR 84]. These spectra confirm that good morphology of the organic layers is attained at low substrate temperatures and low growth rates [SO 90/1]. For comparison the transmission spectrum of a NTCDA layer is shown in Fig. 3.2.

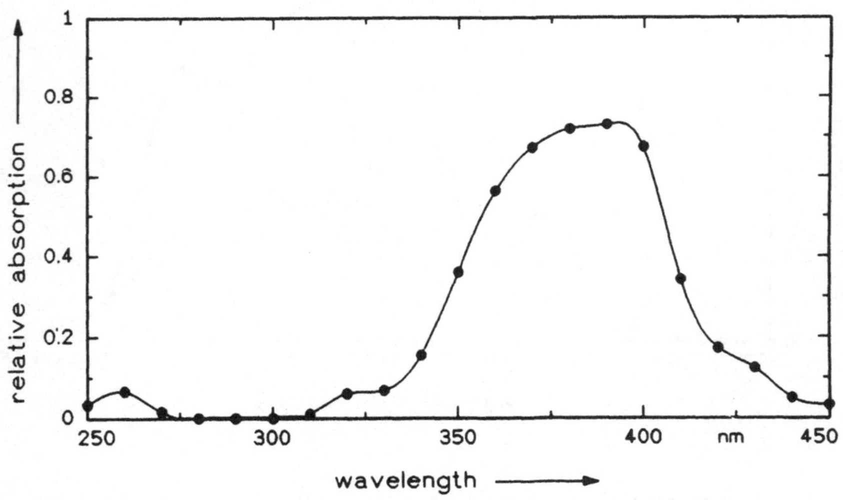

Fig. 3.2: Absorption spectrum of a NTCDA layer.

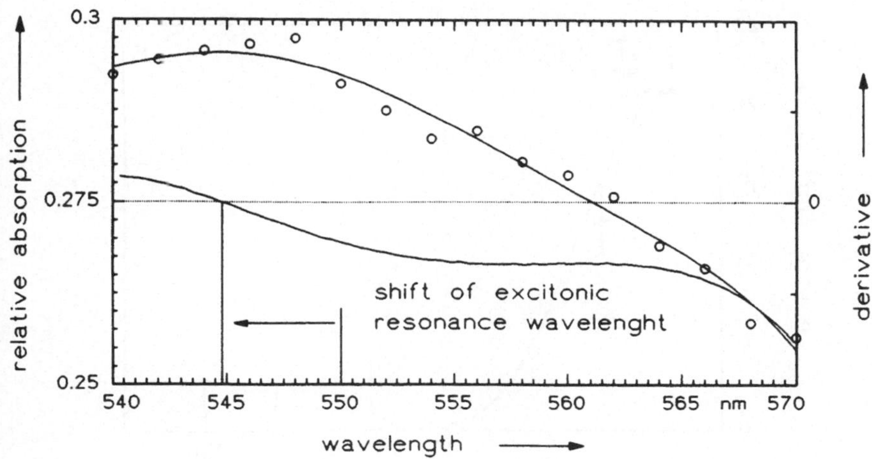

Fig. 3.3: Absorption spectrum of a PTCDA/NTCDA
quantum film.

Encouraged by these results growth of quantum well structures has been investigated. A thin PTCDA layer (W_a = 2.2 eV) of only a few nm thickness is sandwiched between two NTCDA layers (W_a = 3.1 eV). The NTCDA wells confine the optically generated excitons in the PTCDA layer as demonstrated by F.F. So [SO 90/2]. This results in a slight energy shift of the absorption resonance to shorter wavelength. Fig. 3.3 shows the interesting detail of the transmission spectrum and its derivative. The thickness of the PTCDA layer is about 1.5 nm. Because of the short interaction length only a slight absorption change is expected. The resonance wavelength derived from Fig. 3.3 is shifted to about 545 nm. The wavelength difference of 5 nm corrresponds to an energy shift of 20 meV. This result is in good agreement with previous works and theoretical calculations shown in Fig. 3.4 [SO 90/2].

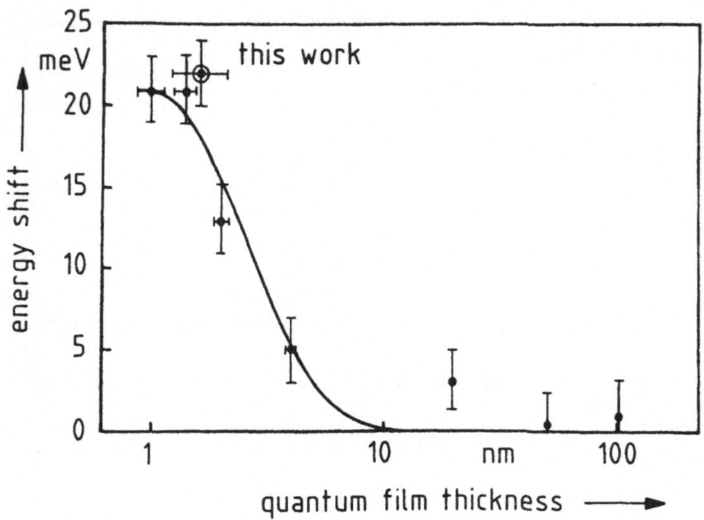

Fig. 3.4: Exciton line shift as a function of
quantum film thickness.

3.2 Electrical Properties

To allow an electrical characterisation ohmic contacts
are indispensible. We obtained low resistance contacts
by evaporation of In layers. Schottky contacts can be
realized too. Fig. 3.5 shows the current-voltage
characteristics of a Au electrode on a NTCDA layer. A
reverse breakdown voltage exceeding 55 V is attained.
This result is promising for electrooptical modulators
based on quantum confined Stark shift in PTCDA-NTCDA
MQW structures.

Furthermore organic-inorganic semiconductor heterostructures are of particular interest for optoelectronic device applications [FOR 84, FOR 90]. The
heterojunction between organic (PTCDA, NTCDA) and
inorganic (Si, GaAs, InP, InGaAlAs) semiconductors
forms a rectifying energy barrier. High reverse
breakdown voltages and low dark currents are attained.
As an example Fig. 3.6 shows the current-voltage
characteristics of a n⁻-GaAs-PTCDA heterostructure. The

breakdown voltage of about -40 V is limited by carrier multiplication in the inorganic semiconductor. The dark current density at V = -10 V is less than 10^{-5} A/cm^2. These results are comparable with high quality Schottky electrodes.

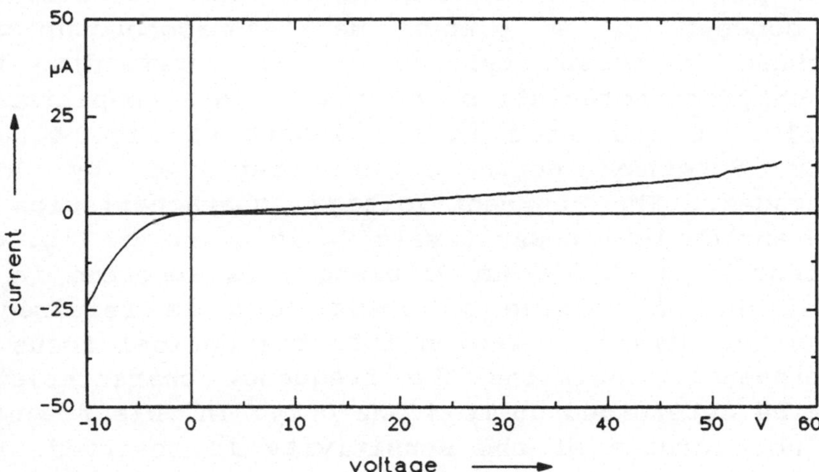

Fig. 3.5: Current-voltage characteristics of a Au contact on NTCDA.

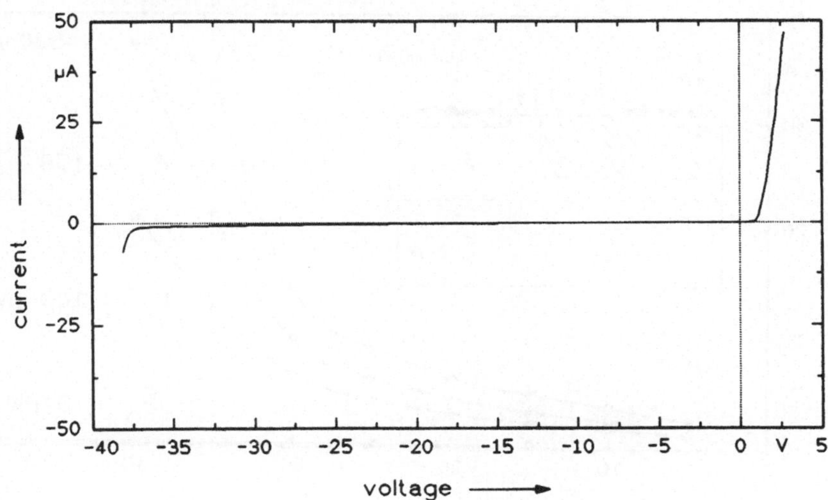

Fig. 3.6: Current-voltage characteristics of a GaAs-PTCDA heterojunction.

4. PTCDA-GaAs Photodetector

The rectifying current-voltage characteristics of organic-inorganic heterostructures with high breakdown voltages and low dark currents are well suited for application in optoelectronic devices. As a first example a photodetector is fabricated and characterized. The concept of a planar metal-semiconductur-metal structure is chosen because of its outstanding high frequency characteristics and its process compatibility to FETs. As indicated in the insert of Fig. 4.1 the planar Schottky contacts are replaced by PTCDA electrodes. The current-voltage characteristics at different optical power levels P_O is given in Fig. 4.1. The test beam at 830 nm wavelength is absorbed in the GaAs layer. A voltage of about 30 V is required to extend the depletion region into the optical focus. By experimental conditions the frequency characteristics could be determined up to 4 GHz only. In this frequency span no decrease of the sensitivity is observed. From the results obtained at MSM-structures the frequency limit should exceed 30 GHz.

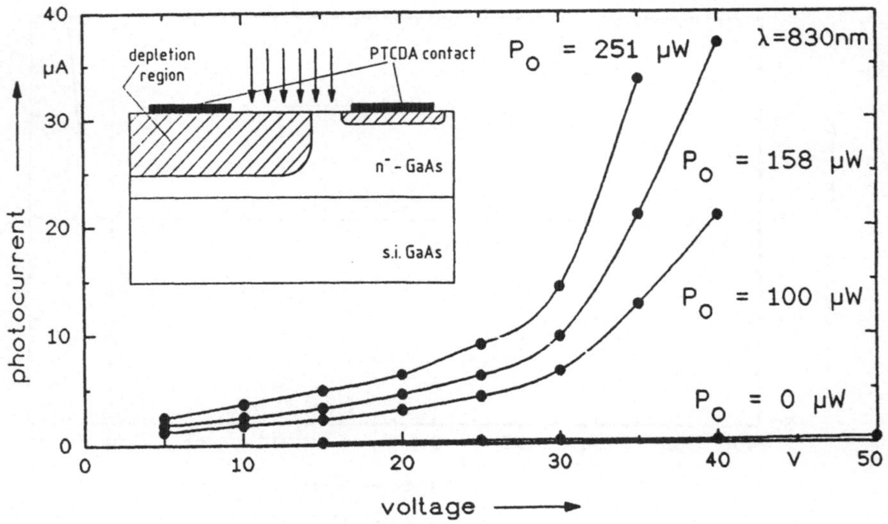

Fig. 4.1: Photocurrent-voltage characteristics of a planar PTCDA-GaAs-PTCDA structure.

5. Discussion and Conclusions

To summarize, we have reported on layer deposition and electrical and optical characterization of crystalline organic semiconductor films. In these experiments we used the well established aromatic compounds PTCDA and NTCDA. As can be derived from the absorption spectra good morphology is attained at low growth rates and substrate temperatures of 77 K. PTCDA layers grown under these conditions show a distinct excitonic absorption resonance at 550 nm close to the gap wavelength of 560 nm. A spectral resonance shift of 5 nm to shorter wavelength corresponding to an energy increase of 20 meV is observed in a NTCDA-PTCDA-NTCDA quantum well structure.

In view of device applications and electrical characterization metal contacts to these organic semiconductors are indispensible. Ohmic current-voltage characteristics are attained using In electrodes whereas deposition of Au electrodes results in Schottky characteristics with high breakdown voltages. The rectifying current-voltage characteristics of inorganic-organic semiconductor heterostructures is used for photodetection. Based on the concept of a planar metal-semiconductor-metal structure a planar organic-inorganic-organic (PTCDA-GaAs-PTCDA) photodetector is prepared.

So far no extensive ageing test have been carried out. But no deterioration of organic layer or device characteristics is observed after temperature cycles up to T = 200 °C for several days. These results are promising for applications of these organic semiconductors in photonic devices.

We gratefully acknowledge the generous financial support by the Volkswagen Stiftung.

References

[BOR 90] Born, L.; Heywang, G.: Crystal structure of 1,4,5,8-naphthalene-tetracarboxylic-dianhydride (NTDA). Zeitschrift für Kristallographie 190 (1990) 147-152

[FOR 84] Forrest, S. R.; Kaplan, M. L.; Schmidt, P. H.: Organic-on-inorganic semiconductor contact barrier diodes. I. Theory with applications to organic thin films and prototype devices. J. Appl. Phys. 55 (1984) 1492-1507

[FOR 90] Forrest, S. R.; So, F. F., Zang, D. Y.: Organic-on-inorganic semiconductor optoelectronic devices. 16th Euro-pean Conference on Optical Communication ECOC '90 (1990) 841-847

[MÖB 89] Möbus, M.; Schreck, M.; Karl, N.: Epitaxial growth of thin films of Perylenetetracarboxylic Dian-hydride and Coronene on NaCl(001). Thin Solid Films 175 (1989) 89-93

[SO 90/1] So, F. F.; Leu, L. Y.; Forrest S. R.: Growth and Characterization of Organic Semiconductor Hetero-junctions and Multiple Quantum Wells. SPIE 1285 (1990) 95-103

[SO 90/2] So F. F.; Forrest, S. R.; Shi, Y. Q.; Steier, W. H.: Quasi-epitaxial growth of organic multiple quantum well structures by organic molecular beam deposition. Appl. Phys. Lett. 56 (1990) 674-676

[ZAN 91] Zang, D. Y.; So F. F., Forrest, S. R.; Giant anisotropies in the dielectric properties of quasi-epitaxial crystalline organic semiconductor thin films. Appl. Phys. Lett. 59 (1991) 823-825

Optical Characterization of Ultrathin Polymer Films by Evanescent Light[*]

Wolfgang Knoll[a,b], Werner Hickel[c], and Michael Sawodny[b]

[a] Frontier Research Program, The Institute of Physical and Chemical Research (RIKEN), Wako, Saitama 351–01, Japan
[b] Max–Planck–Institut für Polymerforschung, Ackermannweg 10, W–6500 Mainz, Germany
[c] Hoechst AG, Angewandte Physik, W–6230 Frankfurt 80, Germany

Summary:
This paper describes the use of evanescent light for the optical characterization of polymer thin films and interfaces. Firstly, a few basic concepts of evanescent wave phenomena, including total internal reflection, plasmon surface polaritons and guided optical modes, are reviewed. It is shown that the excitation of these waves allows for a sensitive determination of the optical architecture of the interface(s) involved. This "surface light" can then be used for the same broad range of optical techniques as it is known from experimental set–ups designed for the investigation of various optical properties of polymer samples using plane electromagnetic waves, i.e. "normal" photons. This is demonstrated for diffraction experiments and microscopic investigations. The examples given include thin polymer films prepared by spin–coating or by the Langmuir–Blodgett–Kuhn technique.

---

[*] This contribution is part of a review published in Makromol. Chem. 192 (1991) 2827

1. Introduction

The technical use of polymeric materials in many cases involves thin film preparations: coatings with protective (against radiation or chemical attack), coloring, isolating/conductive (anti–static), lubricating, wetting/dewetting, etc. properties are just a few examples for the wide range of applications where thin polymer films play an important role. Within the broad spectrum of characterization techniques necessary to optimize these various properties, optical methods have their well–established place. However, as far as thinner and thinner samples are concerned, eventually only consisting of one monomolecular layer, some of them lack the sensitivity necessary to be useful.

It is the aim of this report to demonstrate for a number of organic thin film samples and experimental configurations that, instead of working with plane electromagnetic waves, "normal" photons, the use of so–called evanescent modes offers in some cases sensitivity and resolution advantages that allow one to overcome the problems uncountered often in optical thin film characterizations.

The organization of this paper follows the general message that evanescent modes are surface–light [BOA 82] with a particular sensitivity and specificity for interfacial properties and processes. We first review some general concepts of surface–electromagnetic waves starting with the simple total internal reflection (TIR) geometry [MOE 88], then focussing on plasmon surface polaritons [BUR 74, RAE 77] but also including guided optical waves [TIE 69]. Already the examination of the condition for exciting these modes with a conventional light source, typically but not necessarily with a laser, allows one to derive detailed information about the optical architecture of the interface. Once this surface light is turned on one finds all the phenomena described in optics textbooks for plane waves − just transformed into the quasi–two dimensional world of bound electromagnetic waves: transmission, reflection and refraction [ROT 88a] when travelling across the interline between two areas, total internal diffraction at a phase grating [ROT 87a], interferometry [ROT 88a,ROT 88b], surface plasmon [ROT 88c] and optical waveguide microscopy [HIC 90a] are examples for elastic scattering processes of surface waves. Brillouin [LEE 90] and Raman–spectroscopies [KNO 82, DUC 88, KNB 89, KNB 91] demonstrate the inelastic scattering of this light, but will not be discussed here.

As far as different sample configurations are concerned, thin polymeric films in air, in vacuo or at a solid/solution interface can be investigated. The films used for demonstration purposes in this report are either prepared by the spin–coating technique (ranging in thickness from some ten nm up to a few μm) or are laid–down, layer by layer, by the Langmuir-Blodgett-Kuhn (LBK) technique [KUH 72]. The resulting multilayer assemblies range in thickness from single monolayer (ca. 2.5 nm) up to about 1 μm.

2. Evanescent Light

The simplest case for the existence of an evanescent wave is the well–known total internal reflection of a plane electromagnetic wave at the base of a glass prism (index of refraction n_1), in contact with an optically less dense medium (with $n_2 < n_1$), e.g. air ($n_2 = 1$). This geometry is schematically sketched in Fig. 2.1, upper part. If the reflected light intensity is recorded as a function of the angle of incidence, Θ, the reflectivity, R, reaches unity as one approaches the critical ange Θ_c for total reflection (Fig. 2.1, lower part). A closer inspection of the E–field distribution in the immediate vicinity of the interface shows that above Θ_c the light intensity does not fall abruptly to zero in air but that instead one is still dealing with a harmonic wave travelling, however, parallelly to the surface with an amplitude which is exponentially decaying normal to the interface. The depth of penetration l which is defined by the 1/e attenuation is given by

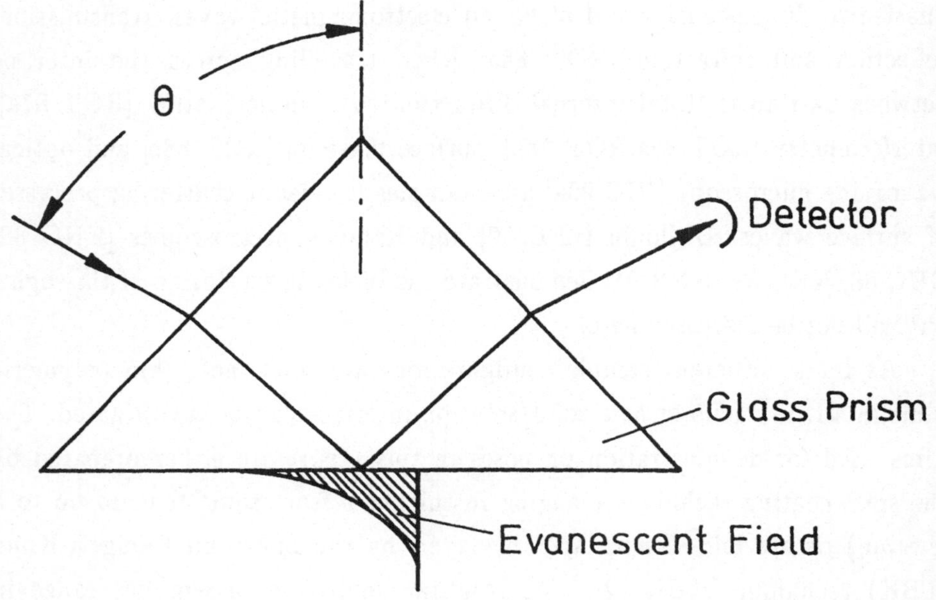

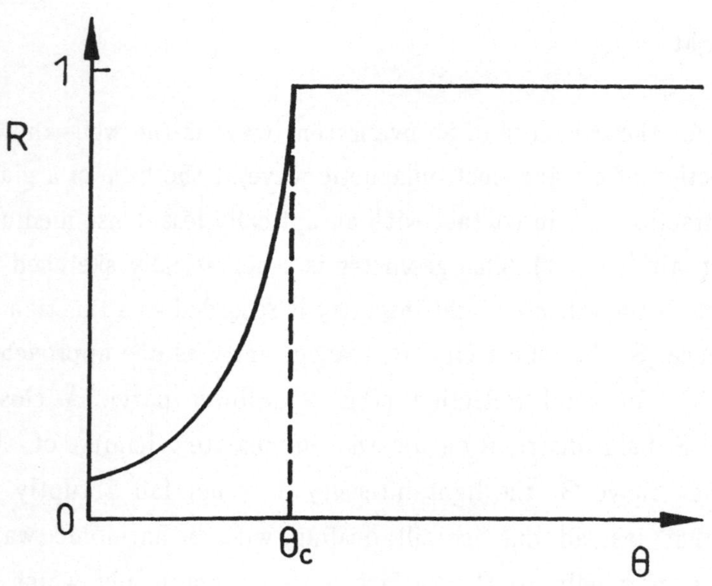

Fig. 2.1

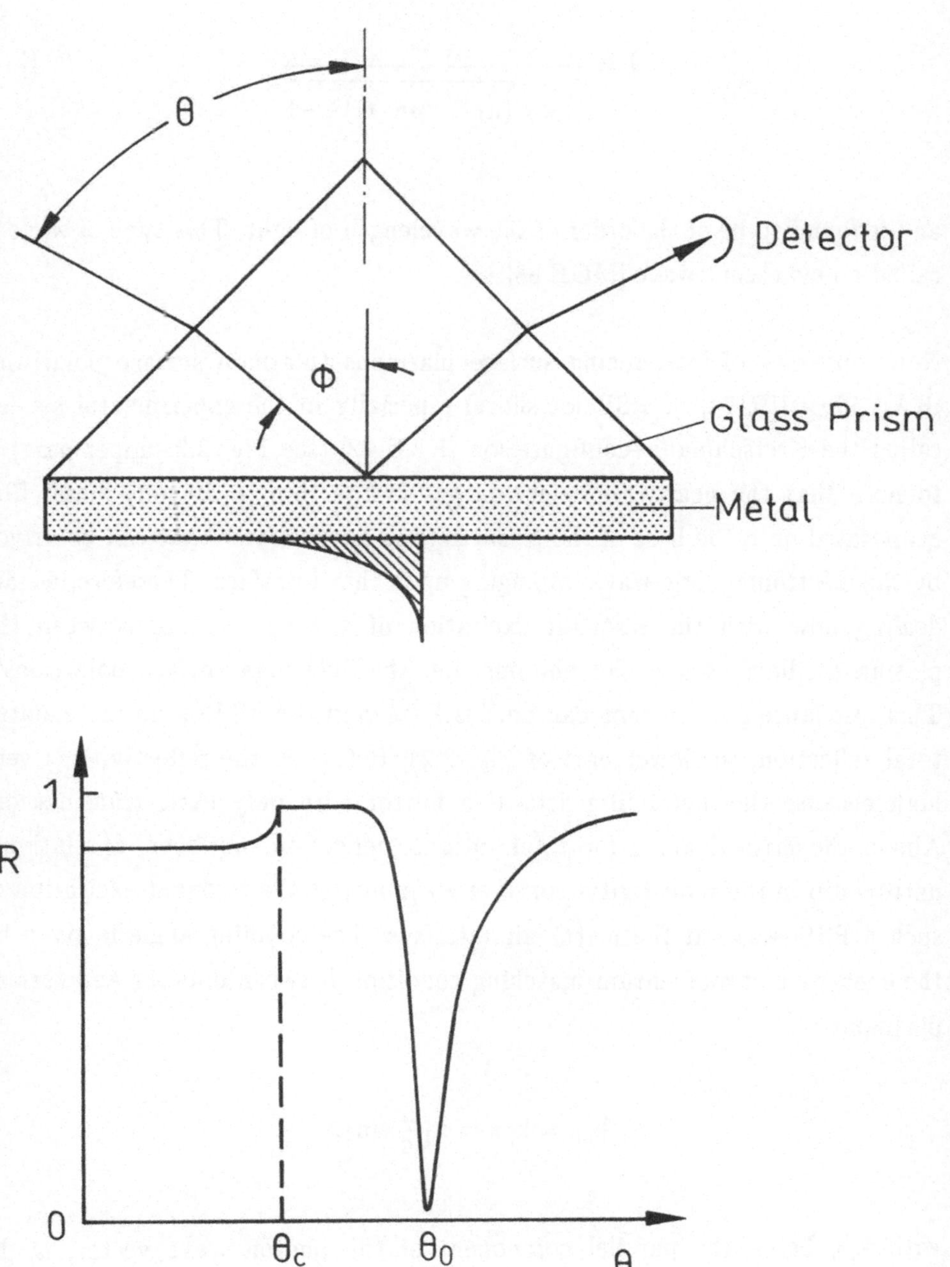

Fig. 2.2

94

$$l = \frac{\lambda}{2\pi\sqrt{(n_1 \cdot \sin \Theta)^2 - 1}} \qquad (2.1)$$

and is found to be of the order of the wavelength of light. This type of wave is called an evanscent wave [MOE 88].

Now, one way of introducing surface plasmons (plasmon surface polaritons [RAE 77, BUR 74] or PSP for short) especially in the experimental set–up called the Kretschmann–configuration [KRE 72] (see Fig. 2.2, upper part) is to note that the nearly free electron gas in the thin ($\approx$ 50 nm) metal film evaporated onto the base of the prism acts as an oscillator that can be driven by the electromagnetic wave impinging upon that interface. Therefore, we are dealing now with the resonant excitation of a coupled state between the plasma oscillations and the photons, i.e. the "plasmon surface polaritons". This resonance phenomenon can be clearly seen in the ATR–scan (attenuated total reflection, see lower part of Fig. 2.2): Below Θ_c the reflectivity is very high because the metal film acts as a mirror with only little transmission. Above the critical angle for total internal reflection, however, a relatively narrow dip in the reflectivity–curve at Θ_0 indicates the resonant excitation of such a PSP–wave at the metal–air interface. The coupling angle is given by the energy– and momentum matching condition between photons and surface plasmons

$$k_{sp}^0 = k_{ph}^0 = n_1 \frac{\omega}{c} \sin \phi_0 \qquad (2.2)$$

with k_{ph}^0 being the parallel component of the photon wave vector, ω the photon energy, c the speed of light, and ϕ_0 the (internal) coupling angle (see Fig. 2.2).

Again, we are dealing with an evanescent wave propagating along the interface with a penetration depth into air of the order of the wavelength.

θ
Detector
Glass Prism
Metal
Waveguide
R
1
0
θ_c
θ_m=3
θ_m=2
θ_m=1
θ_0
θ

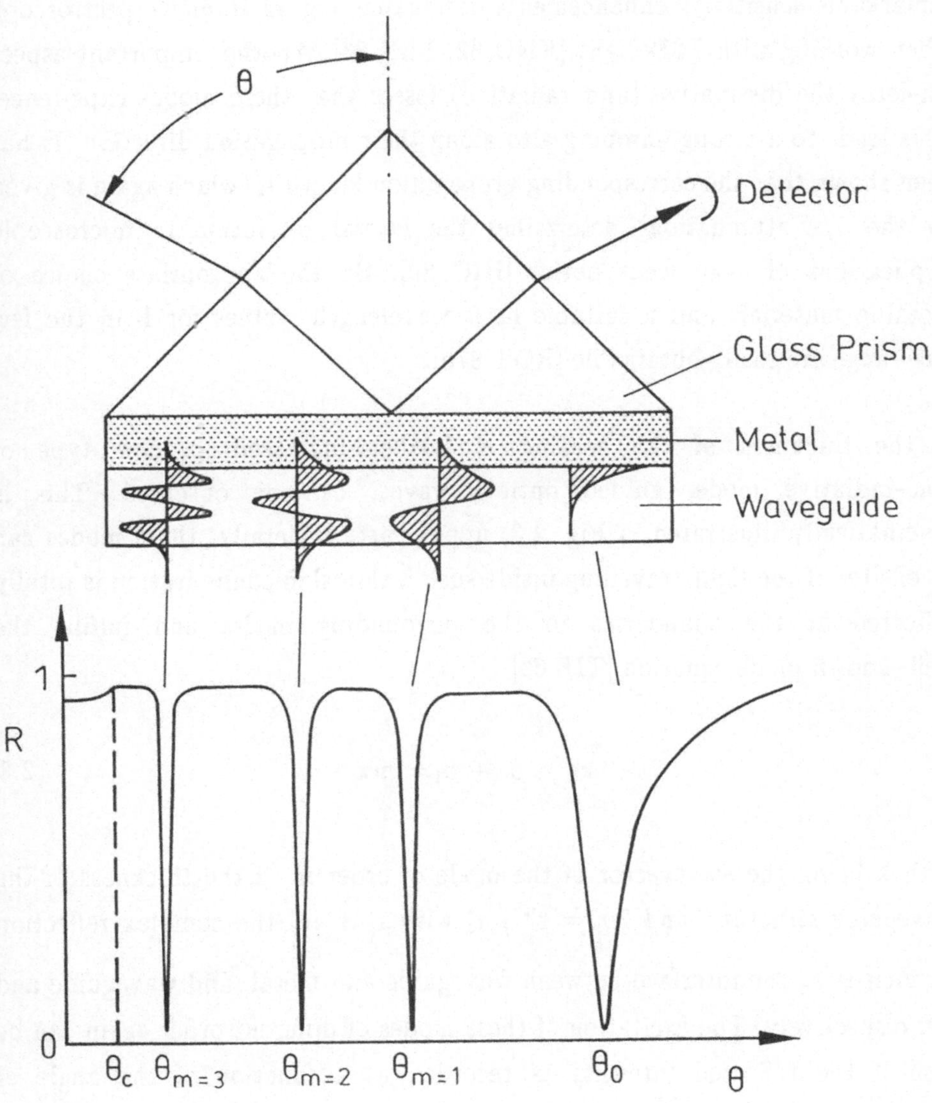

Fig. 2.3

The resonance character of this excitation gives rise to an enhancement of the E—field at the interface by more than a factor of 10 which is the origin of the remarkable sensitivity enhancements obtainable, e.g. in Raman—spectroscopy when working with PSP—light [KNO 82, USH 83]. Another important aspect concerns the dissipative (and radiative) losses that these modes experience. This leads to a strong damping also along their propagation direction. It has been shown that the corresponding propagation length L (which again is given by the 1/e attenuation) determines the lateral resolution in microscopic applications of evanescent optics [HIC 89]. By the appropriate choice of metallic materials and a suitable laser wavelength, values for L in the few μm—range are easily obtainable [ROT 87b].

If the thickness of the coating is further increased, a new type of non—radiative mode, guided optical waves, can be observed. This is schematically illustrated in Fig. 2.3, upper part. Generally, these modes can be excited if the light travelling inside such a thin slab configuration is totally reflected at the boundaries to the surrounding media and fulfills the well—known mode equation [TIE 69]:

$$kd + \beta_0 + \beta_1 = m\pi \qquad (2.3)$$

with k being the wavevector of the mode of order m, d the thickness of the waveguide structure, and $2\beta_i = r_i'' / r_i'$ with $r_i' + ir_i''$ the complex reflection coefficient at the interface between waveguide and metal, and waveguide and air, respectively. The excitation of these modes of different order again can be seen if the reflected intensity is recorded as a function of the angle of incidence, Θ (Fig. 2.3, lower part): narrow dips in the reflectivity curve above Θ_c indicate the existence of the various guided waves. For a given material combination their number and angular position only depend on the waveguide thickness. For illustration purposes we have drawn the field—amplitudes of the first three modes indexed according to their number of nodes inside the

ω
Air
Ag
PSP⁰
PSP¹
Air
Dielectric Layer
Ag
ωL
ksp⁰
ksp¹
ksp
θ0
θ1
θ

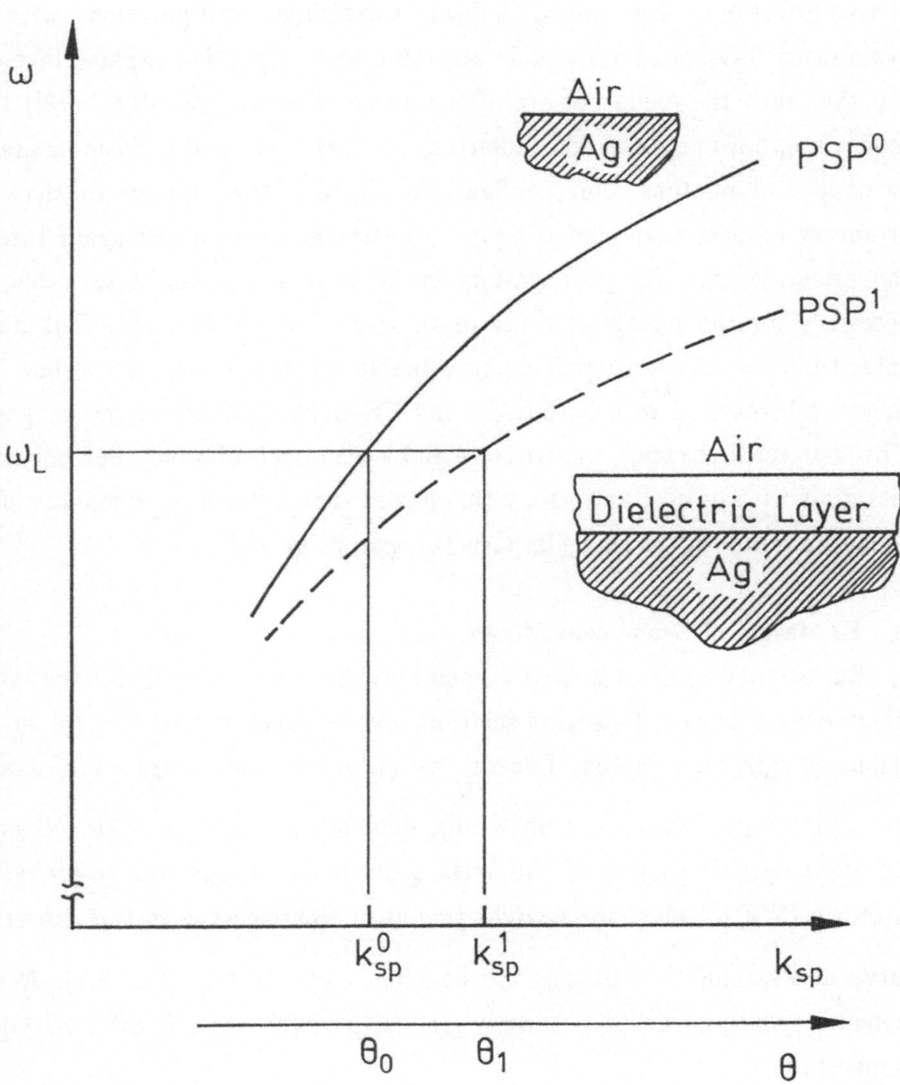

Fig. 3.1

waveguide: m = 1,2,3. Although, strictly speaking, guided optical waves are not evanescent (except for an exponentially decaying tail that reaches out into the air and which can also be used for thin polymeric film characterizations, e.g. in a hybrid waveguide configuration with an evaporated SiO_x main waveguide structure and only a few organic layers on top that shift the mode pattern of the uncoated waveguide [SWA 78]) these modes are, however, also non radiative, i.e. they, too, need a prism or grating to couple to photons. Our configuration is somewhat special in that this geometry ensures that guided light is constantly coupled out again through the prism so that the propagation length L is also reduced to a few μm necessary for the microscopic use of these modes (see below). What makes guided optical waves a particularly valuable diagnostic tool is the fact that they can be excited with both TM– and TE–light (TM: transverse magnetic; TE: transverse electric), i.e. with p– and s–polarized photons. For polymeric materials with optical anisotropy this means that different components of the dielectric tensor of the thin film structure can be probed.

3. Excitation of Evanescent Modes

Surface plasmons as well as guided optical waves are well–defined modes which obey a known dispersion relation ω vs. k. This means, e.g., for surface plasmons that each photon of energy $\hbar\omega_L$ allows for the excitation of exactly one PSP mode. This is schematically depicted in Fig. 3.1. The full curve represents the dispersion of the surface plasmons at a metal (e.g. Ag)/air interface (PSP^0). The thin straight line at ω_L intersects with this dispersion curve at k_{sp}^0 and thus defines the coupling angle θ_c (see Fig. 2.1). A thin dielectric coating causes a shift of the dispersion curve (PSP_1) to higher momentum

$$k_{sp}^1 = k_{sp}^0 + \Delta k_{sp} \qquad (3.1)$$

Hydrocarbon
Polyglutamate
Hydrocarbon
Silver
Glass

Fig. 3.2

100
50
0
reflected intensity /%
40
55
θ/deg
● : 0
▲ : 2
■ : 4
× : 6
▽ : 10
◇ : 20

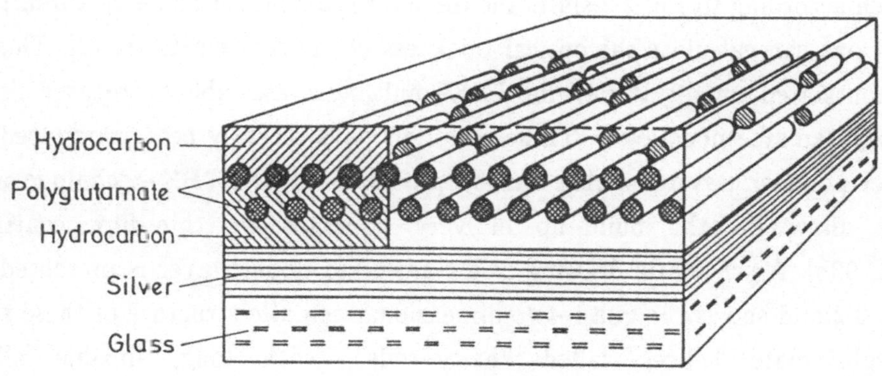

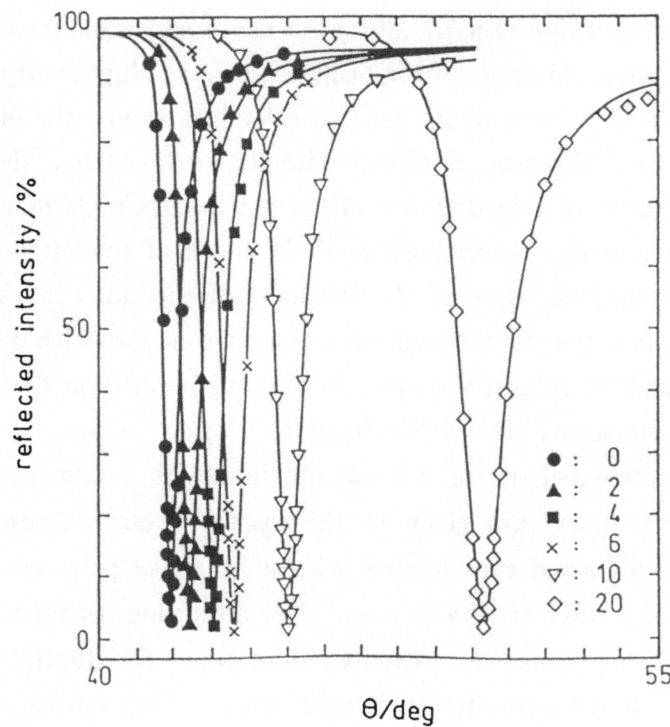

Fig. 3.3

100

which according to Eq. 2.2 shifts the resonance to a higher angle θ_1. From this shift one can calculate the optical thickness of the coating [GOR 77]. This is demonstrated quantitatively for LBK–multilayer assemblies prepared from polyglutamate–monolayers. These materials have recently been introduced as novel polymeric systems that can be processed by the LBK–technique and that allow for the build–up of very homogeneous thin–film coatings [HIC 90b]. A schematic drawing of a transferred double layer is presented in Fig. 3.2 and shows the quasi–two dimensional nematic structure of these stiff polyglutamate helices called "hairy–rods", with long, flexible alkyl side–chains. Now, Fig. 3.3 gives the results of ATR–scans obtained from the bare metal and 2,4,6,10 and 20 layers of polyglutamate, respectively. The symbols are the experimental data points, the full lines are Fresnel calculations. Given the thickness per monolayer (d_0=1.75 nm) from X–ray measurements one finds that all reflectivity data can be described by a constant index of refraction of n_z=1.486. As in any other technique capable of resolving the optical thickness of a monolayer, e.g., in ellipsometry [JOII 90], no independent evaluation of the geometrical thickness and the index of refraction is possible. However, if thicker samples are available, then the simultaneous excitation of guided optical modes and a surface plasmon mode allows for a rather precise determination of the index of refraction of the material and – independently – of the thickness of the film. In cases of optically anisotropic polymers, the excitation of several modes, with different polarizations, s– and p–, respectively, and propagating in different directions, allows for the determination of the full indicatrix.

The example presented in Fig. 3.4 was obtained with a thin layer of a solid polyelectrolyte (structure given in the inset). These ionenes–called materials recently attracted considerable interest as glassy polymers which can be tuned in their index of refraction and their dispersion behavior over a wide range [SIM 89] which is extremely important for future device applications, e.g., in (non–linear) integrated optics. The various guided modes and the surface plasmon resonance at high angles can be described to a

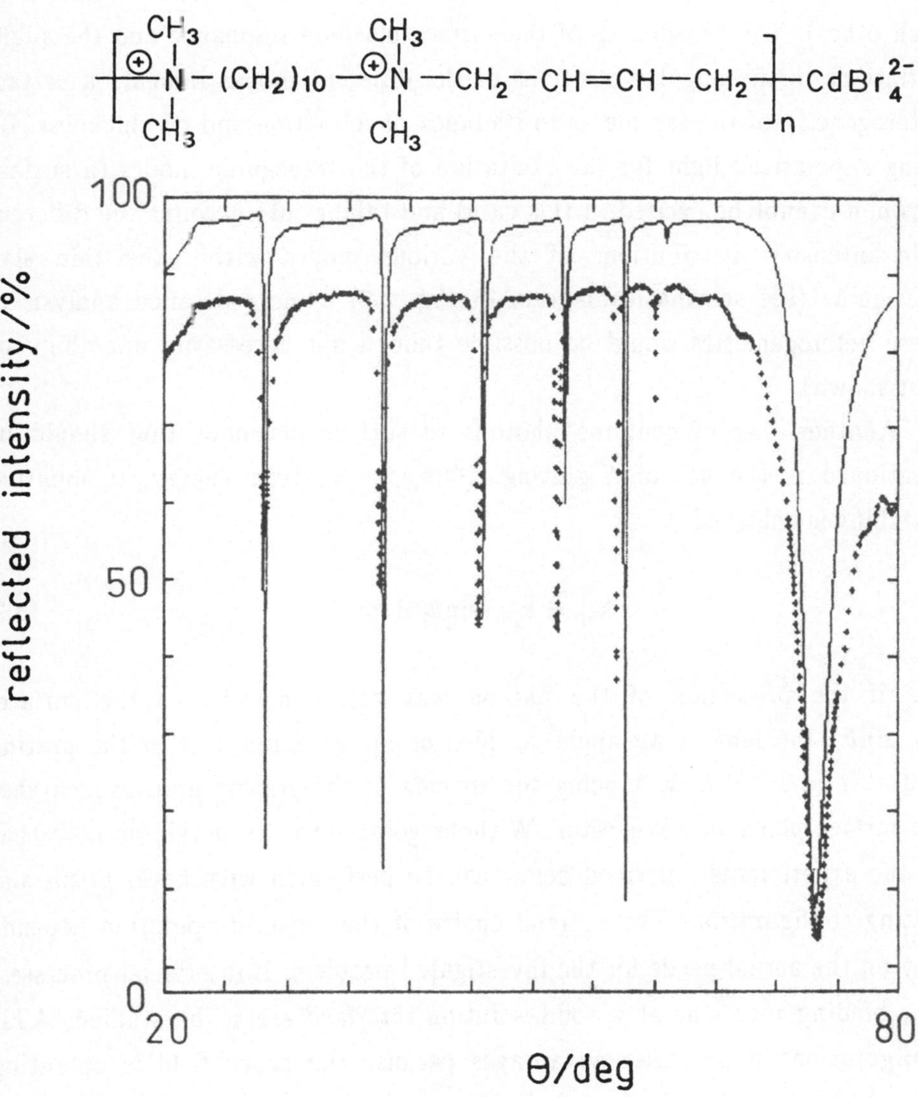

Fig. 3.4

first approximation by a box model with an index of refraction of n=1.49 and a thickness of d=1.85 μm. (For better clarity, the experimental data points and the Fresnel fit (full curve) are somewhat shifted vertically relative to each other). The broadening of the surface plasmon resonance and the slight variations in the angular positions of the different modes indicate a certain heterogeneity of the sample as to its index of refraction and its thickness. By using s—polarized light for the excitation of the waveguide modes (a surface plasmon cannot be excited in this case) and taking into account the different field intensity distributions of the various modes within the thin slab waveguide (see schematic diagram in Fig. 2.3) a more detailed analysis of these heterogeneities would be possible though not necessarily unambiguous (not shown).

Another way of coupling photons to surface plasmons that should be mentioned is the use of a grating. Here, for a given energy, momentum matching is achieved if

$$k_{sp} = k_{ph} \cdot \sin\theta_0 \pm m \cdot G \tag{3.2}$$

i.e., if the projection of the photon wavevector parallel to the surface, $k_{ph} \cdot \sin\theta_0$, incident at an angle θ_0, plus or minus a multiple of the grating vector $G = 2\pi/\Lambda$ with Λ being the spacing of the grating grooves, matches the surface plasmon wavevector. Without going into any detail we note that all the experiments described below can be performed with both, prism and grating configuration. The optimal choice of the mode of operation depends only on the actual needs for the investigated problem: If interfacial processes, e.g., binding reactions at a solid—solution interface are to be studied, ATR configurations might offer advantages because the probe field is operating from the back so that the front half—space is free for on—line manipulations.

4. Diffraction of evanescent waves

So far, we were concerned only with thin–film samples that were laterally homogeneous, i.e. in the plane of the interface with the only relevant parameter being their optical architecture (thickness, index of refraction, etc.) normal to the surface. The question arises, however, to what extent evanescent optics allow also for the characterization of ultrathin coatings with lateral heterogeneities either being caused by preparative artefacts, e.g. thickness variations of planar waveguides prepared by spin–coating, or structures prepared purposefully, e.g. by photo–reactive lateral pattern writing (channel waveguides, etc.) for integrated optics devices.

Following the general idea that evanescent waves are surface–light, it is immediately evident that these modes, too, will couple to (optical) material inhomogeneities seen along their propagation direction.

As a first direct manifestation of this coupling we discuss in the following the special case of a strictly periodic variation of the optical property of a thin film, i.e., the diffraction behavior of an evanescent mode at a phase grating. For this purpose we laterally structured a LBK–film of 5 monolayers of cadmium arachidate (CdA) by selectively photo–desorbing parts of it with UV light through the (very regularly spaced) holes of an electron microscopy copper grid used as a mask. In order to be able to compare directly the behavior of an evanescent wave in a mere dielectric total internal reflection geometry (cf. Fig. 2.1) with the resonantly enhanced case of a surface plasmon wave (cf. Fig. 2.2) we coated only one half of the base of the coupling prism with a thin Ag–layer (and a 7 nm thick SiO_x layer to render the whole surface hydrophilic) and prepared on both sides an identical diffraction grating with a periodicity of D=80 μm. This is schematically depicted in Fig. 4.1.

Two sets of measurements were performed with both configurations: In the first run, the reflected intensity was recorded in the usual angular scan. For the TIR set–up the typical behavior with unity reflectivity above θ_c was found (cf. Fig. 2.1, lower part). The ATR experiment, however, showed two

D
Phase Grating
$g = \dfrac{2\pi}{D}$
ϕ
θ
+3
+2
+1
0
-1
-2
Diffraction Order

Fig. 4.1

$\theta_1 = 43.6$ deg
Diffracted Intensity
Diffraction Orders
-3 -2 -1 0 +1 +2
43
44
θ_2 /deg

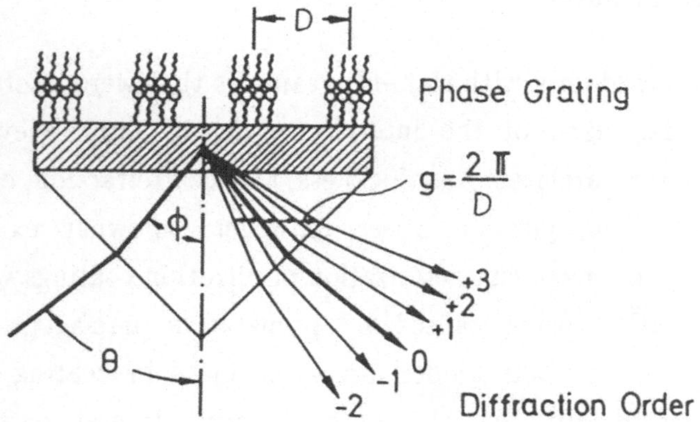

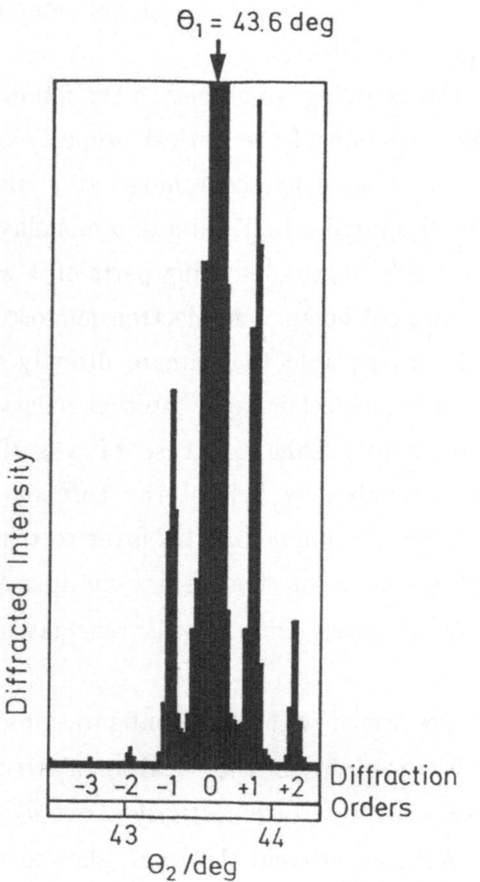

Fig. 4.2

resonances when working with blue light: one at θ_0 for the bare Ag–areas and a second at θ_1 for the coated regions shifted to higher angles according to the optical thickness of the LBK–film.

Now, the periodic alternation between bare metal and coated regions provided additional discrete momentum g $= 2\ \pi/D$ for the reradiating plasmon fields. In addition to the specularly reflected light, the zeroth diffraction order, this gave rise to the multiple diffraction orders seen by scanning only the detector while keeping the angle of incidence fixed at θ_0. Fig. 4.2 gives a direct comparison of the diffracted intensities as obtained for the two configurations. Clearly, the resonance enhancement of the surface plasmon mode shows up as much stronger intensities in the various diffraction orders. We could show that even a thickness modulation of only one monolayer CdA can easily be seen by PSP diffraction in the first order [ROT 87c]. It is straightforward to translate this increased sensitivity for thickness modulations into one for an index of refraction modulation. Such structures are used in kinetic expeiments with transient gratings in holographic techniques, e.g., in forced Rayleigh scattering.

5. Surface plasmon– and optical waveguide microscopies

The above described angular shift of the resonance condition for different coating thicknesses is also the basic mechanism that generates the high contrast achievable in surface plasmon microscopy (SPM) of heterogeneous thin films. Here, the reflected, scattered and diffracted plasmonic light is Fourier–backconverted by a lens to form an image of the interface in real space on a TV–camera in our case (see Fig. 5.1). Only those areas that are at resonance appear dark, whereas all other domains according to their relative shift of the coupling condition are more or less bright. If the laser light is coupled–in at θ_0 the bare metal surface is at resonance for PSP excitation, and hence almost no light is reflected. Any coated region, however, is still off–resonance, and hence reflects the full laser intensity. The different areas, therefore, show maximum contrast in reflection. In previous work we

θ
Laser,
wavelength λ
Screen/Camera
Lens
Prism
Metal
Coating
Evanescent
Field of PSP

Fig. 5.1

Ag
Ag + 3nm SiO2
12 μm

Fig. 5.2

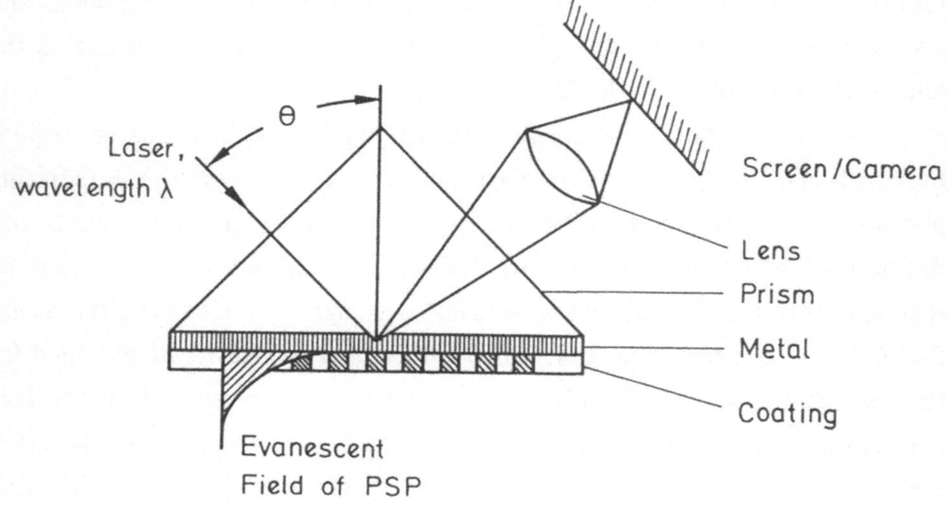

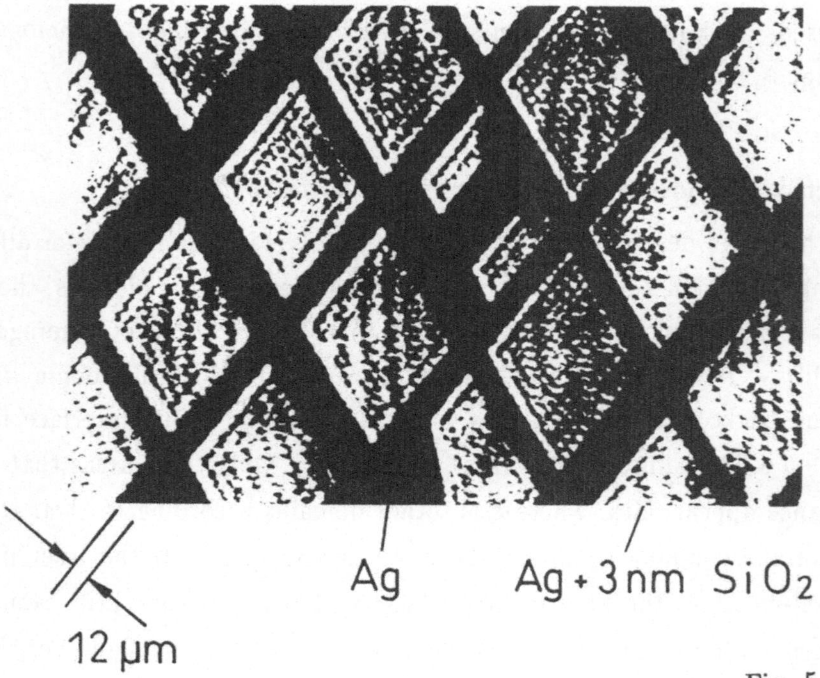

demonstrated that variations of the thickness of different areas as little as a few tenths of a nanometer are enough to generate sufficient contrast for an image [HIC 90c]. Illumination of the sample is typically done with a HeNe laser beam of ca. 1 mm² spot size. The lateral resolution of this microscopic technique which is determined by the propagation length L (see above) has been shown [HIC 90d] to be better than 5 μm. An example is given in Fig. 5.2. An electron microscopy copper grid had been used as a mask for the photo–desorption of parts of a CdA double layer (d=5 nm) laid down by the LBK–technique. The angle of incidence was chosen so as to excite PSP only in the bare metal regions, the CdA–coated areas hence appear bright. The stripes that hold the letter "Z" are only 12 μm wide but can be imaged with good contrast and sharp edges. We should note that the propagation length L could be further reduced by either working in the blue or by using other metals with higher intrinsic dissipation (higher imaginary part, ϵ'', of the dielectric function). The lateral resolution could thus be further increased, but the far–field optics that we use for the imaging is diffraction–limited at about 2 μm anyhow.

Differences in the geometrical thickness of two areas, e.g. by photo–ablation [SWA 91] or by evaporation, are not the only contrast mechanisms in SPM. Any change in the real part of the dielectric function of a thin–film sample could be read–out with equally high contrast: for a single monolayer of a fatty acid (d$\approx$2.5 nm) a change in the index of refraction of Δn = 0.01 would be sufficient to create an image.

A very similar performance characteristic can be obtained with the recently developed optical waveguide microscopy (OWM) [HIC 90a]. The thickness dependence of the angular coupling condition for waveguide modes (see Eq. 2.3) provides us with a contrast–mechanism very similar to the situation encountered in SPM. Different from the latter technique, however, guided light waves can be excited with both, TM– and TE– modes, i.e. with p– and s–polarized light. This can be particularly helpful for the characterization of thin–film samples with an anisotropic index of refraction.

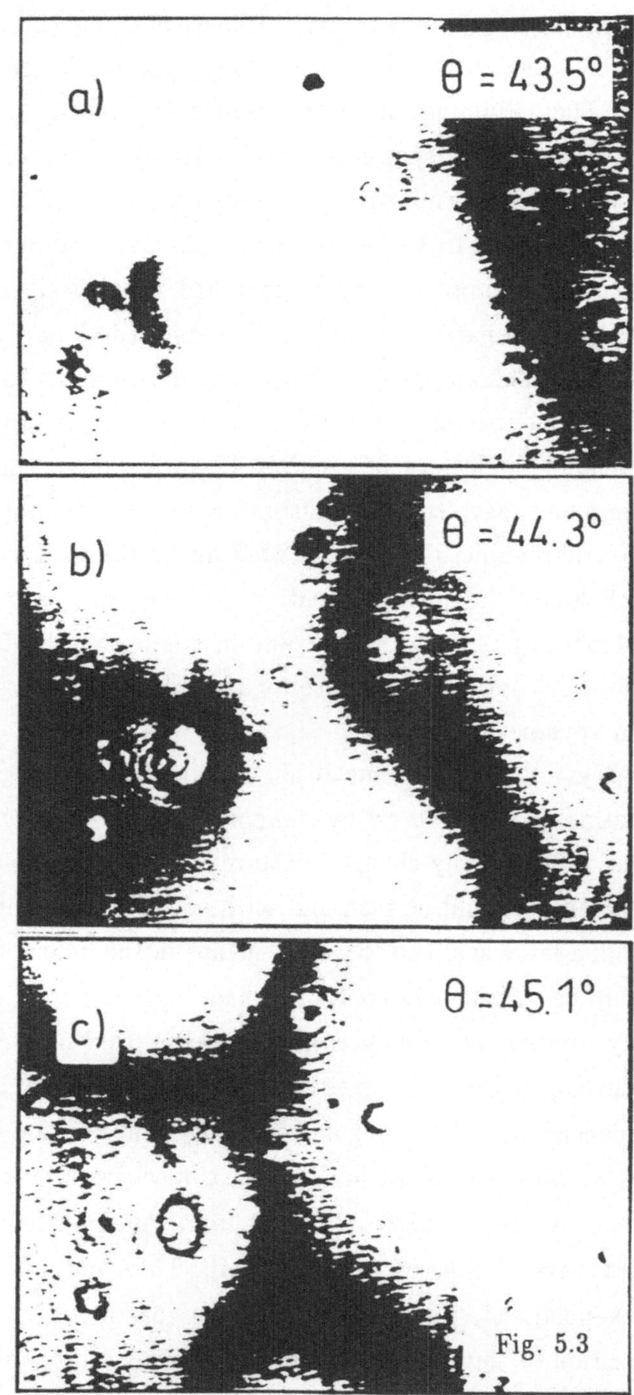

Fig. 5.3

We could demonstrate that our particular lossy thin slab configuration (where the guided lightwave constantly couples out through the prism) allows for OWM investigations with a lateral resolution better than 10 μm and a thickness resolution comparable to SPM, i.e. in the few tenths of a nanometer range [HIC 90a].

The example presented in Fig. 5.3 concerns the application of this novel technique, namely, the microscopic characterization of the lateral thickness heterogeneities of a polymeric planar waveguide structure prepared by the usual technique, i.e., spin–coated from solution. The employed material was the solid polyelectrolyte introduced in Fig. 3.4.

The reflectivity measurements of this preparation taken first (not shown) suggested through an angular broadening of the mode structure again a certain thickness variation of the spun–on film. Optical waveguide microscopic pictures taken with p–polarized light every 0.4⁰ between 43.1⁰

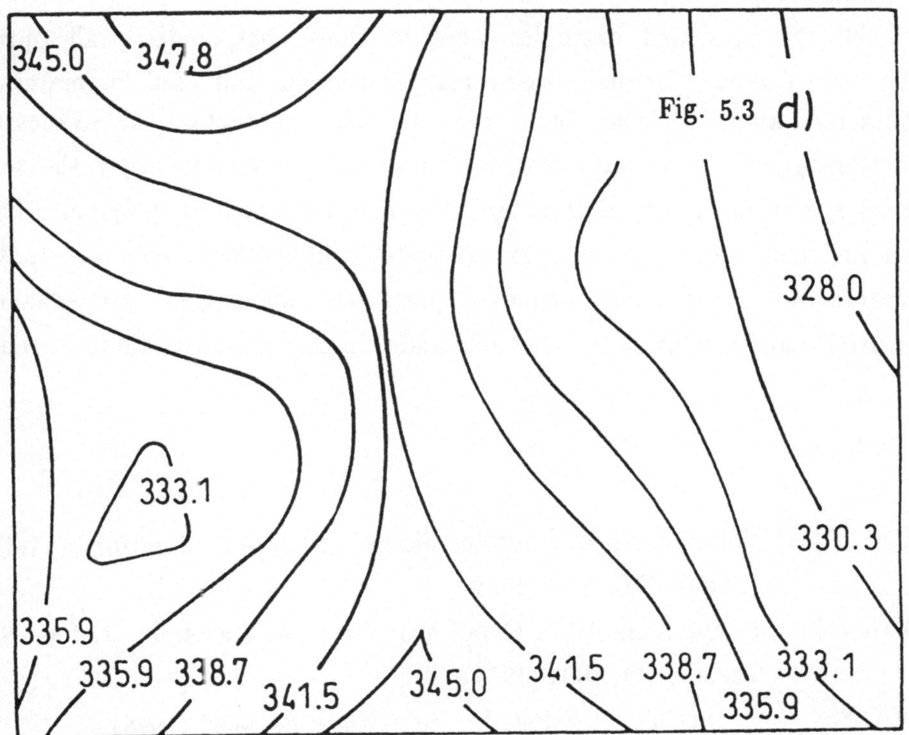

and 45.9⁰ then, indeed, demonstrated these thickness variations by a constantly changing local resonance pattern. This is shown in Fig. 5.3 a–c for three diefferent angles of incidence. Now, with an image–analyzing system it is easy to find for each picture the line(s) of lowest (reflected) intensity defining the local waveguide resonances. These lines are given in Fig. 5.3 d) together with the thicknesses (in nm) they correspond to. These were obtained from Fresnel calculations assuming an isotropic index of refraction of n =1.63. This assumption could be checked independently by OWM with s–polarized light which gave consistent thickness data. One can see that already on a rather local scale (the frame dimensions are 570 x 420 μm^2) substantial thickness variations can lead to a smearing of the waveguide condition (Eq. 2.3) and hence to a reduced device performance, e.g. in integrated optics.

6. Conclusions

All the presented examples were to show that, indeed, all optical phenomena known for plane electromagnetic modes and used for materials' characterization, can be found also for the interaction of evanescent electromagnetic waves with matter. Thus it is possible to apply the same broad spectrum of optical techniques for investigating thin polymeric films and interfaces. Some of these experimental configurations offer substantial advantages over photons whenever particular challenges for enhanced sensitivity and resolution in interfacial and thin film problems are to be met.

7. References

[BOA 82]   "Electromagnetic Surface Modes", e. by A.D. Boardman, Wiley & Sons, New York 1982

[BUR 74]   E. Burstein, W.P. Chen, Y.J. Chen, A. Hartstein, J. Vac. Sci. Technol. **11**, 1004 (1974)

[DUC 88]   C. Duschl, W. Knoll, J. Chem. Phys. **88**, 4062 (1988)

[GOR 77] J.G. Gordon II, J.D. Swalen, Opt. Commun. **22**, 374 (1977)

[HIC 89] W. Hickel, W. Knoll, Acta Metall. **37**, 2141 (1989)

[HIC 90a] W. Hickel, W. Knoll, Appl. Phys. Lett. **57**, 1286 (1990)

[HIC 90b] W. Hickel, G. Duda, M. Jurich, T. Kröhl, K. Rochford, G.I. Stegeman, J.D. Swalen, G. Wegner, W. Knoll, Langmuir 6, 1403 (1990)

[HIC 90c] W. Hickel, W. Knoll, Thin Solid Films **187**, 349 (1990)

[HIC 90d] W. Hickel, W. Knoll, J. Appl. Phys. **67**, 3572 (1990)

[JOH 90] D. Johannsmann, W. Knoll, Prog. Colloid Polym. Sci. **83**, 146 (1990)

[KNB 89] H Knobloch, C. Duschl, W. Knoll, J. Chem. Phys. **91**, 3810 (1989)

[KNB 91] H. Knobloch, W. Knoll, J. Chem. Phys. **94**, 835 (1991)

[KNO 82] W Knoll, M.R. Philpott, J.D. Swalen, A. Girlando, J. Chem. Phys. **77**, 2254 (1982)

[KRE 72] E. Kretschmann, Opt. Commun. **6**, 185 (1972)

[KUH 72] H. Kuhn, D. Möbius, H. Bücher, in "Physical Methods of Chemistry", ed. By A. Weissberger, B.W. Rossiter, Wiley, New York 1972, part III B, chap. VII

[LEE 90] S. Lee, B. Hillebrands, J.R. Dutcher, G.I. Stegeman, W. Knoll, F. Nizzoli, Phys. Rev. B: Condens. Matter **41**, 5382 (1990)

[MOE 88] See any textbood, e.g., "Optics", K.D. Möller, University Science Books, Mill Valley, California 1988

[RAE 77] H. Raether, in "Physics of Thin Films", vol. 9, ed. by G. Hass, M.H. Francombe, R.W. Hoffmann, Wiley, New York 1977, p. 145

[ROT 87a] B. Rothenhäusler, W. Knoll, Appl. Phys. Lett. 51, 783 (1987)

[ROT 87b] B. Rothenhäusler, W. Knoll, Surf. Sci. **191**, 585 (1987)

[ROT 88a] B. Rothenhäusler, W. Knoll, Appl. Phys. Lett. **51**, 783 (1988)

[ROT 88b] B. Rothenhäusler, W. Knoll, J. Opt. Soc. Am. B: **5**, 1401 (1988)

[ROT 88c] B. Rothenhäusler, W. Knoll, Nature **332**, 615 (1988)

[SAW 91]   M. Sawodny, J. Stumpe, W. Knoll, J. Appl. Phys. **69**, 1927 (1991)

[SIM 89]   H.–U. Simmrock, A. Mathy, L. Dominguesz, W.H. Meyer– G. Wegner, Angew. Chem. (Adv. Mater.) **101**, 1148 (1989)

[SWA 78]   J.D. Swalen, K.E. Rieckhoff, M. Tacke, Opt. Commun. **24**, 164 (1978)

[TIE 69]   P.K. Tien, Rev. Mod. Phys. **49**, 361 (1969)

[USH 83]   S. Ushioda, Y. Sasaki, Phys. Rev. B: Condens. Matter **27**, 1401 (1983)

Optimization of Polymer Inverted-Rib-Waveguides by the Finite Element Method

M. Guntau, U. Bartuch, A. Bräuer, W. Karthe

Friedrich-Schiller-University Jena
Institute of Applied Physics

Abstract
The inverted rib-waveguide is a commonly used design of polymer strip waveguides. In order to calculate the optical properties of these devices a computer program on the basis of the finite element method (FEM) was developed. Starting with the calculated optical near fields the attenuation by scattering was estimated.

1. Introduction
The spin-coating technique applied to substrates with prefabricated grooves is a favourable technique for producing polymer strip waveguides. In order to get devices with minimal scattering and fibre - strip coupling losses it is necessary to find out optimal geometrical parameters. This letter gives an example of the application of the finite element method (FEM) to the calculation of the optical near field. The results will be used to get information about the scattering losses.

The basic configuration of the inverted rib-waveguides used in the investigations is shown in Fig. 1.1.

The grooves in the SiO_2-substrate are produced by ion-beam etching through a photolithographical mask, where the walls need not be vertical. The spin-coating of the polymer (PMMA) is the next step. The side walls and the bottom of the groove become roughly during the ion-beam etching, which causes scattering losses.

2. Calculation of the near field

Generally, a vectorial calculation is necessary to simulate the waveguiding process, including polarization effects (e.g. [HAY 86]). However, in the present case the accuracy of attenuation calculated by the scalar approximation is better than the errors of experimental results. Neglecting $\nabla\epsilon/\epsilon$ at material discontinuities the scalar Helmholtz equation

$$\Delta E + k_0^2 n(x,y)^2 E = 0 \qquad (2.1)$$

is used. E is the value of the electric field strength. Assuming propagation in z-direction the eigenvalue equation of the guided mode i can be written as

$$\Delta_{x,y}E + (n(x,y)^2 - n_{eff,i}^2) k_0^2 E = 0 \qquad (2.2)$$

with the effective index n_{eff}.
Now, the FEM shall be applied to the waveguide problem. A special input program was written to subdivide the cross-section into a number of first-order triangular elements (Fig. 2.1) with up to 6561 nodal points.
This computer program allows the reduction of the distance of the nodes at domains, where the greater field strength changes are expected. The variation of geometrical parameters is simple.
Equation (2.2) is converted by the FEM-formalism into a matrix eigenvalue equation, which is solved by inverse simultaneous iteration [PIS 84]. The resulting optical near fields of the modes can be used for further calculations.
Figure 2.2 shows an isoline plot of the calculated field of the basic mode of an inverted rib-waveguide with following parameters:
- groove width = 2 μm,
- groove depth = 0.456 μm,
- PMMA height = 1 μm.

The effective index $n_{eff}=1.4777$ ($\lambda=632,8$ nm) calculated by the FEM is in good agreement with the results, obtained by the effective index method for TE polarization ($n_{eff,TE}=1.4779$, $n_{eff,TM}=1.4771$).

3. Scattering of the waveguides

The fields obtained by the FEM are used to get information about the scattering behaviour of the strip waveguides. The influence of geometrical parameters are investigated.

The exact treatment of the waveguide scattering is very difficult. In this case the coupling of the guided mode into other guided and radiation modes must be considered [LEI 83]. Furthermore, the coupling coefficient of the polarization due to surface roughness must be calculated. Therefore an empirical way was chosen and only the correlation between calculated and measured scattering losses was investigated.

Because of the roughness at the groove walls and the relatively small refractive index step, most of the scattering losses occur at the polymer-substrate boundary. It was assumed that the attenuation due to this scattering is proportional to the integral

$$I_s \sim \int_{\partial polymer} |E|^2 dl. \qquad (3.1)$$

where E is the electrical field strength obtained by the FEM calculations and the integral has to be calculated on the polymer-substrate interface. The proportionality factor has to be derived from an exact theory.

The attenuation of such inverted rib-waveguides has been measured. The loss was determined by scattering. The experimental setup is shown in Fig. 3.1. The track of scattered light of the strip waveguide is imaged on a CCD linear sensor. Knowing the image scale the attenuation constant can be determined.

First results are demonstrated in Fig. 3.2. The width of the groove varied from 1.5 µm up to 4 µm. The FEM-calculated values of I_S from equation (3.1) were fitted with a constant scale factor to the attenuation values obtained by the experiment.

On the basis of this results further numerical investigations were carried out. Keeping other parameters constant, the optical field is more concentrated in the groove and so the loss increases with the depth of the groove (Fig. 3.3).

In another case we changed the slope angle α of the groove side walls. Fig. 3.4 shows that the slope angle has only a small influence on the scattering.

4. Conclusions

The FEM in scalar approach can give useful information about the properties of optical waveguides. This approximation is limited to problems with moderate refractive index steps, were the polarization effects are small. The application of FEM-calculations to the estimation of scattering losses was demonstrated.

References

[HAY 86]   Hayata, K.; Koshiba, M.; Eguchi, M.; Suzuki M.: Vectorial Finite-Element Method Without Any Spurious Solutions for Dielectric Wave-guiding Problems Using Transverse Magnetic-Field Component. IEEE Trans. Microwave Theory Tech., vol. MTT-34, pp. 1120-1124, Nov. 1986.

[LEI 83]   Leine, L.: Theoretische Untersuchungen zur Parameterbestimmung von Schichtwellenleitern und zur Dämpfung der geführten Moden durch statische Grenzflächenrauigkeiten und Brechzahlfluktuationen in der Schicht. Ph.D. thesis, FSU Jena, Jena 1983.

[PIS 84]   Pissanetzky, S.: Sparse Matrix Technology. London: Academic Press 1984.

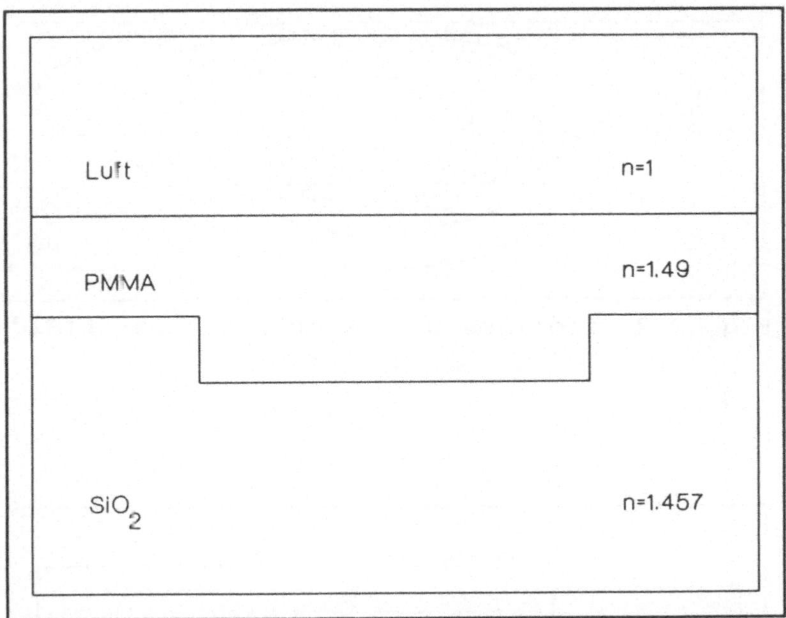

Fig. 1.1   inverted rib-waveguide configuration

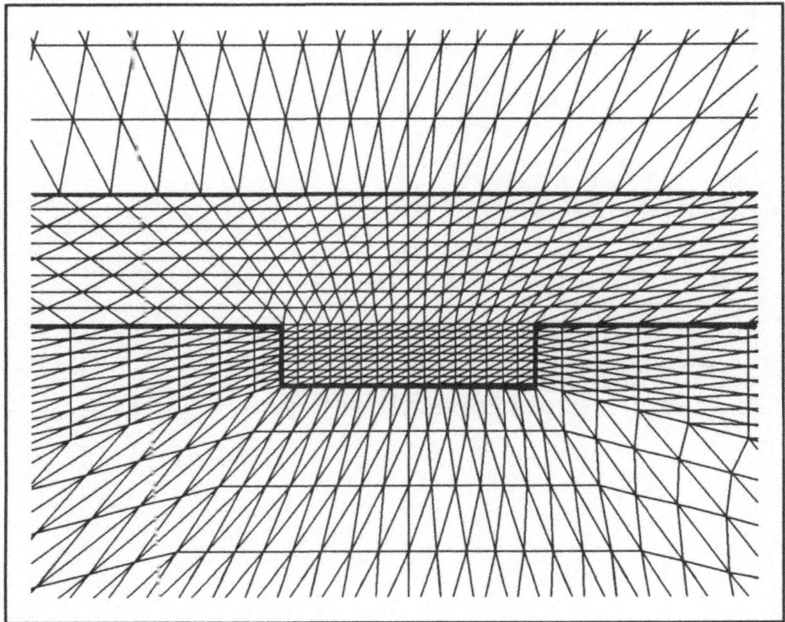

Fig. 2.1   finite element division

118

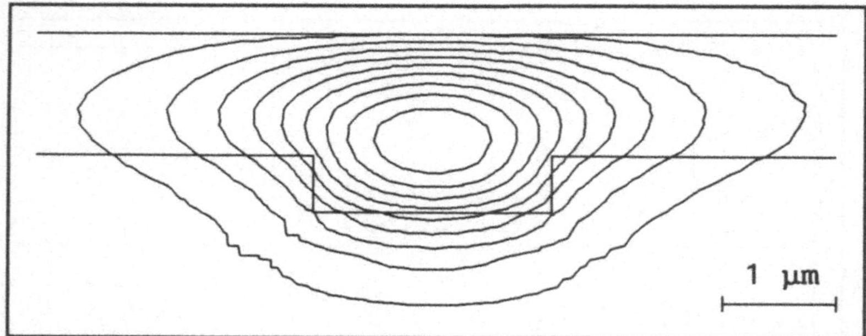

Fig. 2.2   isolines of the optical near field

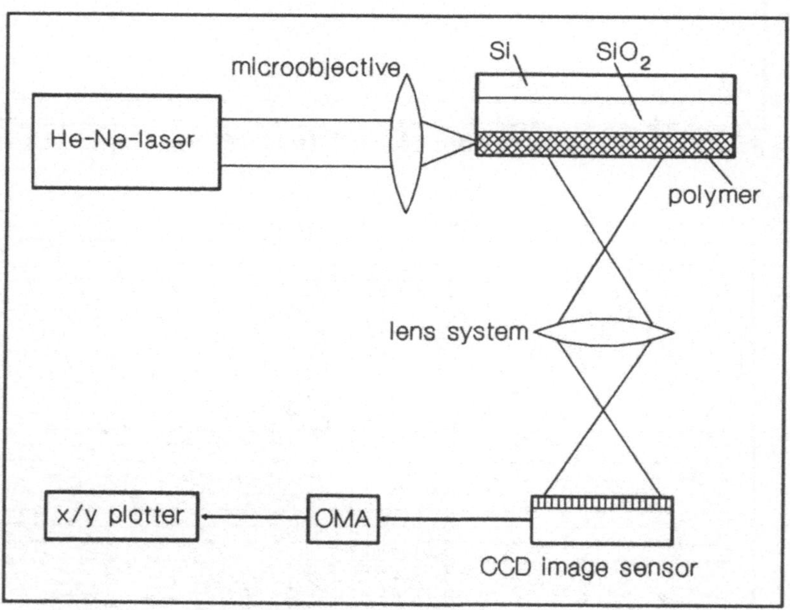

Fig. 3.1   Experimental setup

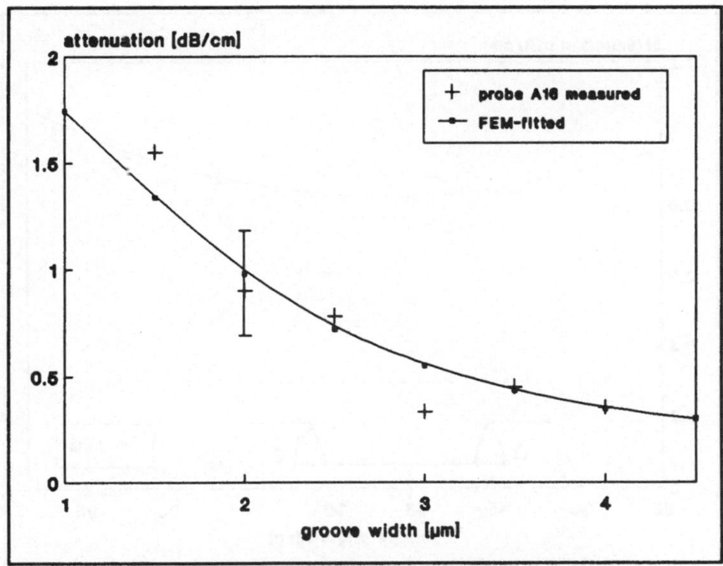

Fig. 3.2   groove depth   = 0.456 µm,
             polymer height = 1.0 µm

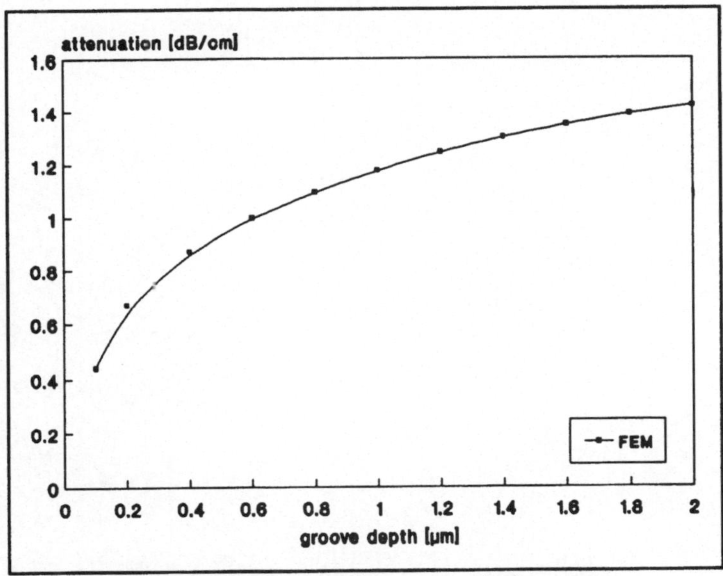

Fig 3.3   groove width   = 2.0 µm,
            polymer height = 1.0 µm

120

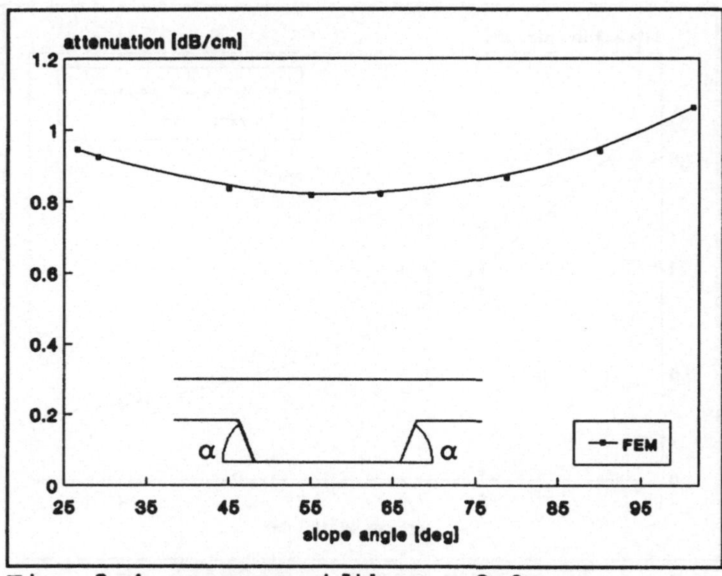

Fig. 3.4   groove width   = 2.0 µm,
            groove depth   = 0.5 µm,
            polymer height = 1.0 µm

$\chi^{(3)}$ - Effects in Polymer Waveguides

A. Bräuer, U. Bartuch, W. Karthe, M.Zeisberger*

Friedrich-Schiller-University Jena
Institute of Applied Physics
*Institute for Physical High Tech, Jena

Abstract

Kerr-like polymers can be used for all-optical switching in waveguides. $\chi^{(3)}$ for the conjugated polymer MP-PPV has been determined (2.7 $\cdot$ 10^{-11} e.s.u.) by prism coupling into the nonlinear waveguide. As an example for all-optical switching in waveguides, the beam-scanner has been realized with a grating coupler in a nonlinear polymer waveguide made from azo-dye doped poly-methyl methacrylate (PMMA). Material trade-offs for opto-optical switching in waveguides are discussed and the currently known nonlinear polymers are evaluated due to their applicability.

1 Introduction

Alongside with the growing application of polymers for optics, their nonlinear optical properties are investigated in a number of groups. Especially, polymers with second order nonlinearities have attracted a great deal of interest for the fabrication of electrooptical modulators and switches. Many industrial research groups work successfully in this field. Third order nonlinearity polymers are still the subject of basic research, but the number of experiments utilizing them is growing. There are two main interests. First, the nonlinear properties of polymers ($\chi^{(3)}$, response time τ) have to be studied, also, in order to compare the results with those found by theoretical calculations. The second direction in this research is the design and fabrication of all-optical waveguide switches for optical communications. The interesting time-scale of function for this elements varies from femtosecond up to millisecond region.
In this paper we report about two waveguide methods for the

122

determination of $\chi^{(3)}$ and, as an example, measurements of electronic nonlinearities in a conjugated polymer. Then, the principle of all-optical switching in waveguides is demonstrated experimentally and demands to the nonlinear material are discussed.

2 $\chi^{(3)}$-Measurements

2.1 Kerr-Effect

In order to realize all-optical switching elements Kerr-materials , i.e. materials having a measurable third-order susceptibility $\chi^{(3)}(\omega;\omega,\omega,-\omega)$ can be exploited. For a single guided wave in an isotropic medium, this susceptibility results in a nonlinear polarization

$$P_i^{NL}=2\varepsilon_o\chi_{1122}^{(3)}E_iE_j^*E_j+\varepsilon_o\chi_{1221}^{(3)}E_i^*E_jE_j \tag{1}$$

where E_i are the electric field components. Far from resonance, the symmetry condition

$$\chi_{1122}^{(3)}=\chi_{1221}^{(3)}=\chi^{(3)} \tag{2}$$

holds. Assuming plane waves, the total polarization of the medium can be written as

$$P_i(\omega)=\varepsilon_o(\chi_{ii}^{(1)}+3\chi^{(3)}|E_j|^2)E_i \tag{3}$$

where it is assumed that all $\chi^{(2)}$-components are zero, which is true for centrosymmetric materials. Formula (3) leads to an intensity dependent refractive index and a respective nonlinear phase shift

$$n=n_o+n_2I$$
$$\varphi=\frac{2\pi}{\lambda}n_2IL \tag{4}$$

where n_o is the linear refractive index, L the length of the waveguide, I $= 1/2\ c\varepsilon_o n_o |E|^2$ the intensity and the nonlinear refractive index parameter n_2 is determined by

$$n_2 = \frac{3\chi^{(3)}}{n^2 \varepsilon_o^2 c} \tag{5}$$

The relation between $\chi^{(3)}$ in e.s.u. and n_2 in m^2/W is

$$n_2[m^2/W] = \frac{1.6 \cdot 10^{-5}}{n_o^2} \chi^{(3)}[e.s.u.]. \tag{6}$$

There are different mechanisms causing Kerr-effect in materials. For application in polymer waveguides, two of them have to be prefered, the nonresonant electronic nonlinearity and the photoinduced structure transition of molecules. Both of them are not connected with strong (nonlinear) absorptions so that phase changes caused by n(I) can accumulate along the waveguide. Typical response times for the Kerr-effect in polymers are in the fs to ps region, whereas the structure transition is much slower (ms to min)

2.2 n_2-Measurements by Waveguide Methods

There are two methods being well adapted for n_2-determination using polymer waveguides: prism coupling (PC) and interferometer (MZI) method. We applied the first one in order to measure n_2 of poly(1,4-phenylene-1-phenylvinylene), MP-PPV, which is a soluble conjugated polymer of the widely studied PPV series [BAR 92]. The experimental set-up is shown in the inset of Fig. 2.1. Light, in this case 30 ps single pulses of a Nd:YAG laser at 1.06 µm wavelength is coupled into the waveguide (E_{in}). The resonance condition for optimum coupling depends among other things on the refractive index of the waveguiding film, which is for a Kerr-medium controlled by the guided wave intensity (cf. equ.(4)). Consequently, the optimum coupling angle varies with changing input fluence. The detection is made by measuring the energy

124

transmitted through the waveguide E_{tr} which leads to exact results in the case of nonresonant nonlinearties, i.e. when the nonlinear absorption in the waveguide need not be considered. The linear losses of the waveguiding film can be measured separately and considered in the E_{tr} determination. For measurements at wavelengths near to the absorption edge of the material where nonlinear absorption has to be expected, the energy E_r reflected from the prism base must be measured separately. In order to do that, a small groove is polished into the prism base. In the measurement described here, nonresonant conditions are obvious because the wavelength of maximum absorption for MP-PPV is at about 0.42 µm, i.e. apart even from half of the laser wavelength so that two-photon-absorption (TPA) does not contribute.

The measured fluence dependence of the optimum coupling angle (cf.Fig.2.1)

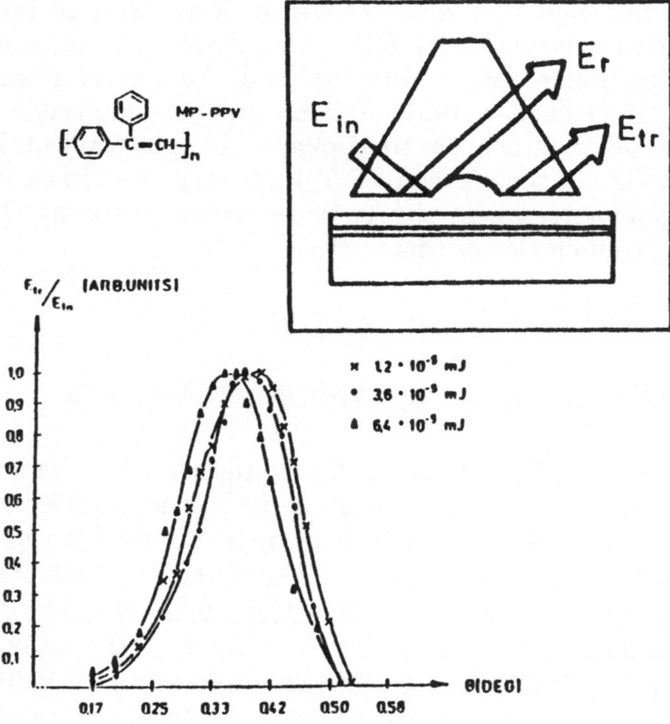

Figure 2.1   Dependence of the transmitted energy on the coupling angle in the nonlinear prism coupling experiment in MP-PPV waveguides for three different values of input energy

leads to a maximum refractive index change of $\Delta n = 4 \cdot 10^{-4}$ and results in $n_2 = 6 \cdot 10^{-17}$ m^2/W.

The prism coupling method is very suitable for n_2-determination and needs less effort than Degenerate Four Wave Mixing (DFWM). Disadvantages are errors in the estimation of the guided intensity caused by the problems of defining the lateral beam width (which can be overcome by using strip waveguides) and uncontrollable leaky modes.

A high stability dual-beam interferometer was invented [GÖR 90] for measurement of small refractive index changes due to the photorefractive effect. It can also be used for n_2-determination [BRÄ 91]. Basically, it is a Mach-Zehnder interferometer (see Fig.2.2). The input beam is split by a double prism into two beams of controllable intensities and directions which are coupled via one and the same

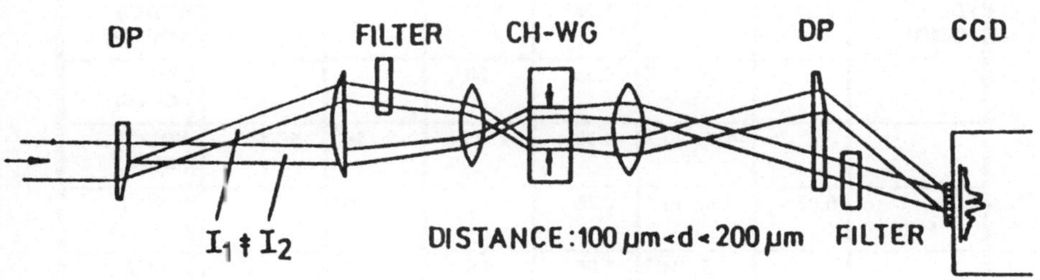

Figure 2.2   Dual beam interferometer for n_2-measurements with double prisms (DP) and channel waveguides (CH-WG) indicated

microlens into two channel waveguides (typical distance 100 to 200 µm), and they leave the waveguides as Gaussian beams. Another microlens and a second double prism are used for imaging the interference pattern of the two beams onto a ccd-line. The intensity of one beam has to be high enough in order to cause the nonlinear Δn due to Eq.4. The resulting phase difference $\Delta\varphi \propto n_2$ between the two beams leads to a

126

shift of the interference fringes, from which the nonlinear refractive parameter n_2 can be determined. This interferometer is very sensitive and stable because thermal drifts are compensated. Refractive index changes of $1 \cdot 10^{-7}$ in one of the waveguides have already been detected.

The high sensitivity of waveguide methods is due to the long interaction length between the optical wave and the material leading to the accumulation of the phase shift along the propagation length. But, of course, there is a need of fabricating low loss waveguides.

Material	λ_{max} in μm	$\chi^{(3)}$ in e.s.u.	λ μm	$\alpha[cm^{-1}]$ β [cm/GW]	Δn	$\Delta n/\alpha\lambda$ $-n_2/\beta\lambda$	Method [Ref.]
polysilane	0.34	$2 \cdot 10^{-12}$	0.53				OKG [YAN 88]
		$2 \cdot 10^{-12}$	0.53				DFWM [YAN 88]
PDA		$7 \cdot 10^{-12}$	0.53	6.9			DFWM [DEN 85]
PDA-4BCMU	0.50	$7 \cdot 10^{-11}$	1.06	0.23			WG(PC) [TOW 88]
		$3 \cdot 10^{-10}$	0.53	50			OKG [HO 86]
		$-1.4 \cdot 10^{-12}$	0.63	<100	$-3 \cdot 10^{-4}$	$5 \cdot 10^{-4}$	WG(PC) [SIN 88]
PDA-PTS	0.67	$1.5 \cdot 10^{-11}$	0.75		$1 \cdot 10^{-3}$		WG(GC) [CAR 83]
		$1.5 \cdot 10^{-9}$	1.06	0.5	$2 \cdot 10^{-4}$	4	WG(MZI) [THA 90]
PPV	0.49	$2 \cdot 10^{-11}$	0.65	1	$2 \cdot 10^{-3}$	30(?)	DFWM [BUB 89]
		$8 \cdot 10^{-11}$	0.57				DFWM [NEH 89]
MP-PPV	0.42	$+2.7 \cdot 10^{-11}$	1.06	1	$4 \cdot 10^{-4}$	3.7	WG(PC) [BAR 91]
DANS	0.43	$2.3 \cdot 10^{-12}$	1.06	2.8 (2)		0.3	WG(GC) [MAR 91]
DAN2	0.44	$3.2 \cdot 10^{-12}$	1.06	4			WG(GC) [MAR 91]
MQW		$5 \cdot 10^{-7}$	0.74	30	$2 \cdot 10^{-4}$	0.9	
A_2B_6-glass	0.49	$1.4 \cdot 10^{-9}$	0.53	1	$5 \cdot 10^{-5}$	0.9	

Table 1   Measured Kerr-nonlinearities $\chi^{(3)}(\omega;\omega,\omega,-\omega)$ for polymers

Other methods for $\chi^{(3)}(\omega;\omega,\omega,-\omega)$-measurements are using the optical Kerr-gate (OKG) (e.g. [YAN 88]) or the Degenerate Four Wave Mixing (DFWM) ([DEN 85]), which are usually not performed in waveguide configuration.

The results of Kerr-nonlinearity measurements are listed in Table 1, and also the wavelength of maximum absorption of the material, λ_{max}, and the wavelength at which the measurement was performed (columns 1 to 3). The wavelength dependence of $\chi^{(3)}$ can clearly be seen, especially the enhancement near to resonance. Note, that only a few authors report about the sign of $\chi^{(3)}$, what is indicated in brackets.

3 All-Optical Switching in Waveguides

3.1 Waveguide Dispersion Relation

There are a number of pecularities for all-optical switching in waveguides. First of all, one can choose in principle, whether the waveguide or the cover or the substrate is made from the nonlinear material for which $n = n_0 + n_2 I$ holds. For all configurations the nonlinear refractive index change $\Delta n = n_2 I$ can be obtained by a similar

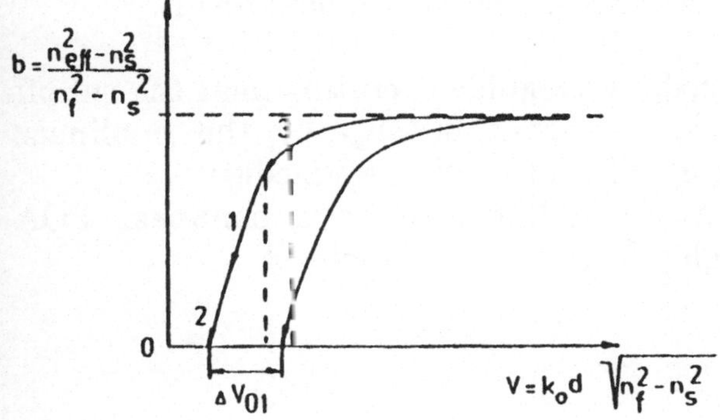

Figure 3.1 Dispersion relation with three possible sections of operation indicated

equation which is basically the nonlinearity averaged over the transverse intensity distribution, provided that the field distributions are not a function of the guided wave power. The latter condition is true for $\Delta n < n_f - n_s$ which holds for most of the practical applications. In our experiments we have used nonlinear polymer films as waveguides.

In a waveguide device one has to take care about the operating point, i.e. in which section of the dispersion curve (BV-diagram see Fig.3.1) one intends to run the nonlinear experiment. Because of $n_f = n_o + n_2 I$ the V-Parameter is a function of the input intensity

$$\Delta V_{NL} = V \frac{2n_o}{n_o^2 - n_s^2} n_2 I \tag{7}$$

For weak nonlinearities ($|n_2 I| \ll |n_o^2 - n_s^2|$) and $n_2 < 0$, three interesting sections of the dispersion curve do exist (cf. Fig. 3.1):

section 1:   there is a monomode waveguide, the nonlinearity shifts n_{eff}
to   lower values, however, the monomode region is preservedin the waveguide, for constant guided power the phase shift is $\Delta\varphi = (2\pi/\lambda)\Delta n_{eff} L$ (L: length of the waveguide)

section 2:   the monomode waveguide is used near to cut-off, if the nonlinear shift can be increased to $|\Delta n_{eff}| > |n_{eff} - n_s|$, a "cut-off"modulator can be made in principle, but the adjustment of the operating point will be difficult.

section 3:   there is a two-modal waveguide operating near the cut-off the second mode, the shift of Δn_{eff} by the nonlinear refractive index is different for the two modes: $|\Delta n_{eff,2}| \gg |\Delta n_{eff,1}|$; in this manner, a nonlinear BOA coupler with high efficiency can be realized

3.2 All-Optical Beamscanner

The nonlinear all-optical beamscanner was analyzed first in [ASS 89]. The configuration we have used is somewhat different (see Fig.3.2). A HeNe-Laser was prism-coupled to a waveguide made from PMMA doped with methylred, which is a molecule possessing a trans-cis transition induced nonlinearity. The SiO_2-substrate is corrugated with a grating of about 600nm grating constant and 150 nm depth. In the region of the grating ,the waveguide condition is disturbed periodically which leads to a partial outcoupling of the guided wave, making an angle Φ with the

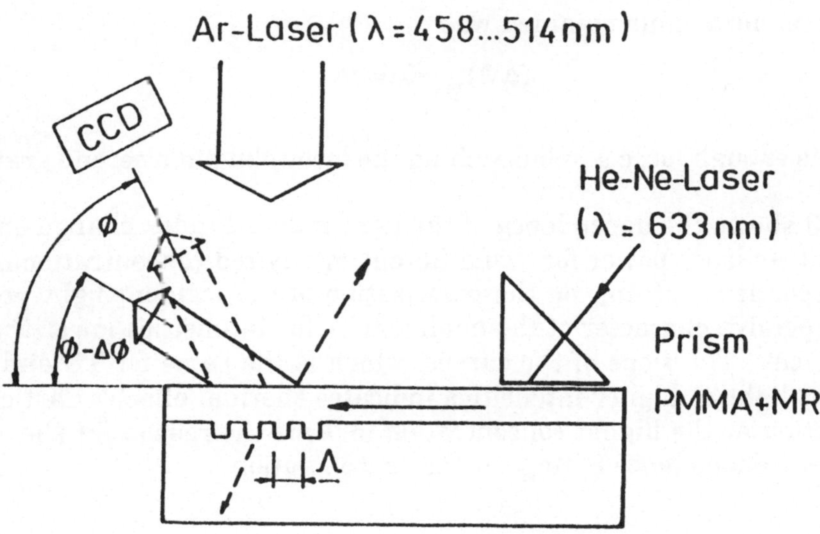

Figure 3.2   All-optical beamscanner using methylred-doped PMMA waveguide and corrugated grating in the SiO_2 substrate (grating constant 600 nm)

waveguide's surface. This angle is determined by the phase matching condition

where n is the index of the waveguiding material
n_{eff} the mode index and

$$ncos\,\Phi = n_{eff} - m\frac{\lambda}{\Lambda} \tag{8}$$

Λ the grating constant.

Illumination of the waveguide grating with Ar-laser light (λ = 458 to 514 nm, $I \leq 20$ mW/cm^2) leads to a light induced change of the refractive index, $\Delta n_{eff} = n_2 I$, and eventually, the angle Φ is changed due to

$$\Delta n_{eff} = -\sin\Phi \cdot \Delta\Phi. \tag{9}$$

The maximum value obtained was

$$(\Delta\Phi)_{max} = 0.6\,mrad \tag{10}$$

which is enough for, e.g. mismatching the incoupling in a second grating.

Fig. 3.3 shows the dependence of the light induced index change on the incident Ar-laser power for two different methylred concentrations. The dependence of Δn_{eff} on the polarization of the incident light proves the dispersive character of the nonlinearity for intensities lower than 10 mW/cm^2. The slope of the curves, which is the same for TE and TM incident light at higher intensities indicates thermal effects. Particle interaction at the higher concentration (6 %) is the reason for the nonlinear dependence of Δn_{eff} on the concentration.

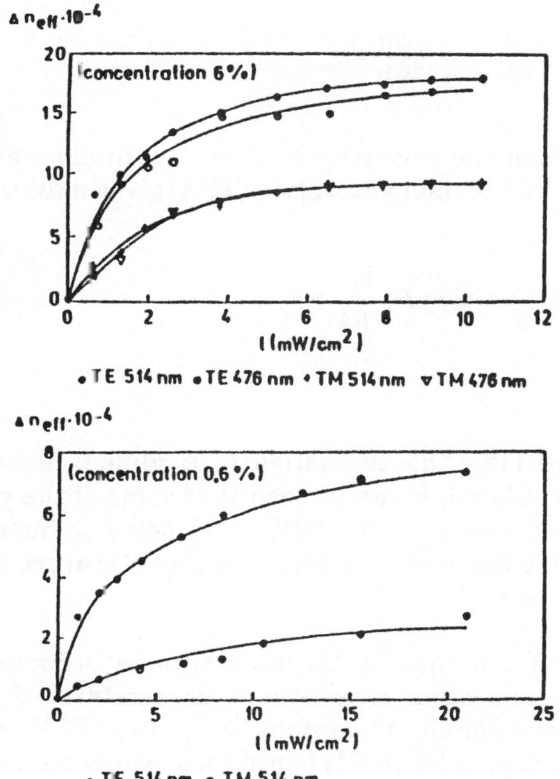

Figure 3.3   Optically induced refractive index change in methylred-doped PMMA waveguides   for two different concentrations

3.3 Material Trade-offs for Opto-Optical Switching in Waveguides

The main advantage in applying waveguides for opto-optical switching is that a wave with high intensity can be guided over a long distance, so that optically induced phase changes (equ. (4)) can accumulate and reach multiple of π, which is a necessary condition for all-optical switching. This advantage is lost, if losses (including scattering in the waveguide) play an important role. Therefore, the figure of merit for the material includes, besides of the nonlinear refractive index change also the absorption constant α. For any optical switching system, the condition has to be fulfilled, where c_{sw} is a numerical constant of the order of unity, whose precise value depends on the exact switching scheme. For, e.g. the nonlinear directional coupler, $c_{sw} = 2$ is obtained ([MIZ 89]). For

132

$$\frac{|\Delta n|}{\alpha \lambda} > c_{sw} \tag{11}$$

wavelengths greater than the wavelength of maximum absorption λ_{max} of the polymer, mainly two-photon absorption (TPA) gives a contribution. Then,

$$\alpha \approx \beta I \rightarrow \frac{|n_2|}{\beta \lambda} > c_{sw} \tag{12}$$

is obtained.

Following eq. (11) and (12), the evaluation of nonlinear polymers in Table 1 has to be reconsidered. It can be seen that some of the polymers have better figures of merit than MQW and semiconductor-doped glasses. But note , that the α-values given in the literature are only estimations in many cases.

There are a number of attempts in the literature in order to find a relation between $\chi^{(3)}$ and other measurable parameters of a given material, e.g. the absorption maximum λ_{max} (e.g.[FLY 87]) for nonresonant $\chi^{(3)}$). Recently, in [SHE 91] based on a two-band model, the dispersion of $\chi^{(3)}$ was theoretically predicted by calculating the nonlinear nondegenerate absorption and using Kramers-Kronig relation. The conformity with the measurements in semiconductor materials is excellent. The observed change of sign of n_2 midway between the two-photon absorption edge and the fundamental absorption is predicted as well as the n_2-peak at frequencies just above the TPA-edge. As pointed out in [HEE 86], the conjugated polymers can to some extend considered as semiconductors. In Fig. 3.4 the n_2-values found from Tab. 1 are represented versus the ratio between the maximum absorption wavelength and the wavelength of the measurement, neglecting the uncertainties of the sign of n_2. This empirical curve is in good qualitative

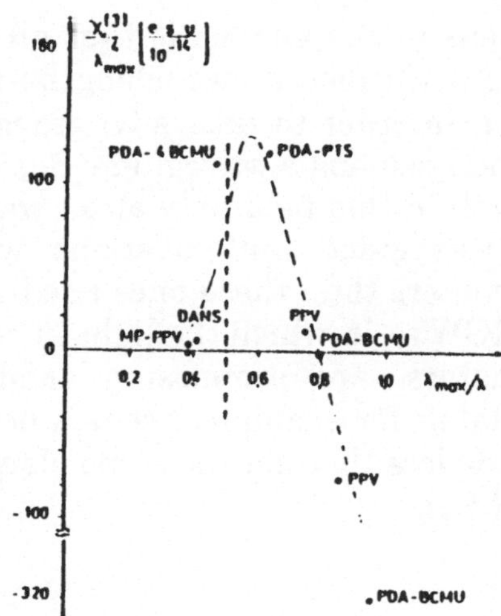

Figure 3.4   Measured dispersion of parameters n_2 for polymers (from Tab. 1)

agreement with the one given in [SHE 91].

The conclusion of the figure of merit discussion is that one has to run an all-optical waveguide element at wavelengths just above the TPA-edge, where low absorption and high n_2 is expected. It was done in [AIT 91], and an excellent functioning nonlinear directional coupler of 6.25 mm length in AlGaAs was realized, although $\chi^{(3)}$ for this material is not extremely high $(5.3 \cdot 10^{-12} \text{e.s.u.})$.

4   Conclusions for Further Development in All-optical Switching in Polymer Waveguides

Of course, the quality of an all-optical device depends on the material parameters. The research in many chemistry laboratories in order to synthesize polymers with higher $\chi^{(3)}$ values must be highly appreciated. Also, the possibility of "tailoring" of polymers should lead to materials with absorption maxima away from wavelengths of interest for, e.g. communication purposes.

134

The consequence of the discussions in chapter 3.3 is that all nonlinear waveguide experiments must be performed at wavelengths where the material has very low absorption in order to take advantage from the accumulating nonlinear phase change along a waveguide. So, it seems to be not very senseful to do research in this field only at on wavelength. Another approach is to find waveguide configurations with lower demands for the nonlinear parameters than those ones used up to now. Resonance waveguides (e.g. ARROW's), in which the light in confined by mismatched Fabry-Perot resonators, are promissing candidates. A nonlinear ARROW-cut-off modulator, for example, needs a nonlinearity which is by one order of magnitude less than for the same element made from classical waveguides [MAN 91].

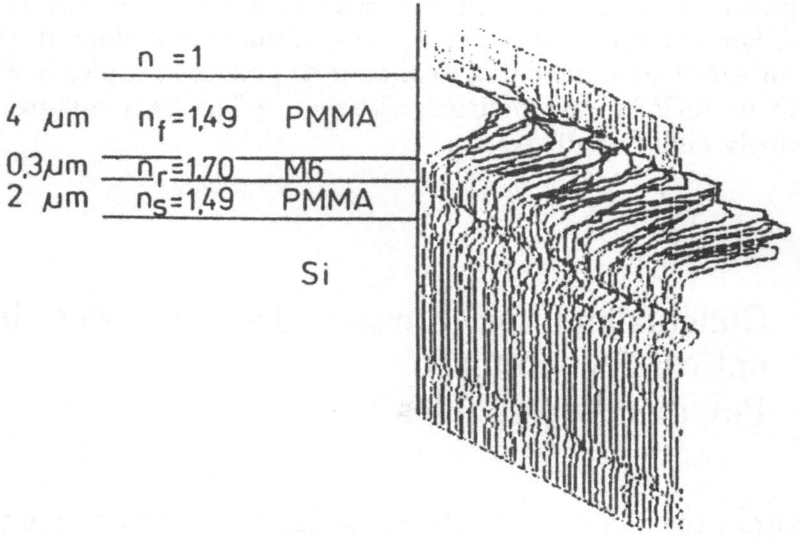

Figure 4   Planar ARROW made from polymers and measured near field intensity

As a first step we have fabricated a planar ARROW using two different polymers on a Si-wafer. As it can be seen in Fig. 4, the main intensity of light is, as expected, in the low index polymer. Optimization of the configuration and fabrication of laterally confined ARROW's are underway.

References

[AIT 91]   Aitchison, J.S., Kean, A.H., Ironside, C.N., Villeneuve, A., Stegeman, G.I.: Nonlinear Directional Coupler in $Al_{C.18}Ga_{0.82}As$ Near Half the Band Gap. PAPER Pd 2-1 on OSA Conference on Nonlinear Guided Wave Phenomena in Cambridge, UK, Sept. 1991

[ASS 90]   Assanto, G., Stegeman, G.I.: Nonlinear all-optical beam scanner. J. Appl. Phys. 67 (1990) 1188

[BAR 92]   Bartuch, U.; Bräuer, A.; Dannberg, P.; Hörhold, H.-H.; Raabe, D.: Measurement of high nonresonant third-order nonlinearity in MP-PPV waveguides. Int. J. Optoelectronics 7 (1992) 275

[BRÄ 91]   Bräuer, A., Bartuch, U., Dannberg, P., Zeisberger, M., Bauer, Th.: Linear and Nonlinear Effects in Polymer Optical Waveguides. SPIE Proc. 1559 (1991) 470-477

[BUB 89]   Bubeck, C., Kaltbeitzel, A., Neher, D., Stenger-Smith, J.D.,

[BRÄ 91]   Bräuer, A., Bartuch, U., Dannberg, P., Zeisberger, M., Bauer, Th.: Linear and Nonlinear Effects in Polymer Optical Waveguides. SPIE Proc. 1559 (1991) 470-477

[BUB 89]   Bubeck, C., Kaltbeitzel, A., Neher, D., Stenger-Smith, J.D., Wegner, G., Wolf, A.: Nonlinear Optical Properties of Thin Films of Polymers with a One-Dimensional Conjugated Electron System. In: H. Kuzmany, M. Mehring, S. Roth (ed.): Electronic Properties of Conjugated Polymers III, Springer Series in Solid-State Sci. 91 (1989) 214

[CAR 83]   Carter, G.M., Chen, Y.J., Tripathy, S.K.: Intensity-dependent index of refraction in multilayers of polydiacetylene. Appl. Phys. Lett. 43 (1983) 891

[FLY 87]   Flytzanis, C. (1987) in: D.S. Chemla, J. Zyss (ed.): Nonlinear Optical Properties of Organic Molecules and Crystals, Vol.2, Academic Press, 121

[GÖR 90]   Göring, R., Rasch, A., Karthe, W.: Quantitative Investigations of Photorefractive Effect in $LiNbO_3$ Channel Waveguides. SPIE, 1274 (1990), 19

[HEE 86]   Heeger, A.J., Moses, D., Sinclair, M.: Semiconducting Polymers: Fast Response Non-Linear Optical Materials. Synthetic Metals 15 (1986) 95-104

[HO 86]   Ho, P.P., Yang, N.L., Jimbo, T., Wang, Q.Z., Alfano, R.R.: Ultrafast Resonant Optical Kerr Effect in 4-Butoxy-carbonylmethylurethane Polydiacetylene. J. Opt. Soc. Am. B4 (1987) 1025

[MAN 91]   Mann, M., Trutschel, U., Lederer, F., Wächter, Ch., Leine, L.: Nonlinear Leaky Waveguide Modulator. J. Opt. Soc. Am. B8 (1991) 1612

[MAR 91]   Marques, M.B., Assanto, G., Stegeman, G.I., Möhlmann,G.R., Erdhuisen, E.W.P., Horsthuis, W.H.G.: Large, nonresonant, intensity dependent refractive index of

[MIZ 89]   Mizrahi, V., DeLong, K.W., Stegeman, G.I., Saifi, M.A., Andrejco, M.J.: Two-photon absorption as a limitation to all-optical switching. Opt. Lett. 14 (1989) 1140

[NEH 89]   Neher, D., Kaltbeitzel, A., Wolf, A., Bubeck, C., Wegner, G.:
Nonlinear Optical Properties of Ultrathin Polymer Films.
In:J.L. Bredas, R.R. Chance (ed): Conjugated Polymeric
Materials: Opportunities in Electronics, Opto-Electronics
and Molecular Electronics, NATO ASI Series E 182, Kluwer
Acad. Publ., Dordrecht (1990) 387

[SHE 91]   Sheik-Bahae, M., Hutchings, D.C., Hagan, D.J., Van
Stryland, E.W.: Dispersion of Bound Electronic Nonlinear
Refraction in Solids, IEEE J. Quantum Electronics 27
(1991) 1296

[SIN 88]   Sinclair, M., McBranch, D., Moses, D., Heeger, A.J.: Time-
resolved waveguide modulation of a conjugated polymer.
Appl. Phys. Lett. 53 (1988) 2374

[THA 90]   Thakur, M., Krol, D.M.: Demonstration of all-optical phase
modulation in polydiacetylene waveguides. Appl. Phys. Lett.
56 (1990) 1213

[TOW 88]   Townsend, P.D., Baker, G.L., Schlotter, N.E., Klausner,
C.F., Etemad, S.: Waveguiding in spun films of soluble
polydiacetylenes. Appl. Phys. Lett. 53 (1988) 1782

[YAN 88]   Yang, L. Wang, Q.Z., Ho, P.P., Dorsinvolle, R., Alfano, R.R.,
Zou, W.K., Yang, N.L.: Ultrafast time response of optical
nonlinearity in polysilane polymers. Appl. Phys. Lett. 53
(1988) 1245

Absorption Studies of Optically Nonlinear Side-Chain Polymer and Polymer-Dye Films oriented by Corona Poling

ARMIN WEDEL, RUDI DANZ, WOLFGANG KÜNSTLER[*] and KLAUS TAUER[+]

Fraunhofer-Establishment of Applied Polymer Research, [*]University of Potsdam, [+]Max-Planck-Institut für Kolloid- und Grenzflächenforschung, Kantstrasse 55, O-1530 Teltow-Seehof, F.R.G.

Abstract

Spectroscopic absorption measurements in the wavelength range from 700 to 300 nm are used to study the orientation of optically nonlinear side-chain groups in an acrylic polymer and dye guest molecules in a polymer matrix by corona poling process. The time stability and the annealing behaviour of these films are measured. The poling process produces efficient orientational order at room temperature of the acrylic side-chain polymers. With stepwise annealing up to poling conditions the orientation order was measured. The guest-host systems show a different behaviour.

1   Introduction

The copolymerization of optically nonlinear polymers and the incorporation of optically active dopants into polymer films induce new useful electrical and optical properties of these materials. Polymer glasses doped with optically nonlinear organic dye molecules and nonlinear optical polymers posses reasonably large second order nonlinear optical susceptibilities [SIN 86]. This concept involves electric field poling in the near of the glass transition temperature (T_g) of the sample resulting in an alignment of the dipolar dye molecules or side chains and yielding a noncentrosymmetric structure.

Corona poling is a well situated method for charging of polymers and has been used to pole electrets or polyvinylidene fluoride (PVDF) films, for example. Corona poling by discharge in an inhomogeneous electric field, usually at atmospheric pressure, can be performed at room temperature, but poling at elevated temperature has several advantages for polymers. Applying a temperature in the near of T_g during poling guest-host systems increases the molecular mobility of the guest molecules. Lowering the temperature well below T_g during applied field, the guest molecules can be frozen into their new orientation.

2 Experimental

Copolymer synthesis

The side-chain copolymers were synthesized by a special dispersion polymerization technique with a surface active azo-initiator. The polymerization was carried out in the following manner: 90 g of water as dispersing agent, 0.8 g ($2.008 \cdot 10^{-3}$ moles) of the azo-comonomer (Figure 1), 10 g (0.1 moles) of methylmethacrylat (MMA) and 0.25 g of surface active initiator were well mixed at 80 ^{0}C in a polymerization flask equipped with stirrer, condenser and nitrogen injection [TAU 91]. The polymerization was stopped after a polymerization time of 390 minutes and the final copolymer dispersion was coagulated in methanol/HCL mixture.

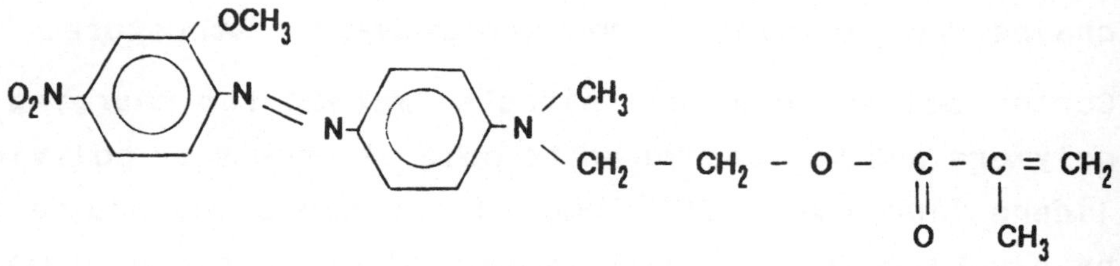

FIGURE 1   Azo-comonomer of the side-chain polymer.

The polymerization conversion was above 95%. The copolymers were washed with methanol and purified by two-fold precipitation from a toluene solution with methanol/HCL. The purified copolymer was characterized by viscosity and elemental composition which gave an intrinsic viscosity number $[\eta]$ of 1.57 dl/g (25 ^{0}C, benzene as solvent, Ubbelohde viscometer) and a dye content of 4.9 % w/w related to MMA units. T_g was measured via DSC (Perkin Elmer DSC 7) as ca. 115 ^{0}C. Dispersion polymerization leads to much higher polymerization rates and molecular weights compared to solution polymerization.

2.1   Sample preparation

The side-chain polymer was dissolved in ethyl-acetoacetate. For preparation of polymer-dye composite films 10% w/w of azoviolet dye (Fluka 73020) and azo-comonomer were dissolved with polymethylmethacrylate (PMMA, $T_g \approx 120$ ^{0}C) in ethylacetoacetate, respectively. The azoviolet dye was also dissolved with polystyrene (PS, $T_g \approx 100$ ^{0}C) in xylene, and with polycarbonate (PC, $T_g \approx 145$ ^{0}C) in chloroform. All polymer solutions were mixed well, filtered through a glass frit filter and spin coated onto indium-tin-oxide (ITO) glass. Spun films were dried in an oven at 60 ^{0}C for 1 hour to reduce the content of solvents. The thickness of dried films ranged from 1.5 to 2.5 μm.

2.2 Corona poling

A cross sectional view of the poling arrangement is given in Figure 2. The corona discharge was generated by a

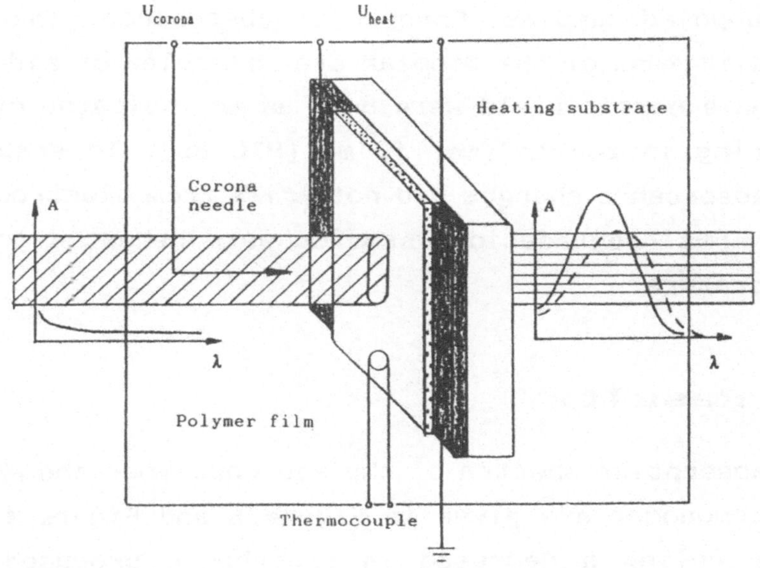

FIGURE 2   Experimental set-up for spectrophotometer in situ-poling and heating

needle electrode biased with –4 kV across a 10 mm air gap normal to a grounded planar electrode. The temperature could be raised by a heating substrate [PAG 90]. The temperature was measured with thermocouple on the top of this electrode. The substrates were placed film side up onto planar electrode. Reaching the poling temperature (T_p) in the near of T_g the poling field was applied and held constant, while after 5 min the temperature of the film was lowered to room temperature (RT). The poling field was applied for approximately 15 min across all films.

2.3 Absorbance measurements

The absorption spectra were measured with the UV/VIS/NIR spectrometer Lambda 19 from Perkin Elmer.

We have compared the optical absorbance of thermopoled and unpoled samples. Changes in absorbance generated by the alignment of the dipolar dye molecules or side chains in the electric field were used as an indicator of frozen ordering in the polymer films [HIL 89]. To ensure that the absorbance changes did not arise from electrochemical or thermal degradation, samples were reheated to poling temperature.

3 Results

The absorption spectra of the azo-copolymer and PMMA with azo-comonomer are given in Figure 3 and Figure 4.

After poling a decrease in absorbance produced by the alignment of the dipolar dye molecules as well as a shift to longer wavelengths for the azo-copolymer and to shorter wavelengths for the PMMA/azo-comonomer composite were observed. The absorbance was measured after 1, 2, 4 and 24 hours to resolve the time dependence of the absorption behaviour at RT. The short-time behaviour of the absorbance shows a slight increase for both the azo-copolymer and the PMMA/azo-comonomer. This increase in absorbance is attributed to a slight relaxation of oriented dye molecules. After annealing at poling temperature T_p the absorbance increased back to the unpoled value for the copolymer and to about 50% of the unpoled value for the PMMA/comonomer composite (Figure 5).

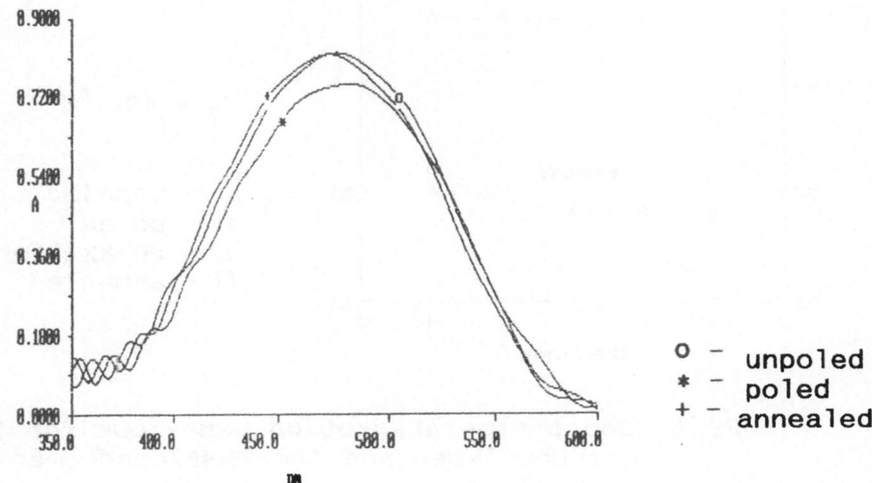

FIGURE 3   Spectra of the side chain polymer.

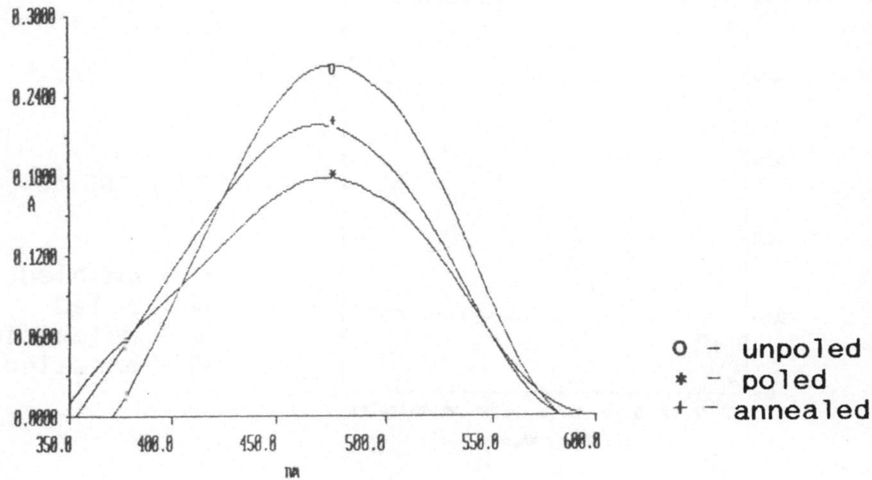

FIGURE 4   Spectra of the PMMA/comonomer composite.

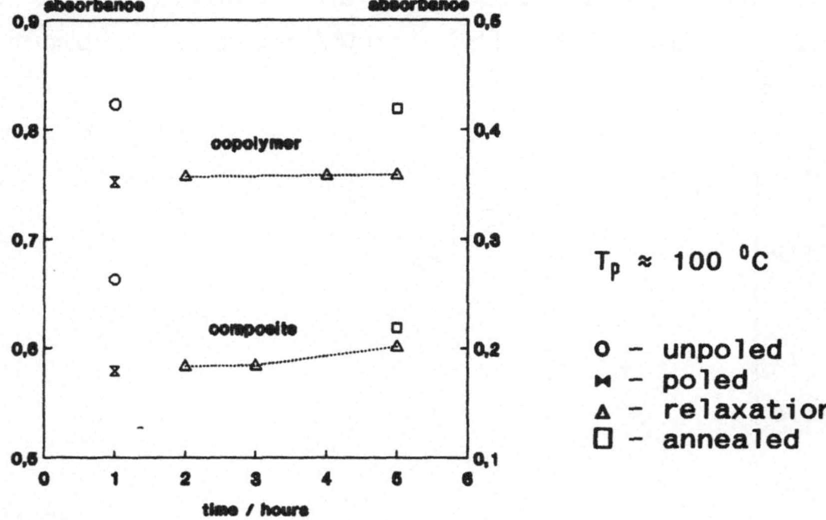

FIGURE 5   Short-time absorption behaviour for the
azo-copolymer and the PMMA/comonomer
composite.

The same behaviour was observed for PMMA/azoviolet dye
composites (Figure 6).

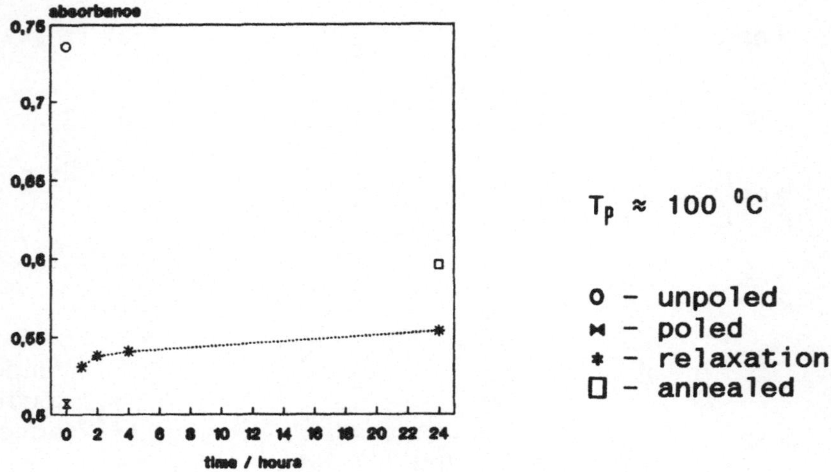

FIGURE 6   Absorption studies of a PMMA/azoviolet
dye composite.

For the other polymer/dye composites the short-time behaviour is the same but the annealing behaviour differs from that of PMMA composites: After annealing a decrease in the absorbance was observed (Figure 7).

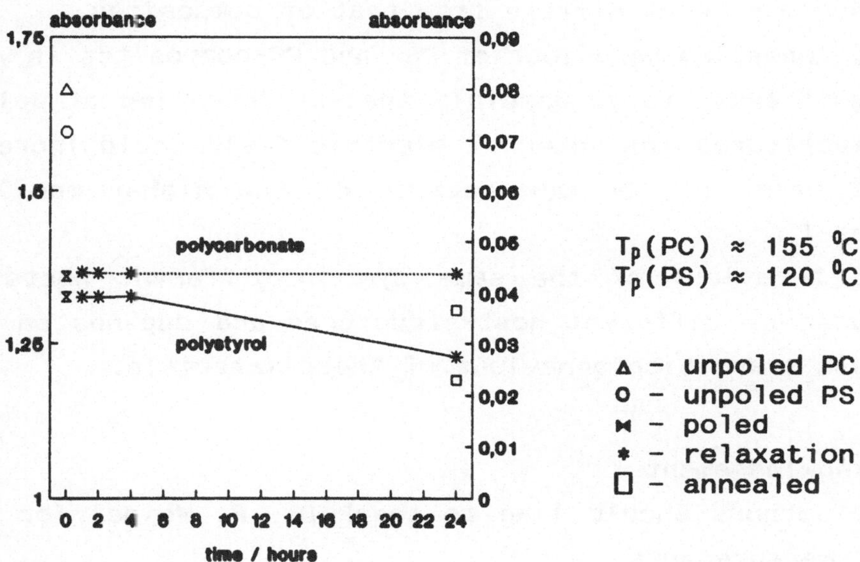

FIGURE 7   Absorption studies of PC/azoviolet and PS/azoviolet dye composite.

4   Conclusions

Corona poling induces an alignment of optically nonlinear dye molecules in an azo-copolymer and in polymer/dye composites. The dipolar orientation after poling remains stable for these materials at RT in comparison to the results of Hampsch et al. [HAM 90], which show a strong relaxation of dipolar orientation by SHG measurements. This orientational order can be described by an order parameter Φ, which is directly proportional to an induced birefringence due to the interaction between dipoles and the internal electric field [MOR 89].

The results of this study indicate that the azo-copolymer has a lower order parameter ($\Phi = 0.1$) in comparison to polymer/dye composites ($\Phi = 0.3$).

After annealing the azo-copolymer shows a reversible behaviour, what differs from that of composites.

The annealing behaviour of PS- and PC-composites is very significant. It is possible that by annealing at poling temperatures the internal electric field could increase the order of the dye because of the higher mobility near T_g.

The behaviour of the same dye in different hosts is caused by different host-structures and depends on the glass transition behaviour of these materials.

Acknowledgement

The authors should like to thank Dr. P. Weigel for the DSC measurements.

5 References

[HAM 90] Hampsch, H.L.; Yang, J.; Wong, G.K.;
Torkelson, J.M.: Dopant orientation dynamics in doped
second order nonlinear optical amorphous polymers.
2. Effects of physical aging on poled films.
Macromolecules, 23, 15, (1990) 3648-3654

[HIL 89] Hill, J. R.; Pantelis, P.; Dunn, P. L.; Davies,
G. J.: Organic polymer films for second order nonlinear
applications, Proc. SPIE, 1147, (1989) 165-176

[MOR 89] Mortazavi, M. A.; Knoesen, A.; Kowel, S. T.;
Higgins, B. G.; Dienes, A.: Second-harmonic generation
and absorption studies of polymer-dye films oriented by
corona-onset poling at elevated temperatures,
J. Opt. Soc. Am. B ,6, 4, (1989) 733-741

[PAG 90] Page, R. H.; Jurich, M. C.; Reck, B.; Sen, A.;
Twieg, R. J.; Swalen J. D.; Bjorklund, G. C.; Willson,
C. G.: Electrochromic and optical waveguide studies of
corona-poled electro-optic polymer films,
J. Opt. Soc. Am. B, 7, 7 (1990) 1239-1250

[SIN 86] Singer, K.D.; Sohn, I. E.; Lalama, S.I.:
Second harmonic generation in poled polymer films,
Appl. Phys. Lett., 49, (1986) 248-250

[TAU 91] Tauer, K.; Morozova, A.; Wedel, A.: Synthese
von nitrogruppenhaltigen Copolymeren durch radikalische
Copolymerisation, submitted to Macromol. Chem.

Electrooptical Measurement of $\chi^{(3)}$ in Polymers with Conjugated π-Electron Systems

W.SCHMID, M.HEROLD, TH.VOGTMANN, M.SCHWOERER

Physikalisches Institut and BIMF
Universität Bayreuth
Postfach 101251
D-8580 Bayreuth

ABSTRACT : Electrooptical measurements of $\chi^{(3)}(\omega;0,0,-\omega)$ have been performed with films of polydiacetylene and poly-p-phenylenevinylene (PPV). In a reflection-technique the electric field induced birefringence is measured, wherefrom $\chi^{(2)}$ and $\chi^{(3)}$ can be calculated. This technique which is a modification of an established $\chi^{(2)}$-method presents some advantages compared to wave-guiding-techniques or complicated THG and DFWM.
In the investigated polymers $\chi^{(3)}$ is caused by fast electronic processes in the conjugated π-electron systems and the measured values should be comparable to those obtained by all-optical methods (THG, DFWM) in the nonresonant area.

1. Introduction

Measuring the third-order nonlinear optical susceptibility $\chi^{(3)}$ of polymers for example by Degenerate Four Wave Mixing (DFWM) , Third Harmonic Generation (THG) or Nonlinear Waveguide Coupling is a rather difficult and expensive task, because of the need of high-power laser systems (normally in the ps-range) and/or because of a complicated alignment procedure.

Recently a new electrooptical reflection-technique has been published [RÖH 91]. Here the nonlinear effect in the material is caused by a high electrostatic field (about $10^4 \dots 10^6$ V/cm) and therefore a low-power cw-laser (e.g. a HeNe-laser) is sufficient. This method is a modification of a $\chi^{(2)}$-method described by Teng and Man [TEN 90].

Via the electric Kerr-effect the applied electrostatic field induces birefringence in the polymer, which can be examined by transmitting a linearly polarized laserbeam through the sample and measuring the change of its polarization.

In order to realize this, the sample must be supplied with a transparent (ITO) and a reflecting electrode (e.g. an evaporated gold layer).

A further advantage of this method is that $\chi^{(3)}$ can be measured even in materials producing too much straylight so that DFWM cannot be applied.

The relaxation of the nonlinear processes cannot be studied by this method. Furthermore it is applicable only to isolating materials.

2. Description of the method

A detailed description of this technique was published by P. Röhl [RÖH 91]. Therefore only a short survey on the physical principle of the method will be given here.

One starts with the common expansion of the nonlinear polarization

$$\vec{P} = \varepsilon_0 \hat{\chi}^{(1)} \vec{E} + \varepsilon_0 \hat{\chi}^{(2)} \vec{E} \vec{E} + \varepsilon_0 \hat{\chi}^{(3)} \vec{E} \vec{E} \vec{E} \tag{2.1}$$

In an electrooptical experiment the electric field is composed of the external static or quasistatic field and the electric field of the propagating light.

$$\vec{E} = \vec{E}_{stat} + \vec{E}_{opt} \tag{2.2}$$

Inserting (2.2) to (2.1) and separating to different orders of E_{opt} the arising physical effects are seen. The electric Kerr-effect in the second line of (2.3) is the basis of the described method. In (2.3) $\chi^{(2)}$ is set to zero (isotropic materials).

150

$$P = \varepsilon_0\, \chi^{(1)} E_{stat} + \varepsilon_0\, \chi^{(3)} E_{stat}^3 \qquad\qquad \text{electrostatic effects}$$

$$+ \varepsilon_0\, (\chi^{(1)} + 3\chi^{(3)} E_{stat}^2)\, E_{opt} \qquad \text{electric Kerr-effect}$$

$$+ \varepsilon_0\, 3\chi^{(3)} E_{stat} E_{opt}^2 \qquad\qquad \text{EFISH}$$

$$+ \varepsilon_0\, \chi^{(3)} E_{opt}^3 \qquad\qquad\qquad \text{all-optical effects (DFWM,THG)}$$

$$(2.3)$$

The organic materials which were investigated consist of linear molecules with conjugated π-electron systems along the molecular chain. Their hyperpolarizability γ is high only in this direction. All other components of the hyperpolarizability-tensor will be neglected. Assuming a special orientational distribution of the molecules in the polymer film the relationships between γ and the tensor-components of $\chi^{(3)}$ can be drawn up. In the absence of a preferential orientation in the film plane and with the electrostatic field perpendicular to the film (this direction will be denoted by the index 3) the relevant components of $\chi^{(3)}$ are :

$$\chi^{(3)}_{3333} \qquad\qquad \sim \; 3\gamma \, \langle \cos^4\Theta \rangle$$

$$\chi^{(3)}_{1133} = \chi^{(3)}_{2233} \; \sim \; \frac{3}{2}\gamma \, \langle \sin^2\Theta \cos^2\Theta \rangle \qquad\qquad (2.4)$$

Θ is the angle between the molecular axis and the 3-direction and the brackets denote spatial averaging.
With an isotropic or nearly isotropic orientational distribution one obtains the relation :

$$\chi^{(3)}_{3333} \approx 3\chi^{(3)}_{1133} \qquad\qquad (2.5)$$

Due to the changes of the refractive indices Δn_1 (polarization *in* the filmplane) and Δn_3 (polarization *perpendicular* to the film) caused by the electrostatic field the analogous relationship to (2.5) is:

$$\Delta n_3 \approx 3\,\Delta n_1 \qquad\qquad (2.6)$$

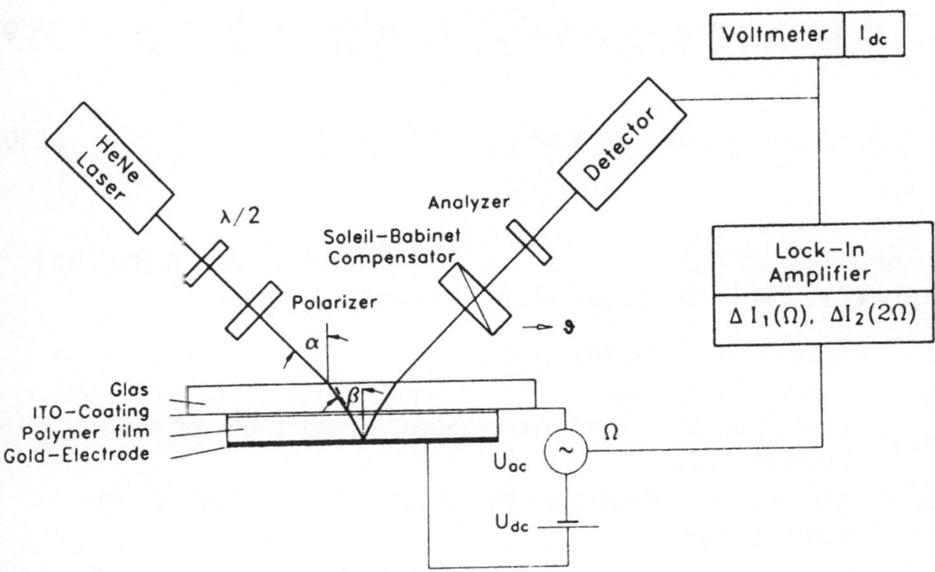

Figure 2.1: Experimental setup

Hence the polymer becomes birefingent in the external field. The s– and the p–component of the initially linearly polarized light experience a different phase shift δ during their propagation through the film and the outcoming light will be elliptically polarized.

In order to measure δ the reflected beam has to pass a Soleil Babinet compensator by which an additional phase shift ϑ can be obtained and an analyzer oriented perpendicular to the polarizer in the input beam (see Fig. 2.1; for a detailled description see [TEN 90]).

One makes use of a modulation technique by composing the electro-"static" field of a dc-component and a low-frequent ac-component in the kHz-range, whereby δ will be modulated.

$$E_{stat} = E_{dc} + E_{ac} \sin \Omega t \tag{2.7}$$

Hence the laser intensity at the detector will also be modulated with the frequency Ω as well as with the double frequency 2Ω.

$$I = I_{dc} + \Delta I_1(\Omega) + \Delta I_2(2\Omega) \tag{2.8}$$

$$\Delta l_1(\Omega) = g(\beta)\,\frac{d}{n_0\lambda}\,\sin\vartheta\,\sqrt{I_s I_p}\,E_{ac}\left(\chi_{333}^{(2)} + 2\,\chi_{3333}^{(3)}\,E_{dc}\right) \qquad (2.9)$$

$$\Delta l_2(2\Omega) = \frac{1}{2}\,g(\beta)\,\frac{d}{n_0\lambda}\,\sin\vartheta\,\sqrt{I_s I_p}\,E_{ac}^2\,\chi_{3333}^{(3)} \qquad (2.10)$$

$\beta = \arcsin(\sin\alpha\,/\,n_0)$: reflection angle (in the film)

$g(\beta) = (\,4\pi/3\,)\,(\,1 - \tan^2\beta\,)\,\sin^2\beta\,/\cos\beta$: geometry factor

d : thickness of the film

n_0 : index of refraction

$I_{s,p}$: intensities of s- and p-component, respectively, of the incident wave

ϑ : phase shift between s- and p-component caused by the compensator

In order to complete the formula the $\chi^{(2)}$-effect is drawn up. If the lock-in amplifier detects at Ω, the plot of Δl_1 versus E_{dc} should show a linear dependence according to (2.9), from which $\chi^{(2)}$ and $\chi^{(3)}$ can be evaluated.

A second possibility to measure only $\chi^{(3)}$ is to detect at 2Ω and plot Δl_2 versus E_{ac}^2 according to (2.10). This may serve as a control .

3. Materials

The investigated materials were unsubstituted poly-p-phenylenevinylene (PPV ; synthesized in the precursor route and thermally eliminated) and poly-4BCMU, a soluble polydiacetylene (Fig. 3.1). The films were prepared by spincoating or cast from solution with a doctor blade machine.

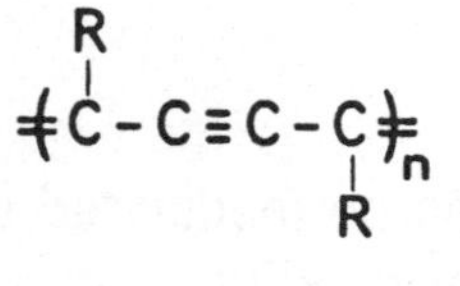

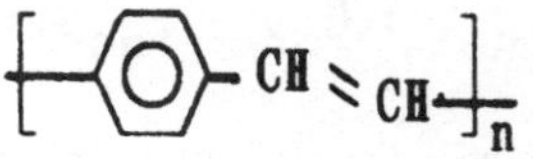

$R = -(CH_2)_4-O-CO-NH-CH_2-CO-O-(C_4H_9)$

Figure 3.1: The structure of poly-4BCMU (a) and PPV (b)

4. Results

4.1. poly-4BCMU

The plots in Fig.4.1 and Fig.4.2 show measurements of poly-4BCMU samples detected at frequency Ω and 2Ω, respectively. The applied electric fields (about 10^5 V/cm) were in the same range as typical fields achieved at DFWM. For these field strengths the curves show a linear behaviour and a good agreement of the $\chi^{(3)}$-values for the two different kinds of detection.
All examined samples yielded values of $\chi^{(3)} = (3\pm1)\,10^{-11}$ esu.
The measurement does not show a $\chi^{(2)}$-effect for this isotropic (and therefore centrosymmetrical) material, as expected.

4.2. PPV

More problems arise with PPV. The reproducibility of the results is worse than for poly-4BCMU. Furthermore the plots of $\Delta I_1(\Omega)$ versus E_{dc} may yield curves with a not-vanishing offset at zero-field, which would indicate a high $\chi^{(2)}$-value (according to Eq. 2.9). As PPV is a centro-symmetrical material this does not seem very likely. A possible prefe-rential orientation caused by the film preparation method (e.g. in the film plane) would influence the validity of eq. (2.6). But this cannot explain the "$\chi^{(2)}$-effect".
The problems may be related to a complicated influence of the contacts to the bulk material. While the current-voltage-characteristic for the poly-4BCMU samples shows a linear dependence (at least for fields below $1.5\cdot10^5$ V/cm), at the PPV samples the current is proportional to the square of the voltage even at low electric fields. This is a hint to charge carrier injection (law of CHILD [HAM 81]) which would lead to space charges and therefore to a modification of the electric field in the material. In order to understand this problem detailled investi-gations remain to be done.
Preliminary measurements yield $\chi^{(3)} = (5\pm3)\,10^{-11}$ esu.

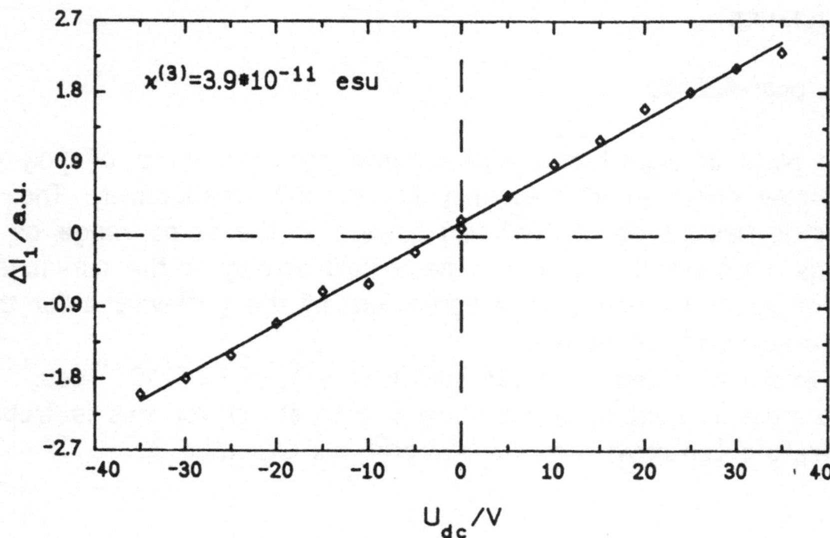

Figure 4.1: Determination of $\chi^{(3)}$ of poly-4BCMU at detection frequency Ω (acc. to eq. (2.9))
$E_{dc} \leq 1.5 \cdot 10^5$ V/cm $E_{ac} = 0.5 \cdot 10^5$ V/cm

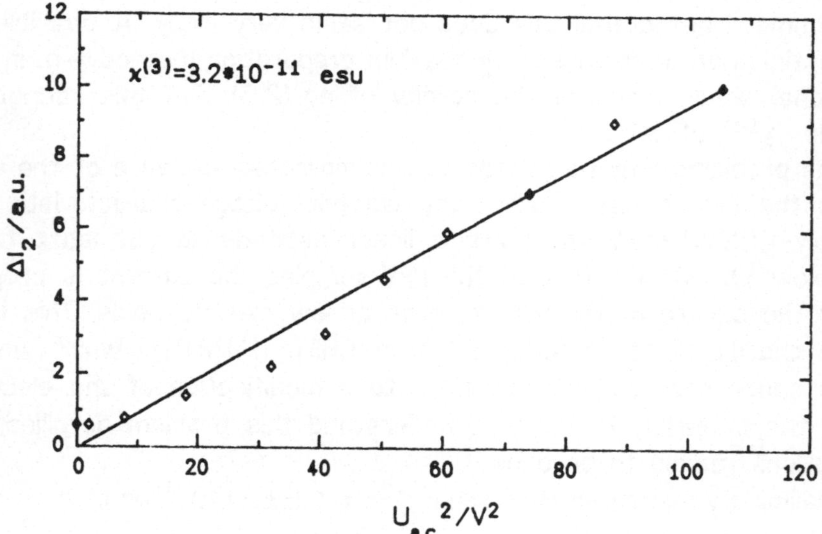

Figure 4.2: Determination of $\chi^{(3)}$ of poly-4BCMU at detection frequency 2Ω (acc. to eq. (2.10))
$E_{ac} \leq 0.5 \cdot 10^5$ V/cm $E_{dc} = 0$ V/cm

5. Comparison of the results with $\chi^{(3)}$-values of DFWM and THG

The $\chi^{(3)}$-effects in these materials are related to a polarization of the conjugated π-electron system. Since this process is very fast, $\chi^{(3)}$-values measured by an electrooptical method ($\to \chi^{(3)}(\omega;0,0,-\omega)$) should be comparable to those obtained by DFWM ($\to \chi^{(3)}(\omega;-\omega,\omega,-\omega)$) or THG ($\to \chi^{(3)}(3\omega;-\omega,-\omega,-\omega)$), provided that one- or multi-photon resonances are not excited.

Table 5.1 and table 5.2 show several $\chi^{(3)}$-values from the literature and from our electrooptical measurement.

Table 5.1: $\chi^{(3)}$-values of poly-4BCMU

Authors		Method	λ/nm	$\chi^{(3)}$/ 10^{-11} esu
Rao, Prasad	[RAO 86]	DFWM	585 605	40
Townsend, Etemad	[TOW 89]	THG	1064	7 ± 3
Berkovic, Shen, Prasad	[BER 87]	THG	1064	3
Mc Branch, Wudl	[BRA 89]	THG	1064	2 ± 1.5
our results – Bayreuth 1992		DFWM	680	4 ± 2
electrooptical method (this work)			633	3 ± 1

Table 5.2: $\chi^{(3)}$-values of PPV

Authors		Method	λ/nm	$\chi^{(3)}$/ 10^{-11} esu
Singh, Prasad	[SIN 88]	DFWM	580 602	≤ 40
Kaltbeitzel	[KAL 89]	DFWM	≈ 600	7
Mc Branch, Wudl	[BRA 89]	THG	1064	2 ± 1
electrooptical method (this work)			633	5 ± 3

With the exception of the $\chi^{(3)}$-values in the first line of table 5.1 and 5.2, respectively (resonant enhancement), the agreement is quite good. In the nonresonant regime electrooptical and all-optical methods measure the same $\chi^{(3)}$-value. In order to confirm this further measurements (also with other materials) will be carried out in future.

6. Possible problems

6.1. Current flow in the samples

It has already been mentioned in (4.2.) that problems may arise due to the current-voltage-behaviour of the samples.

First of all one has to ensure that the electrodes do not produce blocking contacts, because in this case a great amount of the electric voltage would drop at the contacts and not at the material.

Moreover, especially for high electric fields, charge carrier injection from the contacts may occur which builds up a space charge and therefore leads to a modification of the electric field in the polymer.

In order to avoid these disturbing effects one has to use appropriate contact materials. To avoid charge carrier injection one can also try to work with buffer layers (of an appropriate isolating material) between the injecting electrode and the polymer. But this will involve the difficulty to determine the exact fraction of the applied voltage at the NLO-material.

6.2. Electromodulation of optical absorption

High electric fields may cause a remarkable electromodulation of the optical absorption at the absorption edge of the polymer [BUR 91]. Though this effect is proportional to the square of the electric field, it produces a modulation of the transmitted optical intensity at Ω as well as at 2Ω in the case of an applied ac-field (with frequency Ω) with a dc-offset. Thus it superposes the $\chi^{(3)}$-effect and one has to ensure that this disturbing effect is small.

In order to avoid this one must use a laser wavelength sufficiently distant from the absorption edge of the polymer.

7. Discussion and outlook

So far measurements have been carried out solely with a HeNe-laser. At the wavelength of 633 nm electromodulation of optical absorption yields a non-neglectable contribution to the detected signal both for poly-4BCMU and PPV. Measurements at higher wavelengths will be necessary. Beyond that it could be helpful to check the predicted dependence of ΔI_1 and ΔI_2 on the the phase shift ϑ (see (2.9) and (2.10)) caused by the Soleil Babinet compensator.

The question is raised, whether the physical effects that lead to electromodulation of optical absorption also contribute to the $\chi^{(3)}$-value evaluated by DFWM (and even THG).

Furthermore one can try to separate the absorption effect from the "pure" $\chi^{(3)}$-effect (real part of $\chi^{(3)}$) by use of the electrooptical technique. This may help to learn more about the physical origin of optical nonlinearities in conjugated polymers.

8. Conclusion

$\chi^{(3)}$-measurements on polymer films of poly-4BCMU and PPV have been carried out with a new electrooptical technique. This method is simple and cheap concerning the experimental setup and can also be applied to materials producing much straylight. Comparing our preliminary results with those obtained by DFWM and THG leads to the question which different physical effects contribute to the $\chi^{(3)}$-values measured by the various methods, respectively.

158

Acknowledgement

We thank Dr. Peter Röhl for giving us the idea to set up this experiment and for some helpful discussions.
The preparation of the materials has been carried out by Irene Müller and Jürgen Gmeiner.

References

BER 87   G.Berkovic, Y.R.Shen, P.N.Prasad
         J.Chem.Phys. **87**, 1897 (1987)

BRA 89   D.McBranch, M.Sinclair, A.J.Heeger, A.O.Patil, S.Shi, S.Askari,
         F.Wudl   Synth.Met. **29**, E85 (1989)

BUR 91   J.H.Burroughes, R.H.Friend
         "Conjugated Polymers" (eds J.L.Brédas, R.Silbey)
         Kluwer Academic Publishers 1991 , p.571f

HAM 81   C.Hamann, J.Heim, H.Burghardt
         "Organische Leiter, Halbleiter und Photoleiter"
         Braunschweig/Wiesbaden : Vieweg, 1981 , p.83

KAL 89   A.Kaltbeitzel   Ph.D. thesis (1989) – Mainz (D)

RAO 86   D.N.Rao, P.Chopra, K.Suniti, J.Swiatkiewicz, P.N.Prasad
         J.Chem.Phys. **84**, 7049 (1986)

RÖH 91   P.Röhl, B.Andress, J.Nordmann
         Appl.Phys.Lett **59**, 2793 (1991)

SIN 88   B.P.Singh, P.N.Prasad, F.E.Karasz
         Polymer **29**, 1940 (1988)

TEN 90   C.C.Teng, H.T.Man   Appl.Phys.Lett. **56**, 1734 (1990)

TOW 89   P.D.Townsend, G.L.Baker, N.E.Schlotter, S.Etemad
         Synth.Met. **28**, D633 (1989)

Deep Proton Irradiation of PMMA for a 3D Integration of Micro-Optical Components

K.-H. Brenner, M. Frank, M. Kufner, S. Kufner, A. Müller

University of Erlangen-Nuremberg
Physics Institute
Tel. +49-9131-858376
FAX +49-9131-13508
W-8520 Erlangen
Federal Republic of Germany

Abstract

An integration of microoptical components to threedimensional systems is of interest for optical information processing as well as for optical interconnections. Polymethyl methacrylate (PMMA) is a suitable substrate material, because microprisms, beamsplitters and microlenses can be integrated monolithically in the same substrate by deep proton irradiation. The microcomponents can be combined to compact microoptical imaging systems which are easy to align. The fabrication method and the imaging properties of this system are discussed and applications are outlined.

1. Introduction

The term 'microoptics' is mostly used in the context of twodimensional waveguide integration. Twodimensional integration methods have been developed since the 70s [NIS 89]. The advantage of twodimensional optical microintegration is that the design methods and the mask technologies can be easily transferred from electronic to optic component manufacturing. But the same topological restrictions existing for twodimensional electronic layouts arise also for optical circuits. The advantages of optical interconnections cannot be fully exploited with a planar (twodimensional) integration. By making use of threedimensional optical imaging the packing density and thus the degree of parallelism can be increased considerably. With free space light propagation the volume can also be used more efficiently because different beams of light can pass through each other without interference.

One approach to threedimensional integrated microoptics is based on diffractive optical components realized by etching in glass because this technique allows to use the same fabrication processes for all kinds of passive components [JAH 89]. Refractive optical components such as lenses, prisms and beamsplitters usually require different fabrication processes for different components. The most well developed technologies exist for the case of microlens fabrication [IGA 84, HUT 90]. A complete system combining heterogenous technologies (for example microlenses with diffusion techniques [IGA 84], semiconductor microlasers [JEW 91], gratings made of photoresist [LOH 88], etch techniques [HEI 90] or dichromatic gelatine [BRE 88]) requires special provisions for alignment because the positional tolerances of microelements are very small [CAN 89].

This paper describes an integration method for passive microoptical components like lenses, prisms, beam splitters in one substrate. The fabrication methods for these components are all based on deep proton irradiation [BRE 90]. The use of refractive optics offers large numeric apertures, is very wavelength and angle insensitive and thus makes it possible to operate a system even with incoherent illumination. The architecture is based on the 'stacked system' approach proposed by [IGA 82] and can be considered as an extension of the system proposed in [BRE 91] by including monolithic assembly and means for alignment.

2. Basic building blocks

The basic building blocks for microoptical systems are components for light collimation, light deflection and beam splitting. These components can all be fabricated in PMMA with one single technology.

2.1. Fabrication process

Microcomponents for light deflection and for light collimation can both be fabricated by an irradiation of PMMA with a high energy proton beam (fig. 2.1.a). The effect of the radiation is a reduction of the average chain-length and thus a reduction of the molecular weight [FRA 92b]. Depending on the subsequent process different optical functions can be achieved.

If the irradiated domains are removed with a special solvent it is possible to generate deflecting elements (fig. 2.1.c) [GHI 80]. In [BRE 90] a new fabricating method for creating deep structures in PMMA was introduced. The fabrication is based on the irradiation of PMMA with protons and a subsequent removal of the

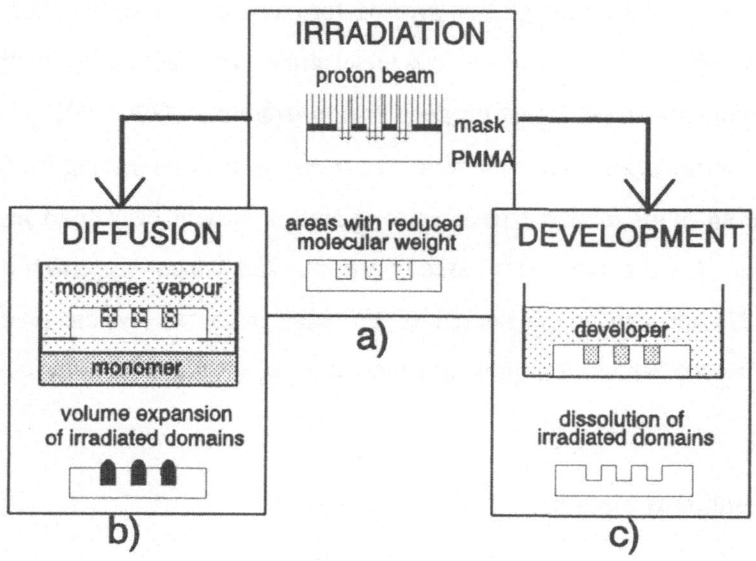

Fig. 2.1: a) Reduction of the molecular weight by irradiation b) Fabriaction of microlenses by monomer diffusion c) Fabrication of microprisms by development

irradiated domains. Using this technology it is possible to generate slits in PMMA with a maximum depth of 1500 μm and a width of down to 10 μm. The accuracy of the slit position depends on the positioning stages and can be up to 0.1 μm. Depending on the angle of irradiation the slits can be produced with an orientation between 0° and +/- 90° relative to the substrate surface. For slits with an angle of +/- 45° the tilted surfaces can be used for light deflection by total internal reflection. In this case a light wave at vertical incidence is deflected by 90° due to total internal reflection [BRE 90].

By diffusion of monomer vapour into the irradiated domains also microlenses can be produced (fig. 2.1.b) [FRA 91]. The irradiated domains show a much larger diffusion constant than the surrounding PMMA-matrix. Thus monomers can diffuse selectively into the irradiated regions and generate a volume growth there. With this technique surface profile microlenses with diameters between 80 μm and 1 mm and relative apertures as high as 1 were fabricated [FRA 92a].

The threedimensional integration method proposed here results from a combination of these two techniques. By selective processing after irradiation both, microlenses and microprisms can be integrated monolithically in one substrate. Beamsplitters can be produced by dielectric coating techniques applied to the microprism surfaces. Combining these components compact and robust threedimensional microsystems can be constructed. Due to the monolithic integration no alignment between the lenses and the prisms is required.

2.2. Optical properties

For the optical characterization of the prisms and the lenses different measuring techniques have been used. The surface of the prisms was analysed with a piezo

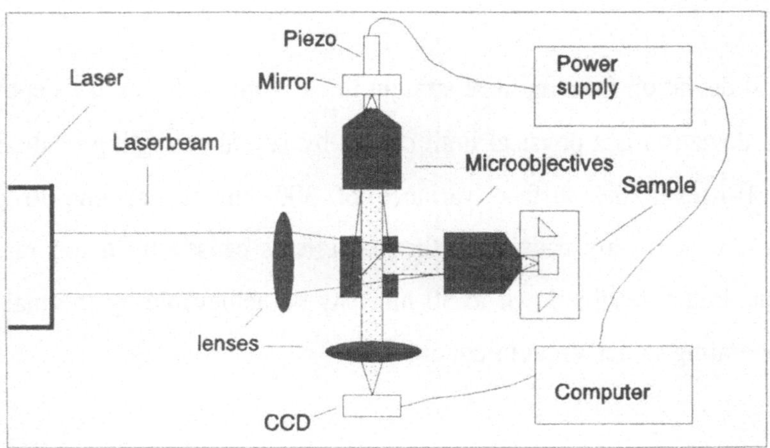

Fig. 2.2: Setup of a Linnik interferometer.

164

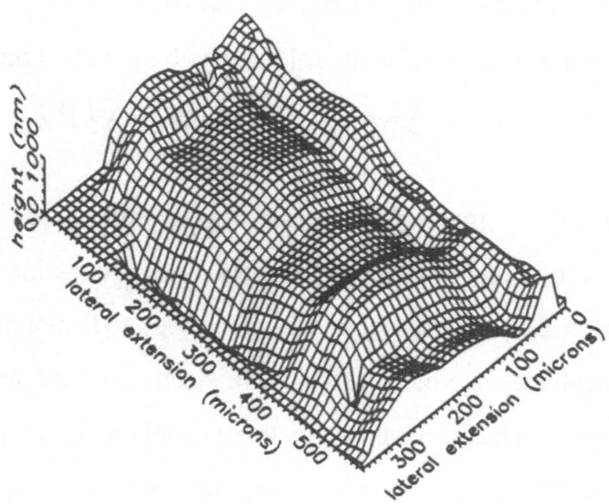

Fig. 2.3: Surface profil of a microprism with a height of 350 μm. The average roughness is about 200 nm.

controlled Linnik interferometer (fig. 2.2) using the phase shifting algorithm for evaluation. The threedimensional phase plot in fig. 2.3 shows the profile of a prism surface. Along the direction of the irradiation from top to bottom, 75% of the area show a variance of approx. 300 nm. Near the bottom of the structure the resulting surfaces bend due to lateral straggling of the protons. Fig. 2.4 shows a side view of a 700 μm deep edge.

The lateral deviation from an ideal straight line is approx. 5 μm in a depth of 700 μm. This deviation is a physical limit caused by lateral straggling of the incident protons. However the surface variance of 300 nm is not due to physical limitations, it is mainly caused by thermal effects caused by a too rapid dose deposition. Better results down to 50 nm may be achievable by thermal control and by operating under vacuum conditions.

The microlenses were tested in an image formation setup shown in fig. 2.5. A

Fig. 2.4: Side view of an edge with a depth of 700 μm.

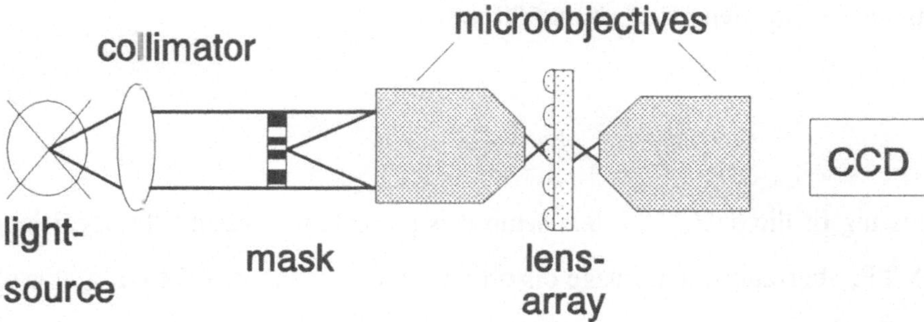

Fig. 2.5: Test setup for microlenses.

microscope objective is used in reverse and forms the demagnified image of a mask. The demagnifaction factor was 20x. The microlens images this demagnified mask in a 2f-imaging configuration. The subsequent image of the mask formed by the microlens is observed with a second microscope objective according to fig. 2.5. This image is recorded by a CCD camera and sent to a computer for further

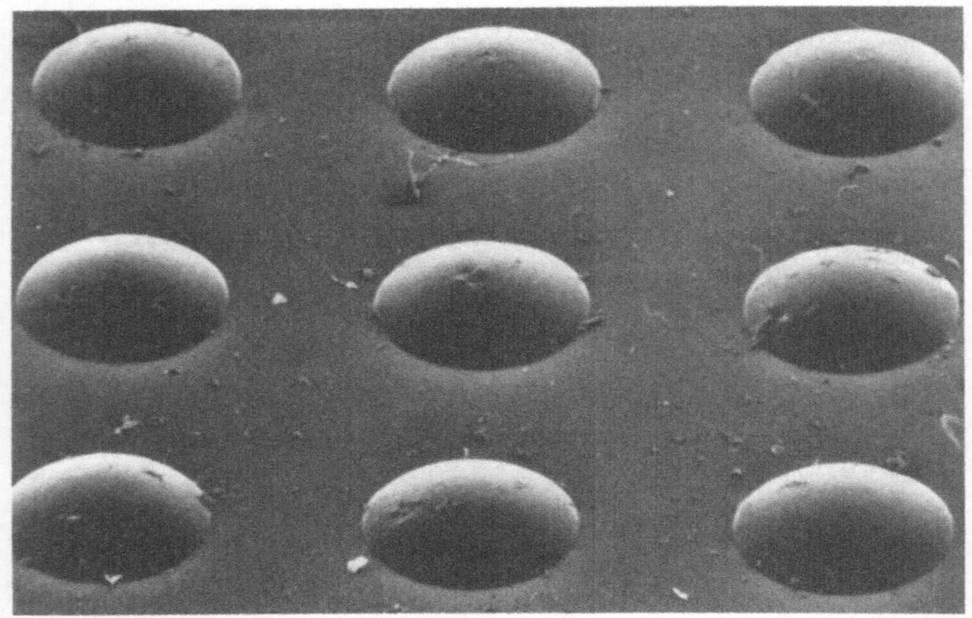

Fig. 2.6: SEM picture of a microlens array. The lenses have a diameter of 500 μm and a focal length of about 620 μm.

processing of the data. With this setup it is possible to measure the focal length, the MTF, aberrations and image distortions. Fig. 2.6 shows a foto of a microlens array fabricated with the process described in fig. 2.1 b. The lenses have a diameter of 500 μm and a focal length of about 620 μm. In fig. 2.7 an example of a 1:1 image formed by a 500 μm diameter lens is shown. The highest grating frequency in this image is 256 line pairs per mm.

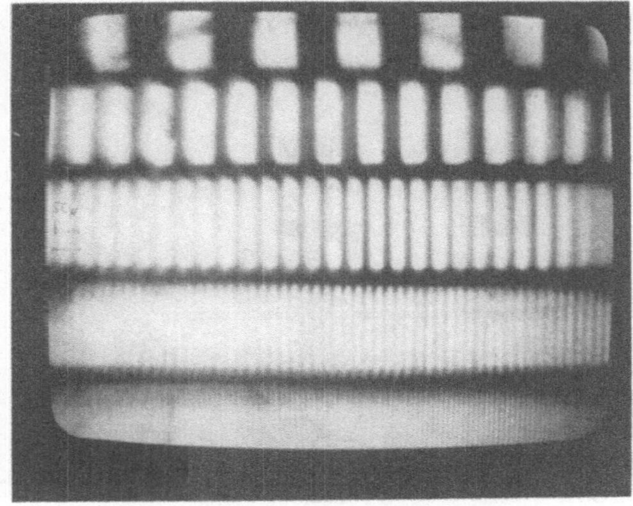

Fig. 2.7: Image formation of a grating with 16, 32, 64, 128, and 256 line pairs/mm.

3. Compact optical microsystems

3.1. Concept of modularity

The concept of dividing a system into a small number of simple and flexibly combinable modules is successfully used in electronics in the form of integrated circuits.

Modules exist at the low level (NAND- or NOR-gates) and also at higher levels of complexity (CPUs). With passive microoptics the need for modules is not so obvious. But especially for the construction of complex free space optical systems with numerous lenses, prisms, beam splitters and also active devices a modular design can simplify the assembly significantly. Lens arrays or surface emitting lasers [IGA 84, JEW 91] are a step in this direction.

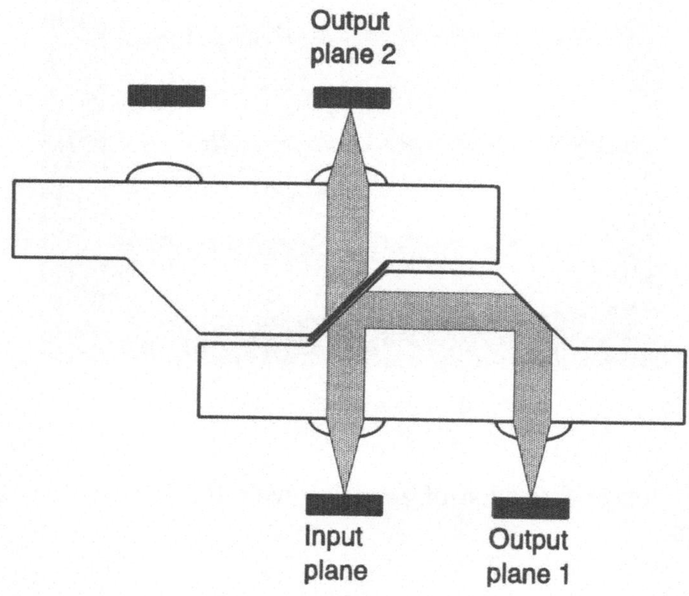

Fig. 3.1: Modular architecture with miniaturized passive optical elements.

Fig. 3.1 shows one type of component for a modular architecture which makes use of the regular structure of optical systems [BRE 91]. Here the distances between the deflecting and between the collimating elements are the same. This component can be repeated periodically in each direction. As a generic 4f-system it can be used for 1:1-imaging, spatial filtering and beam splitting. In the following section a modular microoptical system is constructed step by step. The used components consist of deep proton irradiation microelements integrated monolithically in one PMMA substrate.

3.2. Fabrication of a modular integrated microoptical system

A basic component for a modular system as shown in fig. 3.1 is formed by prism arrays arranged equidistantly with an altering prism angle of +/- 45° with respect to the substrate surface. Fig. 3.2 shows a photo of a prism array fabricated in this way.

Fig. 3.2: Foto of a microprism array. The length of the 45° edge is 500 μm.

On the backside of this prism array a microlens array is formed in a second step. Since the pitch for the lens array and the prism array is the same only one mask alignment for the lens irradiation process is necessary. A photo of the extended component with the lens array opposite to the prism array is shown in fig. 3.3. In this component the prisms have a cathete of 350 μm and the lenses have a diameter of 500 μm. Fig. 3.4 shows a beam splitting experiment by aligning two prism arrays as indicated in fig. 3.1. A 35 nm thick gold layer was used to split the beam.

Fig. 3.3: Foto of a monolithic integrated system of microprisms and microlenses. The length of the 45° edges is 500 μm. The diameter of the lenses is 500 μm, too.

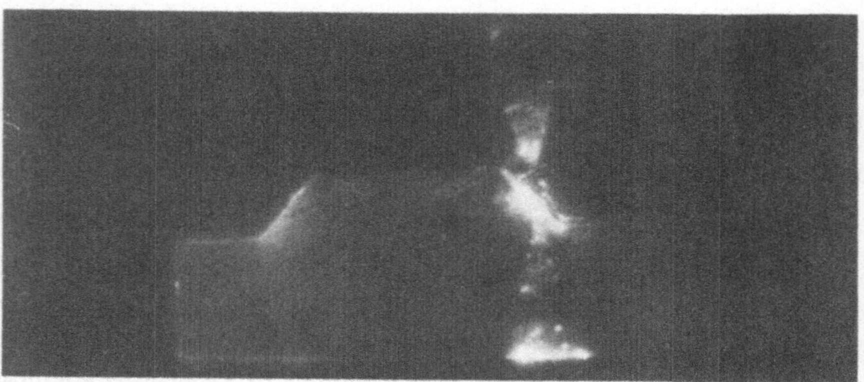

Fig. 3.4: Beam splitting experiment with two microprism arrays. As a coating layer 35 nm gold was used. The length of the 45° edges is 500 μm.

4. Optical experiments

The performance of a microoptical system is limited by the quality of each single element. Therefore in the previous chapter test methods for prisms and lenses were described. But another aspect of a system is also very important. It is how all the single elements perform together. Vignetting is one example of a system property. For threedimensional microoptic systems the imaging properties are an important criterion. In the system described above 4f-imaging is the mostly used operation. The image formation can be tested with the basic setup shown in fig. 4.1. A mask is projected in the front focal plane of a microlens. An adjacent lens generates the image of the mask after deflection at a +45° prism and a -45° prism.

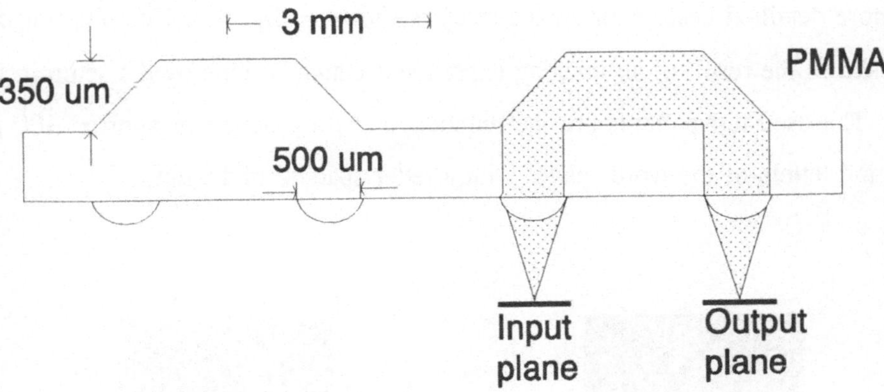

Fig. 4.1: 4f setup with microlenses and microprisms.

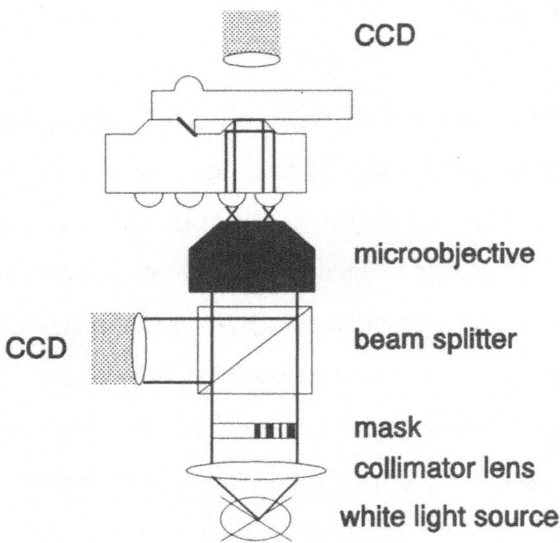

Fig. 4.2: Test setup for image formation in a system of microlenses, microprisms and microbeamsplitters.

A more detailed description of the setup is shown in fig. 4.2. The photo in fig. 4.3 shows the result of an imaging experiment which is done by the setup in fig. 4.2. The used test patterns are the number '3' with a heigth of approx. 100 μm and the letters of the word 'plate' with a letter spacing of 15 μm.

Fig. 4.3: Image formation by use of a monolithic integrated 4f system.

5. Summary

It is shown that the three most important passive optical elements, lenses, prisms and beam splitters, can be miniaturized and integrated in one substrat using only refractive optics. The integration is monolithically for prisms and lenses. Beam splitters can be fabricated by coating and aligning two prism arrays. The alignment is reduced to only one degree of freedom. Compact and robust systems can be constructed from modular building blocks. The next aim is to include active devices into the system in order to realize integrated optical interconnections between optoelectronic chips.

Parts of this work are funded by the german ministry of research and technology (BMFT) under grant 0585/5.

174

References:

[BRE 88] K.-H. Brenner, F.Sauer: "Diffractive-reflective optical interconnects", Appl. Optics 27 (20), 4251-4254, 1988.

[BRE 90] K.-H. Brenner, M. Frank, M. Kufner, S. Kufner: "H^+-Lithography for 3-D integration of optical circuits", Appl. Optics 29 (26), 3723-3724, 1990.

[BRE 91] K.-H. Brenner: "Techniques for integrating 3D-optical systems", Proc. SPIE 1544, 263-270, San Diego, 22-23 July 1991.

[CAN 89] J.W. Cannt, V. Diadnik: Gimbal for aligning Laser and Lenslet arrays for coherent operation in an external cavity", Appl. Optics 28 (9), 1602-1605, 1989.

[FRA 91] M. Frank, M. Kufner, S. Kufner, M. Testorf: "Microlenses in polymethyl methacrylate with high relative aperture", Appl. Optics, Vol 30, No. 19, 2666-2667, 1991.

[FRA 92a] M. Frank, M. Kufner, S. Kufner, J. Moisel, M. Testorf: "Microlenses in PMMA with high relative aperture: A parameter study", submitted to Pure and Applied Optics.

[FRA 92b] M. Frank, J. Göttert, M. Kufner, S. Kufner: "Untersuchungen zur tiefenlithographischen Strukturierung von PMMA", KfK-IMT-Bericht 103/21, 1992.

[GHI 80] V. Ghica, W. Glashauser, "Verfahren zur spannungsfreien Entwicklung von bestrahlten Polymethacrylat-Schichten", Offenlegungsschrift DE 3039110.

[HEI 90] M. Heißmeier, U. Krackhardt, N. Streibl: "A Dammann-grating with diffraction orders of arbitray intensity etched in Al_2O_3", Opt. Comm. 76, 103-106, 1990.

[HUT 90] M.C. Hutley: "Optical techniques for the generation of microlens arrays", J. of Modern Optics 37 (2), 253-265, 1990.

[IGA 82] K.Iga, M. Oikawa, J. Banno, Y. Kokubun: "Stacked Planar Optics; an application of the planar microlens", Appl. Optics 21, 3456-3460, 1982.

[IGA 84] K.Iga, Y. Kokubun, M. Oikawa: "Fundamentals of Microoptics", Academic Press, Tokyo, 1984.

[JAH 89] J. Jahns, A. Huang: "Planar integration of free-space optical components", Appl. Optics 28 (9), 1602-1605, 1989.

[JEW 91] J. Jewell, J.P. Harbison, A. Scherer, Y.H. Lee and L.T. Florez: "Vertical-Cavity Surface-Emitting Lasers: Design, Growth, Fabrication, Characterization", IEEE J. of Quantum Electronics, 27 (6), 1332-1346, 1991.

[LOH 88] A.W. Lohmann, W. Lukosz, J. Schwider, N. Streibl, J. Thomas: "Array illuminators for the optical computer", Proc. SPIE 963, 37-44, Toulon, 1988.

[NIS 89] H. Nishishara, M. Haruna, T. Suhara: "Optical integrated circuits", McGraw-Hill, New York, 1989.

[ZIE 85] J.F. Ziegler, J.P. Biersack, U. Littmark "The stopping of Ions in Solids", Pergamon Press, New York, 1985.

Fabrication of Micro-Optic Elements by UV-initiated Polymerization

G. Bagordo, K.-H. Brenner, T.M. Merklein, A. Rohrbach

Physikalisches Institut der Universität
Staudtstr. 7/B2
8520 Erlangen, Fed. Rep of Germany

Abstract:

UV-Initiated Polymerization (UVIP) is a method for fabrication of diffractive and refractive optical elements e.g. lenses, prisms and other phase-only structures. Flexible function and design is possible. The surface relief structures are realized by UV-initiated polymerization. This process consists of an UV-sensitized PMMA-resist (PolyMethylMeth-Acrylate), which is spin coated and exposed at a wavelength of 365nm. Afterwards the resist is developed in MMA vapour (monomer of PMMA). Due to additional polymerization the exposed regions swell. Due to the linear response of the resist accurate control of the surface profile is possible by lithographic means.

1. Introduction

Diffractive Optical Elements (DOEs) are of interest for different optical applications [DAM 70, JAH 90, SWA 89].Refractive Optical Elements (ROEs), however, are significantly more insensitive to the wavelength of illumination, because the only dependency on wavelength is caused by material dispersion. Therefore inexpensive incoherent light sources e.g. LEDs can be used. ROEs show approximately linear dependence of the diffraction angle a on the wavelength λ ($\sin(a) \sim a = \lambda/g$ (g is the grating period of the refractive element)). Consequently many implementations of optical systems have been demonstrated based on ROEs [BRE 88, BRE 92, ECK 89, JOH 88, LOH 86, MER 89].

There are different methods for realizing DOEs [CES 89, WAL 90, HUR 82]. Especially reactive ion etching [JAH 90, JAH 92], direct electron beam writing [SHI 87, SMI 73] and deposition methods have been demonstrated in the literature. These methods use sophisticated technology for exposure or structuring. Fabrication of multilevel structures by the reactive ion etching process is possible after realigning and iterating the processing scheme. Thus refraction limited optical elements have been demonstrated [JAH 92].Classical ROEs (e.g. lenses and prisms) are fabricated in general by milling and polishing of glass or other transparent material. Aspherical refractive elements can be realized by diamond turning or by other special adapted preparation methods. Demands on mechanical accuracy are comparably high. Arrays of refractive microoptic lenses with small diameters (d < 1mm) have been demonstrated also by different methods based on surface tension or ion exchange [WOL 91]. These methods are favourable for spherically shaped optical elements.

The UVIP processing scheme provides flexible design of DOEs and ROEs, because the surface profile of the optical element is directly determined by exposure. The linear response of the resist reduces sensibility of the processing scheme versus deviations of the exposure e.g. deviations of the density of mask. Also biasing is not necessary.Realization and control of the UVIP process (Fig. 2.1) is comparable to the well known photographic 'black & white' processing scheme. The different ingredients (MMA, PMMA, Ketone) are provided very pure and are not expensive. Consequently it is possible to fabricate inexpensive optical elements by this method.

The optical elements fabricated by the UVIP process can be used like other refractive or diffractive elements. Because of flexibility on design, it is not only possible to produce lenses and prisms as they are used for common imaging systems, but also special adapted optical elements like compensation plates or other complex filter functions.

The special application of the UVIP process we are aiming at is the miniaturization and integration of digital 3D-optical systems we have implemented before [ECK 89, BRE 89, BRE 92]. The UVIP-process is especially capable for tailoring the phase only filters and the arrays of microoptic lenses as shown in fig. 1.1. The arrays of beamsplitters and beam deflection by 90° can be fabricated e.g. by thermal imprinting [BRE 00].

A generalized version of a miniaturized optical system is given in Fig. 1.1 The typical size will be below one centimeter.

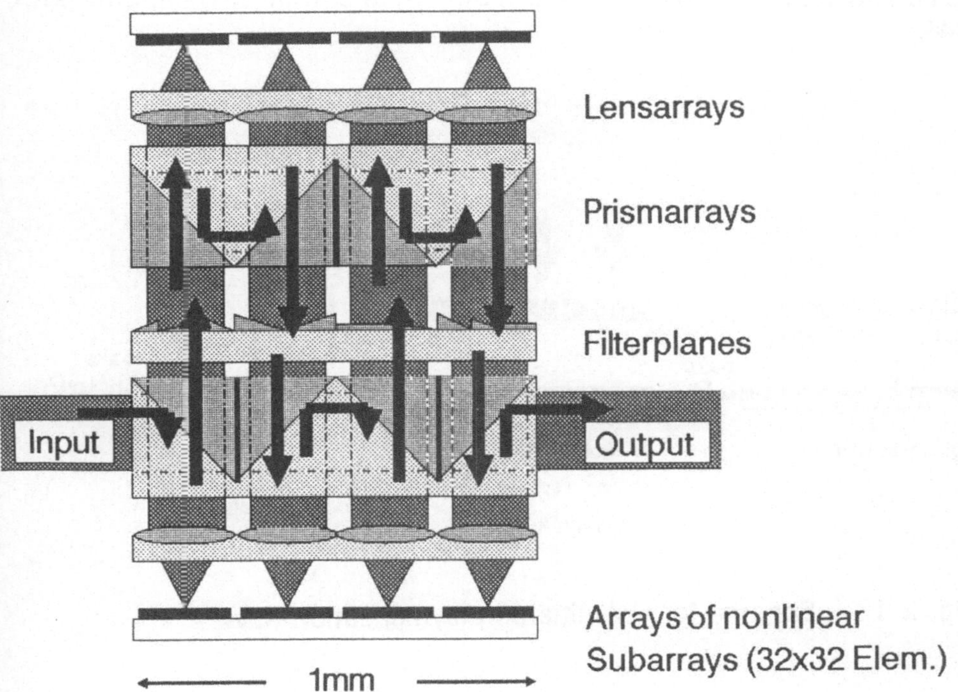

Fig. 1.1   Principle setup of a microoptic system for digital optical data processing.

This conceptual microsystem can be adapted for many of the previously mentioned digital 3D-optical systems. Thus complex and inexpensive optical systems can be realized by combining several arrays of microoptic elements. Hereby the microoptic elements in each array can be preadjusted during fabrication. The stacking of layers enables combining different technologies. The nonlinear arrays at the top and bottom of the stack carry (e.g. 32x32) subarrays of nonlinear devices (e.g. surface light emitting thyristors [HAR 90, OGU 90]). A heatsink can be placed there easily.

2. Process of UV-initiated polymerization

We use spin coated layers (thickness 15-25um), which are sensitized for UV wavelengths by ketone (2,2-Dimethoxy-2-phenylacetophenone) (fig. 2.1). The ketone is most sensitive at 300nm to 350nm wavelengths [FRA 84].

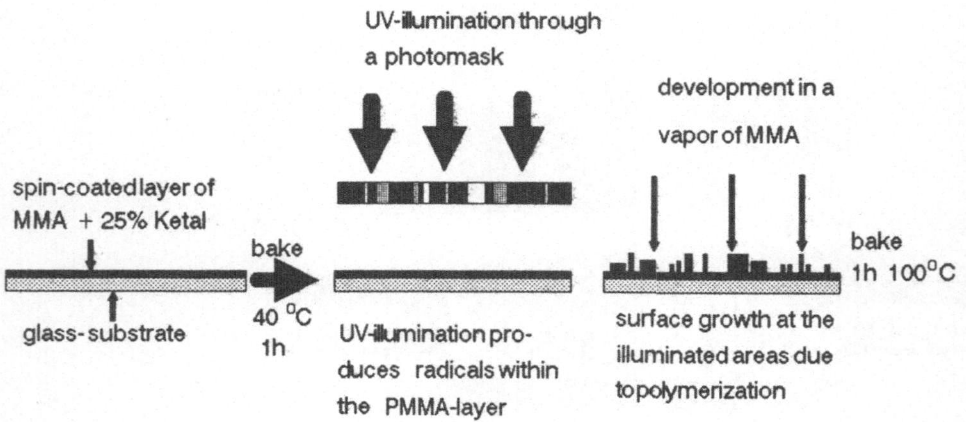

Fig. 2.1   Scheme for UV-initiated polymerization (UVIP)

The UV-radiation causes the photochemical reaction of the ketone described in fig. 2.2. In the presence of monomer additional polymerization is initiated by the exposed ketone.

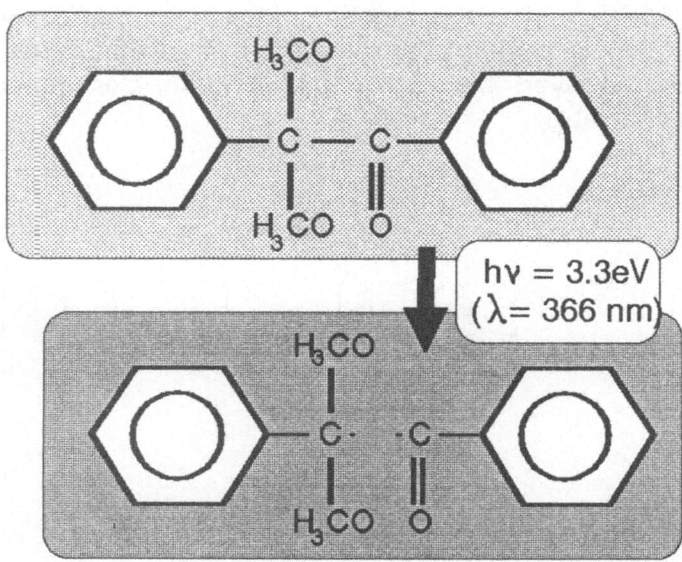

Fig. 2.2   Photochemical   reaction   of   2,2-Dimethoxy-2-phenyl-
acetophenone [FRA 84].

The additional polymerized regions of the resist show an increase of volume (fig. 2.1), as the MMA is now fixed to the oligomer and cannot be removed by the final bake. Thus an intensity pattern generated by a mask is transferred into a surface relief structure of the layer. The resulting change of refractive index is negligible for the proposed processing.

In order to characterize the processing, we separated the exposure and the development of the resist by heating the samples (1h at 40°C (prebake)). Thus the monomer is removed and exposure only causes the photochemical reaction described in fig. 2.2. Polymerization is not possible during illumination due to the lack of MMA. Consequently latent images do not distort exposure. To enable polymerization the samples are developed afterwards by controlled diffusion of MMA vapor into the exposed layer.

The photosensitive resist we are using, consists of the solvent MMA, prepolymerized MMA (oligomer) and the ketone (3:2:1 parts by weight). The resist is spin coated on glass substrates (fig. 2.3). The thickness of the resist (15-30μm) is also a function of the amount of the solvent. The variations (peak to peak) of the thickness in the center area ($\sim$70% of the spin coated layer) are smaller than 0.05μm. Consequently optical quality is achievable.

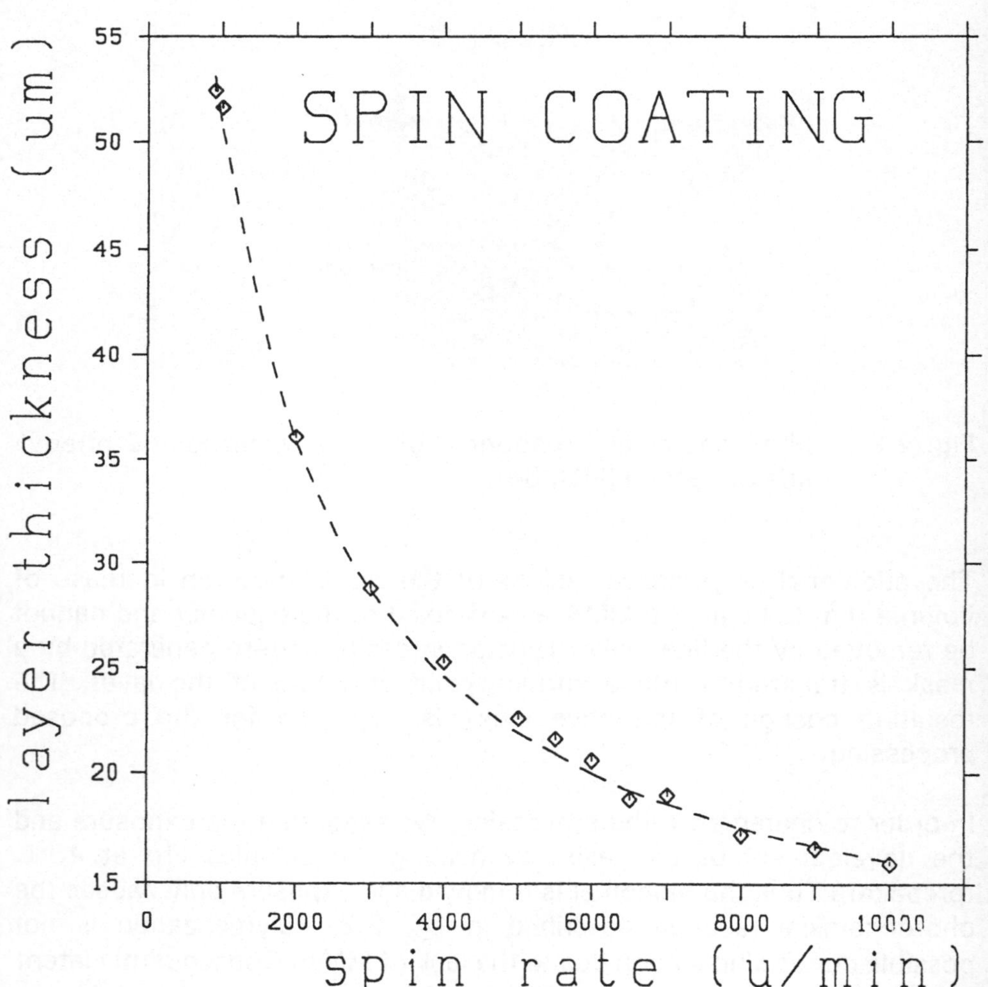

Fig. 2.3   Thickness of the spin coated resist versus rpm.

Before exposure the sensitized layers can be stored in dried air for some weeks. High humidity causes crystallization of the ketone.

The **PMMA**-samples are exposed to UV-light at a wavelength of **365**nm (mercury line). A mask containing the intensity profile is imaged onto the sample (reduction of the mask by 10)(fig. 2.4). We are using pulse width modulation for generating intensity distributions (fig. 2.5). The mask consists of a regular array of spots. Due to an appropriate numerical aperture the spots are not resolved by the imaging system. Consequently variations of the size of the spots cause local gray levels in the resist.

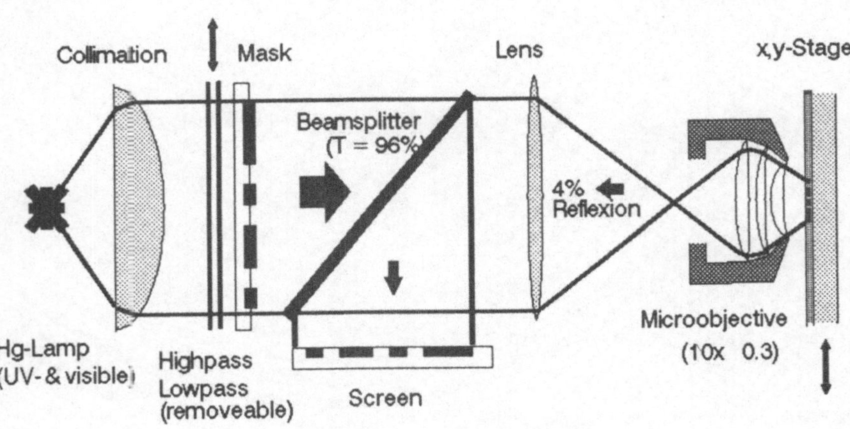

Fig. 2.4   Principle setup for exposure. A UV-stepper (UV-microscope & x- y-stage) is used for writing intensity distributions into the resist.

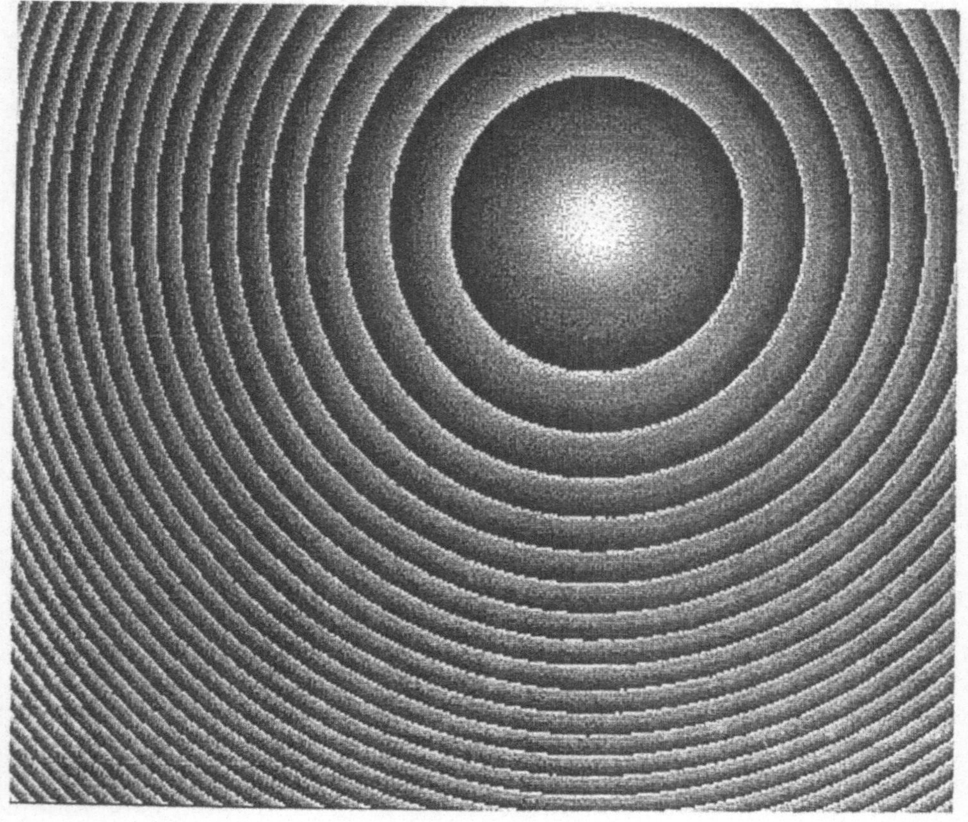

Fig. 2.5   Example for a pulse with modulated mask resulting in a phase distribution of a Fresnel lens.

The angle of illumination and the resolution in the object plane is a function of the numerical aperture of the imaging system. Higher resolution may be obtained by an increased numerical aperture. Due to an increased numerical aperture the depth of focus (region of the beam waist) decreases. Consequently we reduced the numerical aperture (NA) of the objective for adapting the depths of focus Δz to the thickness of the resist d (n = refractive index of the sample, NA = numerical aperture, λ = wavelength of the exposing light):

$$d = \Delta z = n\lambda / (2(NA)^2)$$

$$=> NA = n\lambda/2d$$

The lateral resolution can be calculated by:

$$\Delta x = \lambda/NA$$

A numerical aperture of 0.1 results in a focal depth of $\Delta z \sim 25\mu m$ ($\lambda = 365$nm, $n = 1.5$) and a lateral resolution of $\Delta x \sim 4\mu m$. Higher resolution is possible for thinner resist ($\Delta z \sim 7\mu m \Rightarrow NA \sim 0.2$, $\Delta x \sim 2\mu m$).

Visible light is used for focal control of exposure (fig. 2.4). Some percent of the illuminating intensity ($\sim 4\%$) is reflected back by the surface of the sample and coupled onto a screen by a beamsplitter. Due to chromatic distortions of the imaging lenses it is necessary to adjust the focal control at visible wavelength to the imaging at the wavelength of 350nm.

Development:

Exposure causes no additional polymerization due to the lack of residual monomers. After exposure the PMMA-resist is developed by controlled diffusion of MMA into the resist. The resulting surface profile is not modified, if there is a delay of development for the exposed resist. This delay was tested up to one day.

MMA is vaporized in a chamber (fig. 2.6) for the development of the exposed resist. Vacuum or inert gasses are not required. The sample is stored for approximately one minute inside the chamber. The UV-initiated polymerization in the volume of the resist can be controlled by the partial pressure of MMA and by the duration of diffusion (fig. 2.7). Lateral resolution of the process can be controlled by the partial pressure of MMA. Consequently height and lateral resolution of the surface profile can be controlled independently.

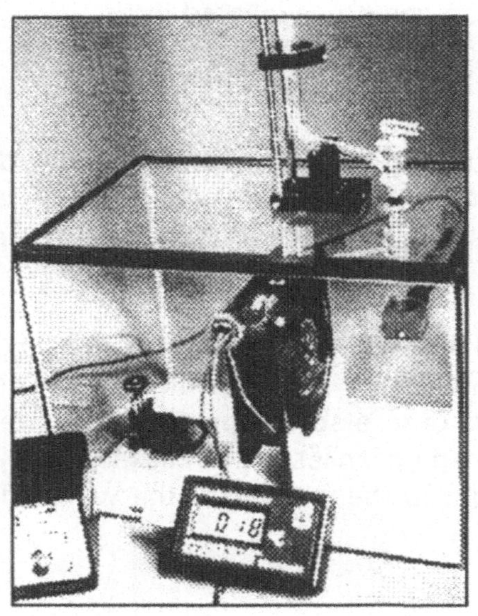

Fig. 2.6   Chamber for developing the exposed resist under controlled partial pressure of MMA

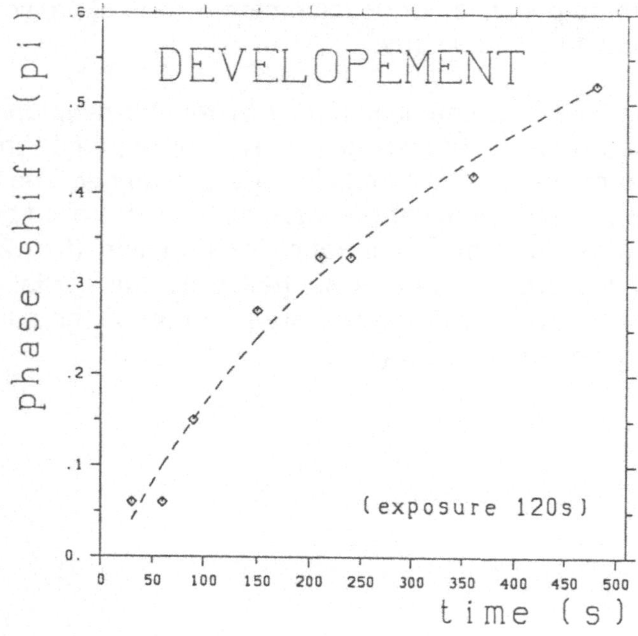

Fig. 2.7   Development of the resist versus time

Postbake:

After development the PMMA-samples are heated again (1h, 90°C) to remove the ketone and residual MMA. The final transparency of the structured layers is equal to the transparency of PMMA [VIE 75] for visible light.

3. Experimental results

The versatility of the processing scheme is demonstrated by experimental results. A linear irradiation profile in the resist results in a prism (fig. 3.1). The resolution and a possible application of the processing scheme is demonstrated in fig. 3.2. The imaged intensity profiles in fig. 3.2a were cosine-shaped stripes. The frequency increases in downward direction. These patterns are used for testing the lateral frequency response of the resist. Fig. 3.2b shows an array of Fresnel lenses.

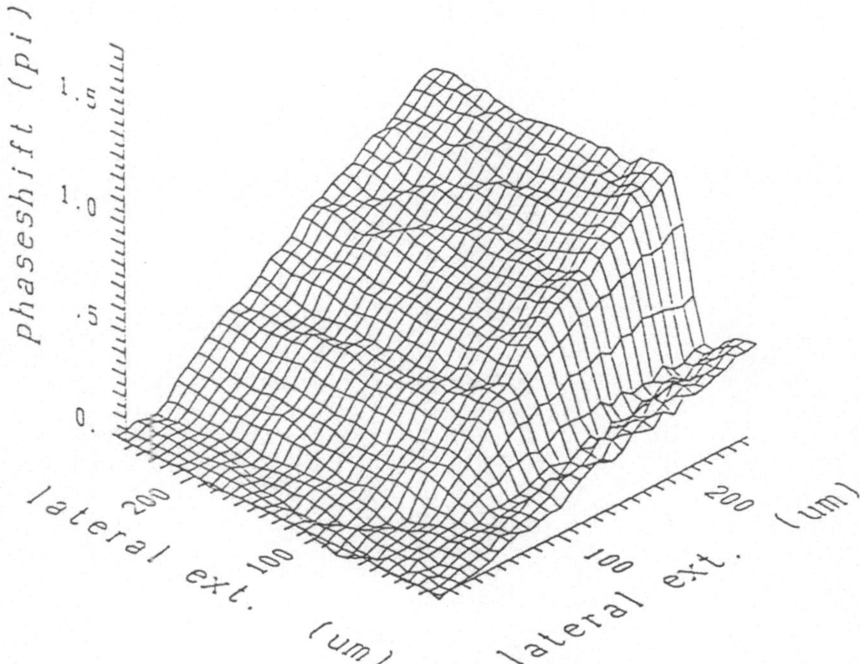

Fig. 3.1   Prism fabricated by linearly increasing illumination. This prism demonstrates the linear response of the resist.

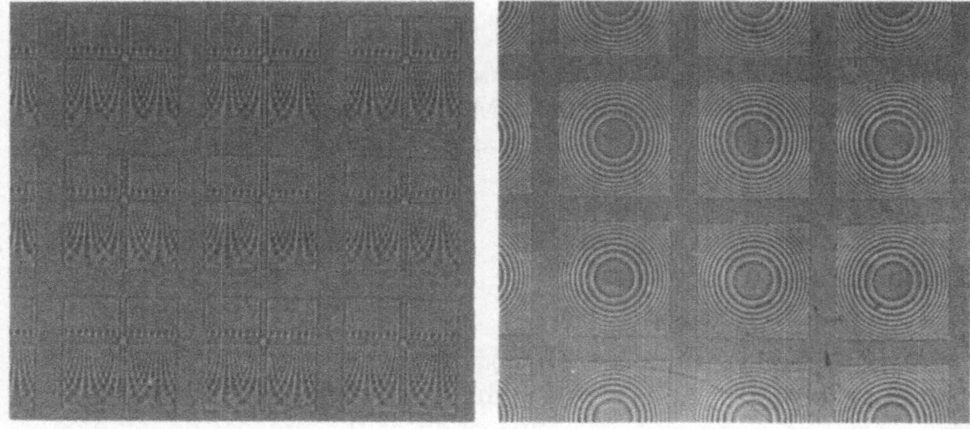

Fig. 3.2   Cosine-shaped gratings (fig. 3.2a) are used for testing the frequency response of the process (pitch 2mm). An array of Fresnel lenses is shown in fig 3.2b (pitch also 2mm).

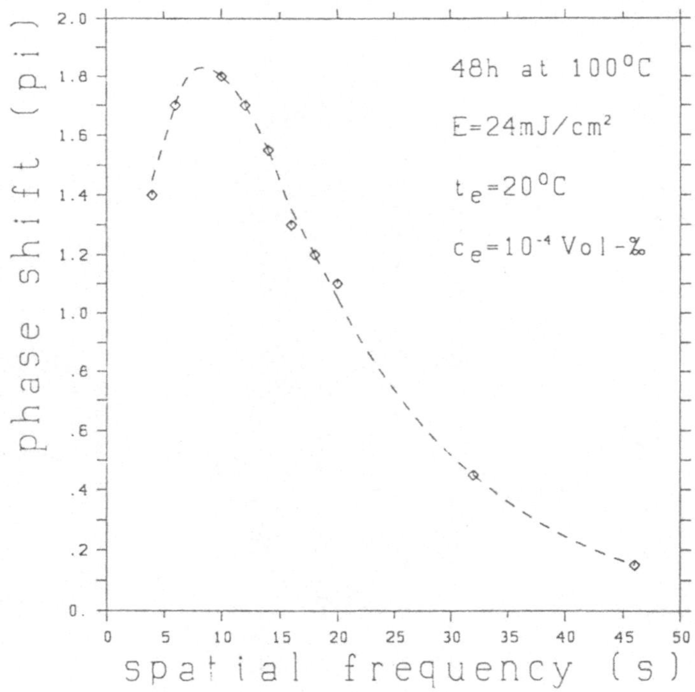

Fig. 3.3   Frequency response of the resist

The resulting modulation of the surface reaches $1.5\mu m$ ($\gg 2\pi$ phase shift at 633nm wavelength). The lateral resolution of the resist reaches $25\mu m$ (fig. 3.3).
Initial experiments with a slightly modified processing scheme demonstrated that an increase of surface modulation ($\sim 6\mu m$) at a lateral resolution of $20\mu m$ is possible.

4. Conclusion

We developed a photoresist for fabricating diffractive and refractive optical elements. The processing of the spin coated UV-sensitive resist involves an exposure at 365 nm wavelength and a development in MMA vapor (monomer of PMMA). Finally the sensitizer (ketone) is removed by heating the sample.Due to the linear response of the resist accurate control of the surface profile is possible by lithographic means. Thus flexible design of phase-only structures is possible. Different optical elements can be integrated on one substrate. The elements can be used from IR to UV-wavelengths, because PMMA shows low absorption and the path of light in the resist is shorter than 1/20mm.

References:

[BAG 00] G. Bagordo, K.-H. Brenner, T.M. Merklein, "Realization of microoptic elements by UV-initiated polymerization", to be publ. in Appl. Opt.

[BRE 00] K.-H. Brenner, C. Doubrava, T.M. Merklein, "Fabrication of microoptic components by thermal imprinting", to be published in Appl. Opt.

[BRE 86] K.-H. Brenner, A. Huang, N. Streibl, "Digital optical computing with symbolic substitution," Appl. Opt. 25 (1986) 3054.

[BRE 86] K. H. Brenner, "New implementation of symbolic substitution logic", Appl. Opt. 25 (1986) 3061-3064.

[BRE 88] K.-H. Brenner, "A programmable optical processor based on symbolic substitution," Appl. Opt. 27 (1988) 1687.

[BRE 88] K.-H. Brenner, "Digital optical computing", Appl. Phys. B46 (1988) 111-120.

[BRE 89] K.-H. Brenner, A.W. Lohmann, T.M. Merklein, "Symbolic substitution implemented by spatial filtering logic", Opt. Eng. 28 (1989) 390.

[BRE 92] K.-H. Brenner, T.M. Merklein, "Implementation of an optical crossbar network based on directional switches", Appl. Opt. 31 (1992) accepted.

[CES 89] L. Cescato, E. Gluch, M. HeiPtmeier, U. Krackhardt, T.M. Merklein, S. Sinzinger, N. Streibl, J. Thomas, "Computer Generated Optical Components in Photoresist", Proc. "Symposium on optics in computing", Toulouse, France, 17/18 Oct. 1989.

[DAM 70] H. Dammann, "Blazed synthetic phase-only holograms", Optik 31 (1970) 95-104.

[ECK 89] W. Eckert, G. Lohman, T.M. Merklein, K. Zürl, K.-H. Brenner, "Optoelectronic implementations of symbolic substitution", Proc. "Symposium on optics in computing", Toulouse, France, 17/18 Oct. 1989.

[FRA 84] H. Franke, "Optical recording of refractive-index patterns in doped poly-(Methyl Metacrylate) Films", Appl. Opt. 23 (1984) 2729-2733.

[HAR 90] K. Hara, K. Kojima, K. Mitsunage, K. Kyuma, "AlGaAs/GaAs pnpn differential optical switch operable with 400 fJ optical input energy", Appl. Phys. Lett. 57 (1990) 1075- 1077.

[HUR 82] Hurtley, M.C., "Diffraction gratings. Techniques of Physics", Academic Press (1982) London.

[JAH 90] J. Jahns, W. Däschner, "Optical cyclic shifter using diffractive lenslet arrays", Opt. Comm. 79 (1990) 407- 410.

[JAH 92] J. Jahns, K.-H. Brenner, W. Däschner, C. Doubrava, T. M. Merklein, "Replication of Diffractive Microcptical Elements Using a PMMA Molding Technique", OPTIK 89 (1992) 98-100.

[JOH 88] K.M. Johnson, M.R. Surette, J. Shamir, "Optical interconnection network using polarization-based ferroelectric liquid crystal gates", Appl. Opt. 27 (1988) 1727.

[LOH 86] A.W. Lohmann, "What classical optics can do for the digital optical computer", Appl. Opt. 25 (1986) 1543.

[LOH 89] A.W. Lohmann, "Scaling laws for lens systems", Appl. Opt. 28 (1989) 4996.

[MER 89] T.M. Merklein, W. Stork, H. Yajima, " An optical full adder" Appl. Opt. 28 (1989) 4313.

[OGU 90] I. Ogura, Y. Tashiro, S. Kawai, K. Yamada, M. Sugimoto, K. Kubota, K. Kasahara, " Reconfigurable cptical interconnection using a two-dimensional vertical to surface transmission electrophotonic device array", Appl. Phys. Lett. 57 (1990) 540-542.

[SHI 87] T. Shiono, K. Setsune, O. Yamazaki, K. Wasa, "Rectangular-apertured micro-Fresnel lens arrays fabricated by electron-beam lithography", Appl. Opt. 26 (1987) 587-591.

[SMI 73] H.I. Smith, "X-Ray lithography: A complementary technique to electron beam lithography", J. Vac. Sci. [SMI 82]P.W. Smith, "On the physical limits of digital optical switching and logic elements", The Bell System Technical Journal 61 (1982) 1975-1993. Technol. 10 (1973) 913.

[SWA 89] G.J. Swanson, W.B. Veldkamp, "Diffractive optical elements for use in infrared systems", Opt. Eng. 28 (1989) 605-608.

[VIE 75] R. Vieweg, F. Esser, "Kunststoff-Handbuch Band 9 'Polymethacrylate'", Carl Hanser Verlag (1975) München.

[WAL 90] S.J. Walker, J. Jahns, "Array generation with multilevel phase gratings", J. Opt. Soc. Am. A7 (1990) 1509-1513.

[WOL 91] B. Wolf, N. Fabricius, W. Foss, A. Dorsel, "Ion exchanged waveguides in glass: simulation and experiments", Proc. SPIE, Hague, Netherlands, 1506 (1991) 40-51.

Micropatterning of Organic-Inorganic Nanocomposites for Micro-Optical Applications

H. Krug, N. Merl, H. Schmidt

Institut für Neue Materialien gem. GmbH, Universitätscampus, Geb. 43, 6600 Saarbrücken

1 Introduction

Microoptic elements like gratings and strip waveguides can be produced by different patterning techniques. Organic polymers and inorganic SiO_2/TiO_2-based sol-gel materials can be used for low-cost mass production of such elements.

Organic polymers have good optical properties and can be easily patterned, but their thermal and mechanical stability is poor [GAL 90a, GAL 90b , ULR 72, WEB 75]. Strip-waveguides are generated by local photopolymerization and gratings are prepared by embossing the polymer heated near T_g. Molecules with non-linear optical properties can be incorporated but relaxation stability after poling is not yet satisfying.

Inorganic films made by sol-gel techniques combine good mechanical and thermal stability with low optical losses. The preparation of crackfree layers with a thickness of more than one micron is extremely difficult [LUK 83]. Densification temperatures near T_g make it impossible to incorporate optical active organic substances as these would be destroyed. Local densification by laser-treatment can generate carbon particles within the structure and thus increase the optical loss to more than 5 dB/cm. In case of embossing techniques,near-net shaping is not possible because of high shrinkage to 70 Vol.% during the densification step. Shrinkage rates can be reduced by the incorporation of polymeric species like PEO, but values better than 30 Vol.% could not yet be obtained [RON 91, TOH 88, MAT 90].

Many of the problems listed above can be overcome by using inorganic-organic nanocomposites of the ORMOCER type. These materials are synthesized through the sol-gel route where the inorganic backbone is formed by a hydrolysis / condensation process and an organic network is built up by polymerization. Crosslinking of the inorganic backbone and the organic network is achieved by covalent or coordinative bonds. The inorganic component increases thermal stability and surface hardness in comparison to pure organic polymers. The organic component allows densification temperatures of less then 150 °C. By the variation of the synthesis parameters many properties can be tailored. For optical applications, the refractive index can be varied by composition. Reduction of mechanical stresses allows to deposit layers of more than 10 μm in thickness in one step by several coating techniques. Optical active compounds can be introduced either as guest-host systems or as network forming units. The low over-all shrinkage allows embossing with a near net shaping quality. The option of curing the materials by UV or VIS irradiation permits patterning by direct laser writing or maskaligner techniques.

2 Experimental and results

2.1 Synthesis and properties of the nanocomposite materials

A system based on methacryl oxypropyl trimethoxy silane (I), zirconium n-propoxide (II) complexed with methacrylic acid (III) was used in this work. The detailed procedure of the synthesis of this type of nanocomposite has been published elsewhere [NAS 90]. Molar compositions between 10:1:1 and 10:6:6 (I:II:III) were used to match the refractive index in the range 1.509 < n < 1.540 at 632.8 nm. For coating, the viscosity was varied in the range of 6 - 7 mP·s by solvent addition to the sol and 1 - 2 wt.-% of a photoinitiator (e.g. Irgacure 369, Ciba Geigy) was added.

Layers with thicknesses of more than 10 μm could be deposited in one step on glass or fused silica substrates by spin-, dip- and a modified flow-coating technique. Heating of the freshly deposited layers to 60°C for some minutes results in a smooth and homogenous surface required for patterning. Layers produced by spin-coating have surface roughness less than 5 nm.

2.2 Embossing procedure

Grating patterns were generated on layers deposited on fused silica using the in-house built device shown in fig. 2.1.

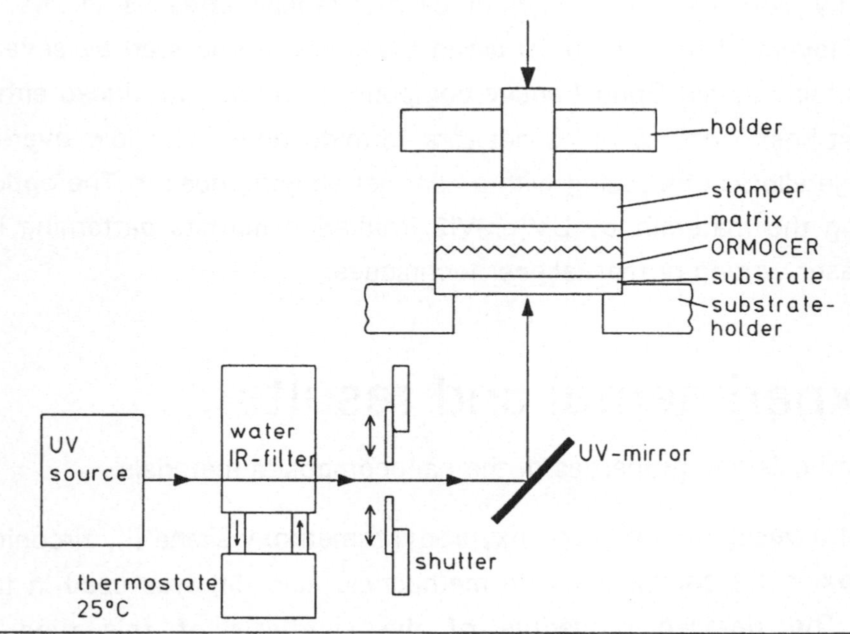

Fig. 2.1. Device for the embossing technique

A black glass stamper with a grating of 2400 parallel lines per millimeter (Fa. Zeiss) was used. It was placed onto the uncured layer and a defined pressure of 10 - 100 $N \cdot cm^{-2}$ was imposed. Photocuring was carried out by irradiation with UV-Light through the silica substrate for 10 minutes. After the curing step the stamper was removed and the layer was heat treated (130°C, 1 h) for final curing. In fig. 2.2 a SEM of a grating fabricated by this procedure is shown.

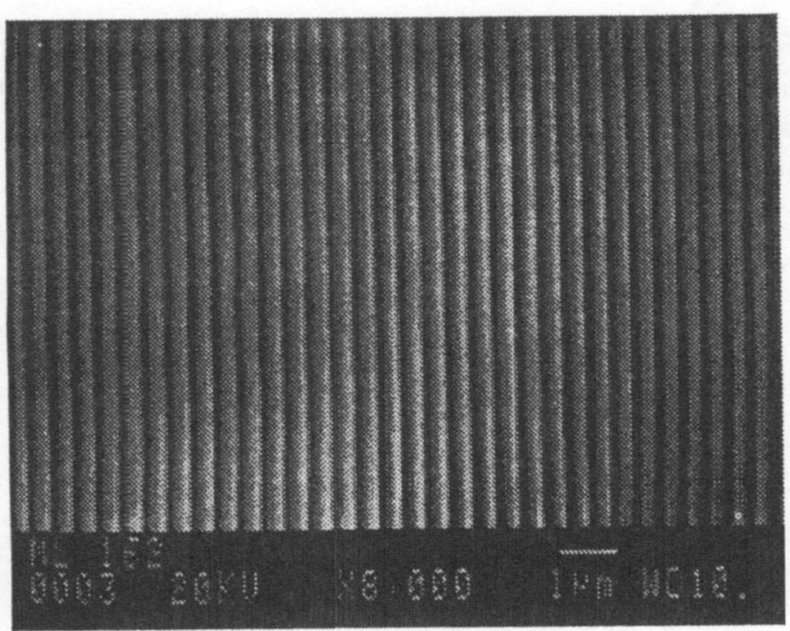

Fig. 2.2. Grating generated by the embossing technique, 2400 lines/mm

The overall shrinkage rate of less then 5 Vol.% results in a near-netshaping quality of the embossed patterns. This low shrinkage rate in combination with high film thickness generates grating amplitudes of several hundred nm, which is essential for high diffraction efficiencies.

Some problems concerning adherence of the material to the stamper still have to be overcome by improving the clean room technology and by insertion of separating thin layers.

2.3 Direct laser writing

Layers deposited in the same manner as described in 2.2. were used for direct laser writing [SCH 91]. The device for this procedure is shown in fig. 2.3.
The substrate is fixed on a unit which can be positioned in x- and y-direction under computer control allowing flexible sample movement. A laser beam focusing unit permits generation of patterns variable in size and geometry.

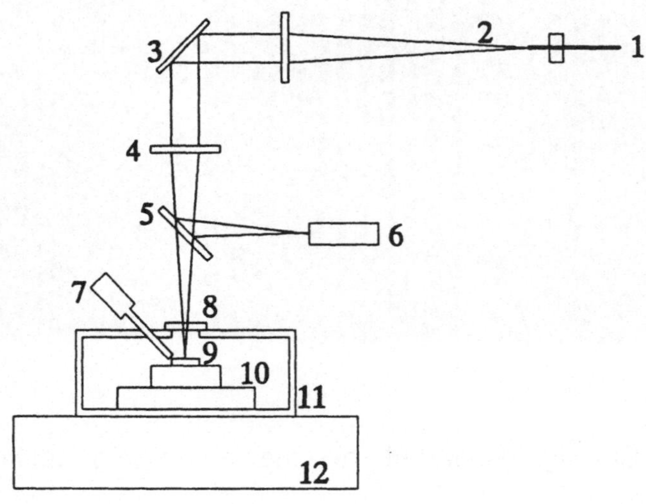

Fig. 2.3: Scheme of direct laser writing equipment. 1: laser beam, 2: beam expander unit, 3: flat mirror, 4: focusing lens (fused silica), 5: fused silica plate, 6: laser beam analyser, 7: TV-microprobe, 8: fused silica window, 9: sample, 10: motor driven xy-stage, 11: reactor housing, 12: optical table

In the areas irradiated by the laser beam, which can be focused down to 10 μm in diameter, the polymer network of the organic component is formed. The material of the non-polymerized areas is then removed by an adequate solvent, e.g. an alcohol or a ketone. The curing of the final patterns is accomplished by heat treatment (130°C, 1h).

A SEM of a strip waveguide which has been evocated by this procedure is shown in fig. 2.4. The optical loss of the waveguide has been determined to 3 dB/cm due to surface roughness and scattering by dust particles. Therefore the patterning process has to be optimized and clean room conditions have to be improved. But nevertheless this is a promising low starting value.

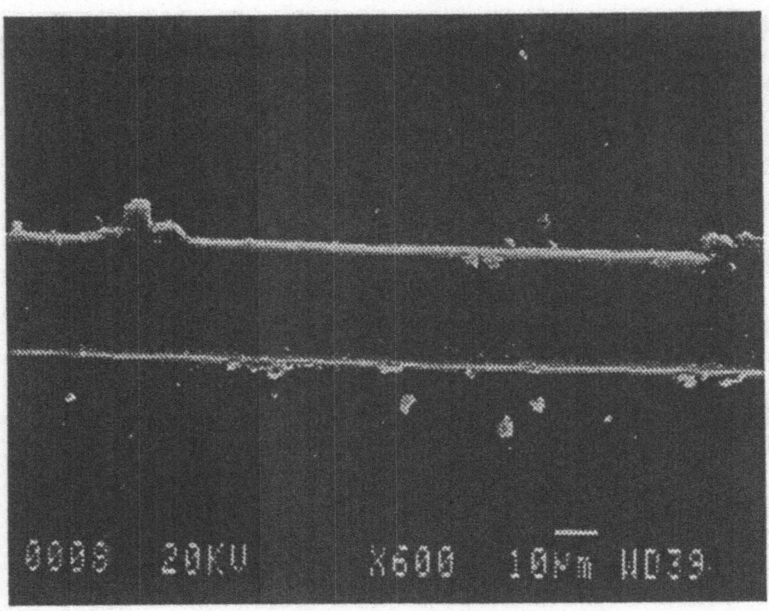

Fig. 2.4. Waveguide patterned by direct laser writing

2.4 Maskaligner technique

The same principle as described for laser writing is used for fabricating patterns by a maskaligner technique (fig. 2.5). The photocuring of a nanocomposite deposited on a substrate is carried out in a maskaligner apparatus providing a parallel arrangement with a defined distance of the coated substrate and a mask, which consists of a metal film deposited on a glass sheet. Through the transmissible areas of the mask irradiation and thus polymerization of the nanocomposite material is achieved. Typical values are 5 - 10 minutes irradiation time and 80 μm distance between mask and layer. The pattern is developed with a suitable solvent to remove the non-polymerized material according to 2.3. The standard heat treatment (130°C, 1h) accomplishes the curing of the nanocomposite and provides maximum hardness and stability.

In fig. 2.6 is shown a SEM of a pattern of parallel lines made by maskaligner technique.

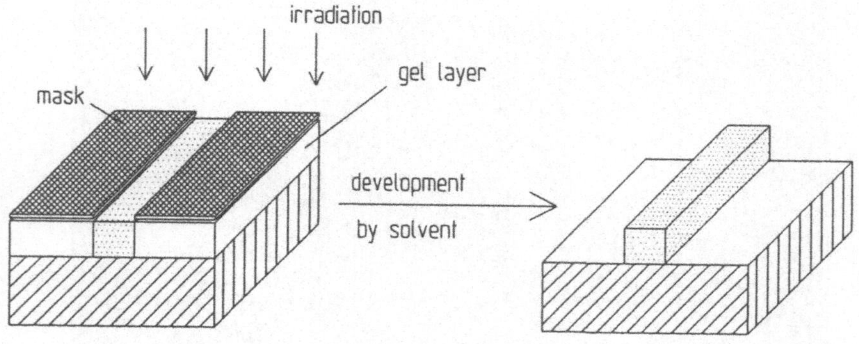

Fig. 2.5. Scheme of the maskaligner device

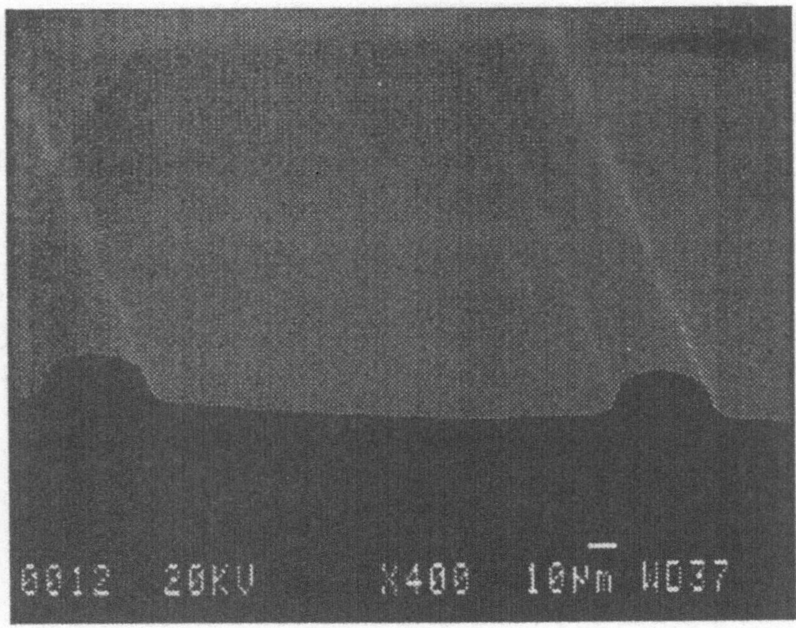

Fig. 2.6 SEM of a pattern of parallel lines made by maskaligner
technique

3 Summary

Organic-inorganic nanocomposites for use in microoptics can be
synthesized by the sol-gel process. Their properties can be taylored by
variation of the materials and the synthesis parameters.
The inorganic backbone is formed during the hydrolysis and
condensation reactions while the polymerization leading to the organic
network is achieved by adding a proper photoinitiator and by light
irradiatiation.

Layers of several μm thicknesses could be obtained showing good optical quality (opt. loss < 1 dB/cm). Their refractive index was adjusted by the appropriate composition.

Gratings were formed by an embossing step with near net shaping. Local photocuring allows micropatterning by flexible direkt laser writing or by maskaligner techniques for low cost mass production.

The special properties of the here presented organic-inorganic nanocomposites make them very interesting for many applications in the field of microoptics.

4 Acknowledgement

The authors want to thank the Minister for Science and Education of the State of Saarland for financial support and Dipl.-Phys. F. Tiefensee for contributions in the field of laser writing and measurement of optical losses.

5 Literature

[GAL 90a]   M. T. Gale, K. Knop, R. Morf, Proc. SPIE 1210, (1990), 83

[GAL 90b]   M. T. Gale, Proc. OPTICS , Nottingham, Engl., (1990), 207

[LUK 83]   W. Lukosz, K. Tiefenthaler, Opt. Lett. 8 (1983), 537

[MAT 90]   A. Matsuda, Y. Matsuno, S. Kataoka, S. Katayama, T. Tsuno, N. Tohge, T. Minami, Proc. SPIE 128 (1990), 71

[NAS 90]   R. Naß, H. Schmidt, E. Arpac, Proc. SPIE 1328 (1990), 258

[RON 91]   R. L. Roncone, L.A. Weller-Brophy, L. Weisenbach, J. J. Zelinski, J. Non. Cryst. Solids, 128 (1991), 111

[SCH 91]   H. Schmidt, H. Krug, R. Kasemann, F. Tiefensee, Proc. SPIE 1590 (1991), 36

[TOH 88]   N. Tohge, J. Non. Cryst. Solids, 100 (1988), 501

[ULR 72]   R. Ulrich, Appl. Phys. Lett. 20 (1972), 213

[WEB 75]   H. P. Weber, Opt. Quantum Electron. 7 (1975), 465

Replication of Microrelief Structures for Diffractive and Integrated Optics

by L. Baraldi and M.T. Gale
Paul Scherrer Institute, Zürich, Switzerland

Summary

The replication of submicron surface relief structures into polymers by hot embossing or moulding techniques is being developed for applications in microoptics, diffractive optics and integrated optics. Microstructures of interest have typical linewidths between 100 nm and 5 µm, and are fabricated by holographic techniques and by direct laser and e-beam writing lithography. Nickel shims are then generated by electroplating and combined to produce replication shims of up to 20x30 cm^2 in size. The deposition of dielectric material onto embossed polymer sheet is being investigated for applications as diffractive security features and in integrated optics. The direct embossing of single-mode polymer waveguide films on glass substrates has also been demonstrated and shows great promise for the low cost fabrication of integrated optical components.

1. Introduction

The replication of surface relief microstructures by embossing and moulding techniques is of great interest since it offers the promise of low cost production of structures for diffractive and integrated optical components. First industrial products include holograms and diffractive foil for security, display and packaging. The additional evaporation of dielectric thin films further opens up a range of new types of optical behaviour for diffractive elements [GAL 90], as well as potential low cost patterning of waveguide structures for integrated optics. Both of these applications are being pursued at the Paul Scherrer Institute in Zürich (PSIZ).

Diffractive optical structures are being developed for applications as visual and machine-readable security features for credit cards, banknotes and valuable documents. Novel optical behaviour is obtained by combining submicron grating structure embossed into polymer sheet with a high refractive index dielectric coating deposited by directional evaporation techniques.

The fabrication of integrated optical components using technology based upon the embossing of polymer substrates and films is being investigated in two forms. The first is the evaporation of high index dielectric waveguides onto structured polymer substrates or sheet. The second is the direct embossing of single mode planar polymer waveguides coated onto glass substrates using an embossing press specially developed for this purpose.

Central to both applications is the fabrication of a replication shim (usually in the form of a nickel sheet of about 100 µm thickness) of the required microstructure and its faithful replication into polymer material by embossing or moulding.

2. Replication shim fabrication

The replication shim is fabricated by electroplating nickel onto an original surface relief structure produced using one or more high resolution lithographic processes (Fig. 2.1a). Submicron structures, grating couplers and stripe waveguides are fabricated in resist by holographic techniques or by direct e-beam writing.

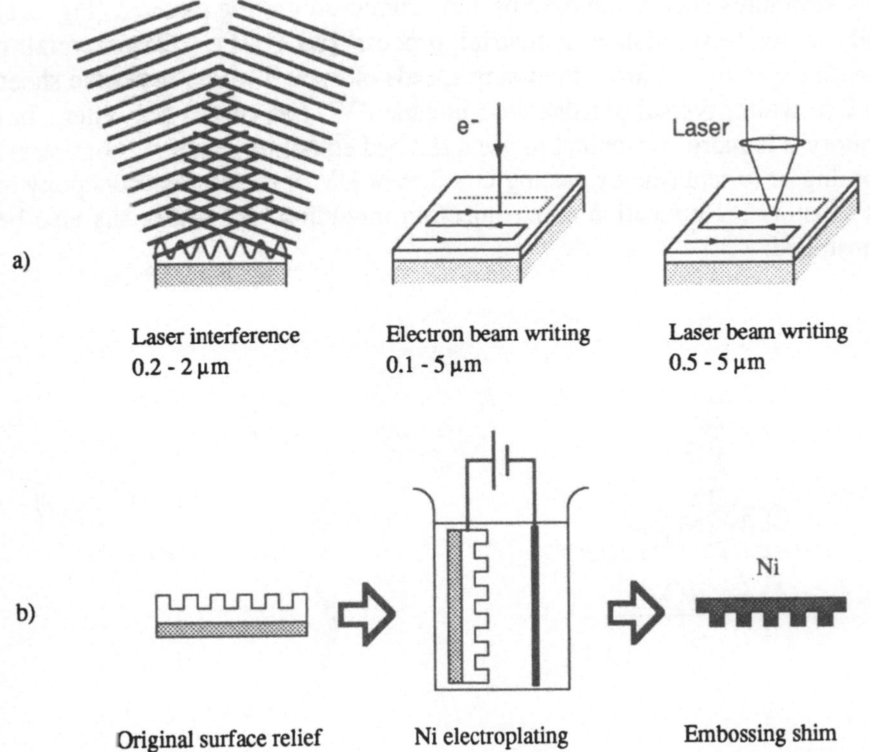

Fig. 2.1: a) Lithographic processes used to fabricate the microrelief structures, with typical achievable dimensions. b) Fabrication of embossing shim by electroplating.

Other semiconductor technology processes, such as dry-etching and lift-off techniques, are also used to produce the required line profiles or to transfer the relief into the underlying material [LEH 89]. Continuous-relief microstructures are typically fabricated by laser beam writing.

The microfabricated surface relief structure is first made conducting by the sputter deposition of a thin silver or gold film or by the electroless deposition of nickel. A nickel foil of about 100 μm thickness is then built up by electroplating and separated from the original relief to give a 1st generation replication shim (Fig. 2.1b). Subsequent generation shims can be fabricated by electroplating copies of this first shim.

3. Replication

Replication of the shim microrelief structure into polymer material can be carried out by embossing, casting or moulding techniques. Hot embossing of diffractive optical structures such as holograms and submicron grating patterns (Fig. 3.1) is already a well-established industrial process [KLU 91]. Commercial roll embossing systems achieve embossing speeds of up to 1 m/sec in plastic sheet of up to 2 m width; typical plastics used include PVC and coated polyester. In the laboratory it is more convenient to use a flat bed embossing system (hot press) for embossing or to replicate by casting in a film of UV or thermal curing epoxy on a glass substrate. Replication using injection moulding equipment has also been demonstrated.

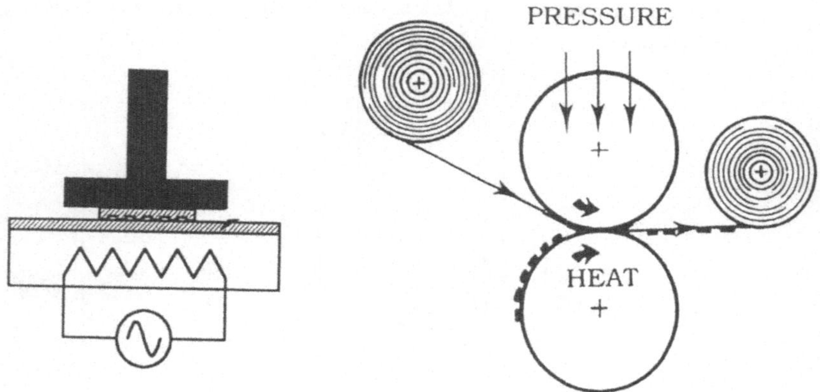

Fig. 3.1: Replication of shim microrelief structure by hot embossing using a flat bed or roll embossing system. Features of dimensions much smaller than 100 nm can be faithfully replicated.

The replication of microstructures for integrated optics is in the laboratory phase. A special press has been constructed for the hot embossing of such structures into single-mode polycarbonate waveguide films on glass substrates (Section 5). The process requires very uniform application of pressure and a well controlled heating and cooling cycle.

4. Diffractive optics

An example of a class of diffractive microstructures produced by embossing is holograms for security applications. At PSIZ a program is underway to develop a new type of security feature based upon zero-order diffractive structures [GAL 90]. Advances in the design and fabrication of very fine grating structures has led to new types of microstructure with optical characteristics which can be optimized for visual and machine identification.

The zero-order microstructures are fabricated by the embossing of submicron grating structure into polymer sheet combined with the directional evaporation of high index dielectric material, both steps being potentially low cost, mass-production processes. After overcoating with a thin polymer film, the resulting microstructure consists of a lamellar dielectric grating with submicron periodicity, fully embedded in plastic. The reflection spectrum is very pronounced, with marked resonance structure and strong polarization behaviour (Fig. 4.1). As the structure is tilted, the main p-polarization reflection peak R_p splits into 2 peaks which shift approximately as a linear function of the tilt angle. Tilting thus produces a marked colour shift in the reflected light.

The advantages of this type of polymer-embedded dielectric microstructure can be summarized as:

(1) *Zero-order read-out:* The structure can be viewed in diffuse or 'poor' illumination under which holograms would be washed out.

(2) *Copy-proof:* The optical behaviour of the diffractive structure is a consequence of the exact physical microstructure with its well-defined materials and refractive index distribution. It cannot be copied by optical techniques because this would not reproduce the identical physical microstructure. Mechanical copying is considerably complicated by the fact that the structure is no longer a simple surface relief.

(3) *Embedded structure:* The microstructure can be fully embedded in a plastic card (or other plastic host), in contrast to current diffractive security features which

are glued onto the surface; it can be inserted at the lamination stage in the production to form an integral part of the card, from it is then extremely difficult to remove.

(3) *Visual appearance:* A colour change on rotation is very difficult to achieve in buried, thin film structures (no alternative structures are currently known).

(4) *Machine identification:* The pronounced polarization and spectral features with fine detail are ideal for secure machine identification.

(5) *Low cost fabrication:* Such microstructures can be mass produced by embossing and evaporation techniques, with production costs only slightly greater than those of current embossed security features.

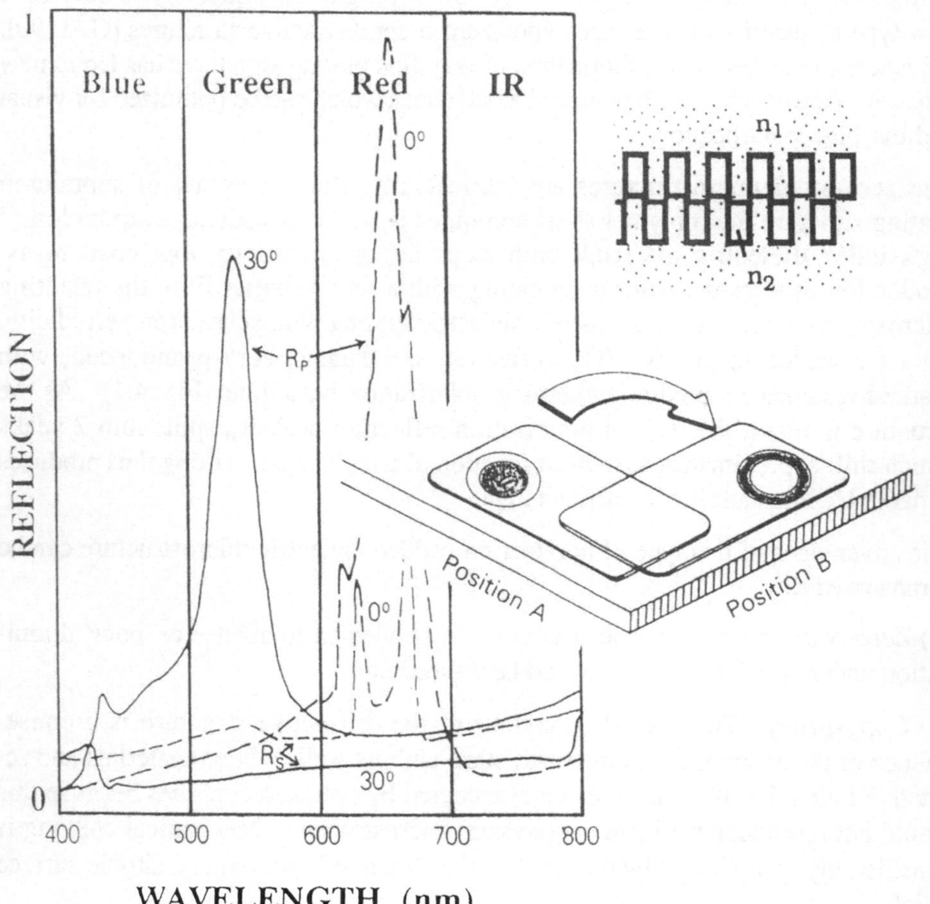

Fig. 4.1: Zero-order embedded dielectric microstructure as a security feature. A red to green colour change is observed on rotation.

5. Integrated Optics

For the low cost fabrication of integrated optical (IO) components by embossing techniques, two principally different approaches are possible (Fig. 5.1). Either the surface of a polymer substrate is microstructured and overcoated with a waveguiding material (left), or a planar polymer waveguide on e.g. glass is directly formed to the final device (right).

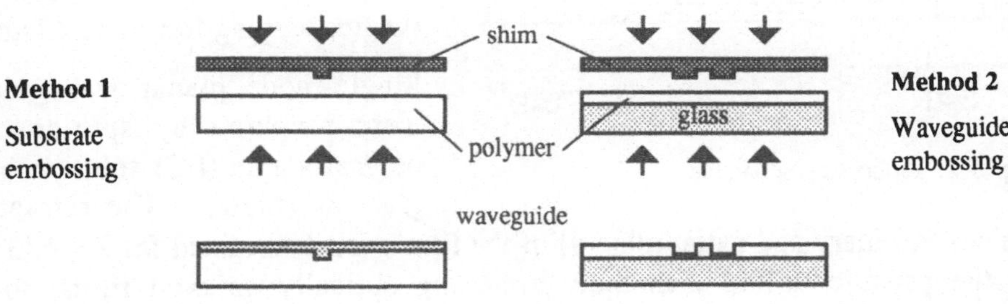

Fig. 5.1: Two methods for the fabrication of IO components based on embossing techniques.

The uniform and reproducible embossing of micrometer thin polymer films on rigid substrates (glass) is more challenging, but offers some substantial advantages compared with the first approach. While thermal expansion of polymer substrate materials typically used in integrated optics (PMMA, polycarbonate) directly affects the final dimensions and stability of device components (grating periods, stripe widths etc.). for the second method the thermal expansion is dominated by the thermally more stable glass substrate. Moreover, for polymer waveguiding materials, the use of a glass substrate allows the choice of the most suitable solvent for the coating process without concern for damaging the substrate surface. This embossing technique is thus suitable for a larger class of thermoplastic polymers, which is especially interesting for those polymers for which great effort has been put into the synthesis (e.g. EO/NLO polymers).

On the other hand, for the first method (overcoating of microstructured substrates), continuous production lines for roll embossing of plastic foil already exist (Section 3) and could be adapted to fulfil requirements for applications in IO. This first approach is thus better suited for large scale production of passive components where less stringent specifications are demanded from the waveguiding polymer but compatibility with the substrate material used is essential.

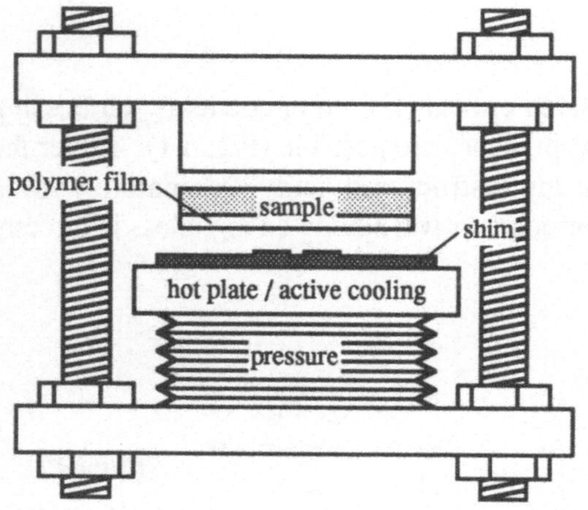

Fig. 5.2: Embossing apparatus.

For the embossing of thin polymer films on glass a special press (schematically illustrated in Fig. 5.2) has been developed. Nickel shims fabricated from an e-beam written mask were used for first tests. The mask contained grating pads (periods of 300, 400 and 600 nm) positioned at the end of groups of stripe waveguides (widths varying from 1 to 10 μm).

Single-mode planar waveguides were prepared by spin-coating polycarbonate (PC) solution onto glass substrates. The refractive indices (ordinary and extraordinary) of the film were determined for $\lambda = 633$ nm by the prism-coupling technique (assuming optically uniaxial films) to be $n_0 = 1.585 \pm 0.002$ and $n_e = 1.576 \pm 0.002$. For a typical sample, the losses for the TE_0-mode were measured to be less than 0.5 dB/cm.

These slab waveguides were embossed at temperatures typically between 180°C and 200°C and a pressure of approx. 25 bar applied over an area of 5x5 cm^2. After an embossing time of 5 to 10 minutes, the temperature was lowered below the glass transition point (for PC to 140°C) before removing the pressure. Light (HeNe laser, $\lambda = 633$ nm) focussed using a combination of two cylindrical lenses could be coupled into ridge waveguides via the grating pads. Although this initial device was not optimized for efficient input-coupling, it served as a first demonstration of waveguiding in embossed polymer films on glass.

Fig. 5.3 shows an atomic force microscope (AFM) measurement carried out on the shim and its corresponding replica in PC on glass. The film thickness is 1.9 μm and the embossing depth (ridge height) approximately 1.1 μm.

6. Conclusions

The application of embossed polymer films in novel diffractive optical elements and in integrated optics has been demonstrated. The technology shows great promise for the low cost fabrication of diffractive and integrated optical devices.

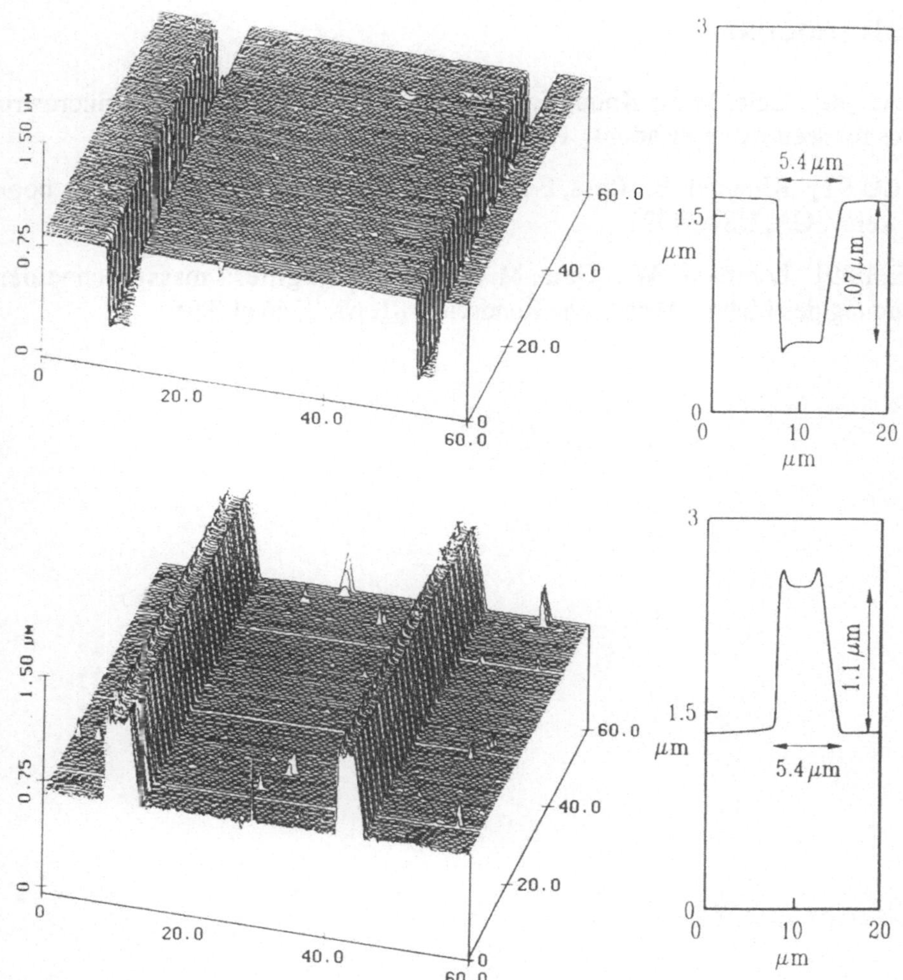

Fig. 5.3: AFM measurements on embossing shim (top) and replica in PC on glass (bottom) of 5 and 4 µm wide ridge waveguide structures.

Acknowledgments

The authors gratefully acknowledge J.S. Pedersen for his work on the electroplating of Ni shims, J.M. Stauffer and Y. Oppliger from *Centre Suisse d'Electronique et de Microtechnique (CSEM)* for writing the e-beam masks and J. Meissner from the *Institute of Polymers, ETH Zürich* for helpful discussions, as well as numerous collegues at PSIZ for advice and help in various fabrication technologies.

References

[GAL 90] Gale, M.T.; Knop, K.; Morf, R.: Zero-order diffractive microstructures for security applications. PROC. SPIE 1201, 83 (1990)

[KLU 91] Kluepfel, B.; Ross, F.; eds.: Holography Market Place. Ross books, Berkeley, CA, USA, 1991

[LEH 89] Lehmann, W.; Gale, M.T.: Submikrongitter: massgeschneiderte Beugung des Lichts. Technische Rundschau 81, Nr. 7, 46 (1989)

Fabrication of Monomode Polymer Waveguides by Replication Technique

P. Dannberg, E.-B. Kley, Th. Knoche*, A. Neyer*
Inst. f. Angewandte Physik, Universität Jena
*Lehrstuhl f. Hochfrequenztechnik, Universität Dortmund

Abstract:

An embossing technique is utilized to press waveguide grooves into thermoplastic substrates. The grooves are subsequently filled with higher refractive index polymer resulting in low loss single mode waveguides for 1.32µm wavelength.

1 Introduction

In a number of papers of this book the potential of polymers for application in Integrated Optics has been demonstrated. Besides the versatility of organic materials (e.g. the possibility of tailoring their properties) there are attractive and well developed processing technologies, first of all replication techniques like molding, injection molding or casting, which might be adapted for Integrated Optics purposes.
There are several obvious advantages of replication techniques over common waveguide fabrication methods: A large number of replica is produced without usual lithographic steps; the technique has mass production capability. There are no demands to the optical properties of the master structure, which can consequently be fabricated by any lithographic process or by a combination of processes. Especially, sophisticated structures like gratings as well as expensive lithographic techniques like deep-etch synchrotron lithography or electron beam writing become justifiable.
Additionally, waveguides and fiber alignment grooves might be produced in one fabrication step as sketched in Fig. 1. This would possibly solve the most serious difficulties (and cost factors) in a mass production of Integrated Optics devices: endface preparation and waveguide-fiber alignment.
Recently, electroplating combined with casting and photopolymerisation

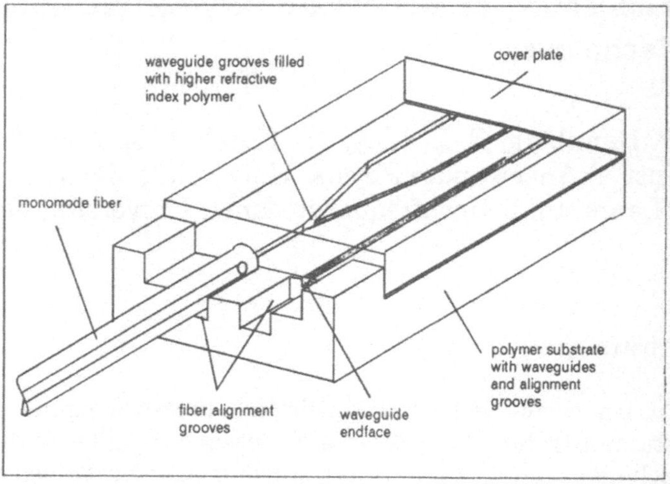

212

Fig. 1 :

Principle of simultaneous polymer waveguide fabrication and fiber alignment

has been used for fabricating polymer microlenses as well as waveguide gratings on glass substrates [HOS 90]. Multimode waveguide structures in poly(methylmethacrylat) (PMMA) have also been generated using the LIGA process (see e.g. [BLE 91], and this book).

In this paper, we used a metal mold (master) to emboss waveguide grooves into thermoplastic substrates, which were subsequently filled with a higher refractive index polymer acting as the waveguide core.

In order to investigate the feasibility of this technique, our goal was to develop passive components as strip waveguides, bends, and Y-junctions, which are monomode at the communications wavelength 1.32µm and their cross section is well matched to the optical fiber. In chapter 2 we will describe details of our technology, wereas first results and conclusions will be represented in chapters 3 and 4, respectively.

2   Technology

2.1   Metal mold fabrication

The fabrication of the metal master is illustrated in Fig. 2: In a first step a resist pattern was produced by standard photolithography. It contains waveguide structures in form of grooves with a cross section of about 6x6 µm^2. Note that the resist pattern defines the final dimensions of our waveguide structure.

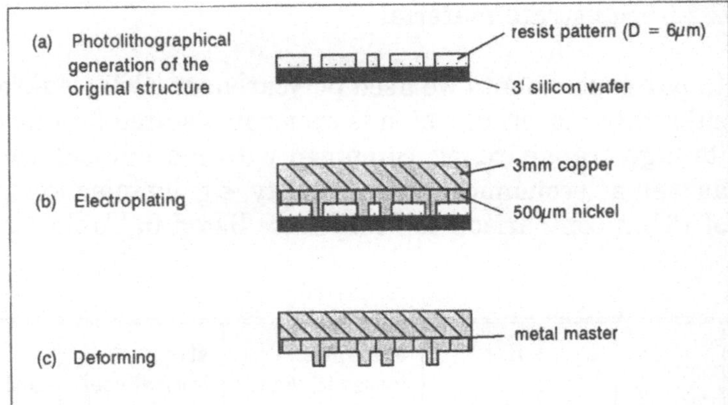

Fig. 2 :

Fabrication
of the metal
master

This pattern was replicated into nickel by an electroplating process. The
resultant nickel slice with a thickness of about 500µm was subsequently
reinforced by 3mm of copper. Fig. 3 shows electron micrographs of the
nickel-surface with a relief pattern of stripes, bends, and Y-junctions.

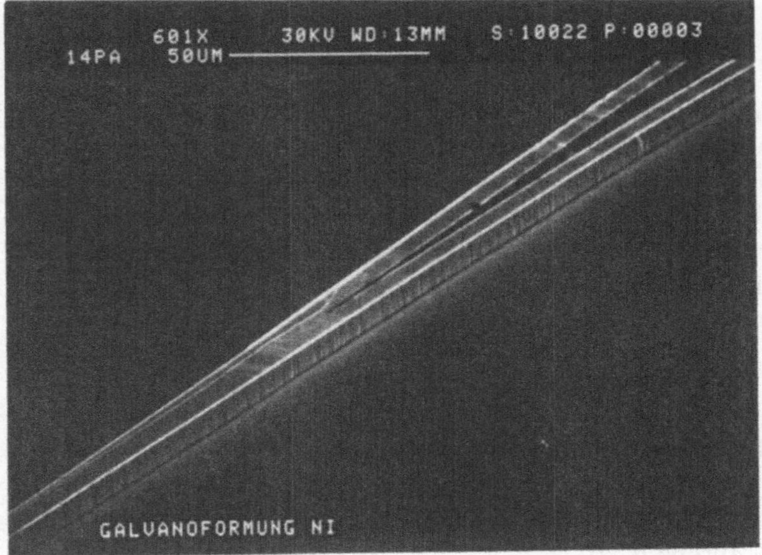

Fig. 3: Electron micrograph of the nickel master with waveguide stuctures.
Y-junction with strip dimensions of 6x6µm^2

The metal master could now be used to emboss the waveguide microstructure into thermoplastic polymer substrates as shown in Fig. 4.

2.2 Substrate material

In our experiments we used polycarbonat (PC) [Makrolon[R]/Bayer] as the substrate material, which is preferably suited for our purpose because of its high transparency, combined with mechanical and thermal stability as well as economical processability, e.g. by injection molding. Properties of PC in comparison to PMMA are listed in Table 1.

	T_g [°C]	n (633nm)	absorption losses [db/cm] at λ [nm] 850　　1320	stress optical coefficient $[10^{-12}m^2/N]$	durability type	impact strength [Ws/m]
PC	149	1.581	0.01 ~ 0.1	100	hard, tough	800
PMMA	95	1.489	< 0.01 < 0.1	5	hard, brittle	27

Tab. 1 :　Properties of substrate polymers (see e.g. [Kre 90])

Additionally, modified PC materials for optical applications (especially for high performance compact disks and magneto-optical disks) are highly perfected and show even better thermal stability (e.g. T_g=238°C for TMC-PC [Bayer]) as well as a substantial reduction of birefringence compared to standard PC. We are going to use these modified PC materials in future experiments.

2.3 Replication of waveguide grooves

Waveguide grooves were produced by an embossing technique sketched in Fig. 4 (steps d, e).
At a temperature of 180°C the metal master was pressed into the PC substrate. By applying a sufficiently high pressure we achieved submicron replication accuracy of the details of our microstructure.
Deforming was performed slightly below the glass transition temperature T_g because otherwise the grooves would have been damaged due to the different thermal expansion of metal and polymer.

2.4 The waveguide core

Up to now fabrication steps well known from plastics processing had been modified and applied to Integrated Optics structures. The crucial step was now to produce the waveguide core (compare Fig. 4 , steps f and g).
For this reason the waveguide grooves had to be filled with liquid prepo-

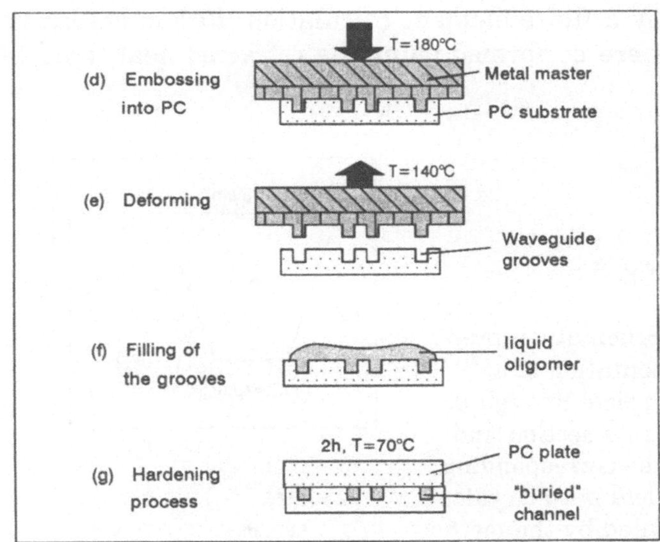

Fig. 4 :

Waveguide
fabrication process

lymer/ oligomer which had to be subsequently hardened. The core material should be highly transparent in the wavelength range of interest, its refractive index had to be slightly higher than that of the the substrate and it should be adjustable with an accurcy of 10^{-3}. Furthermore, we need a good adhesion to the substrate, low shrinkage as well as a certain amount of elasticity to avoid defects during the hardening process. Mechanical and thermal properties should be similar to those of PC, and endface polishing should be feasable.
In our actual experiment, we filled the waveguide grooves with an epoxy polymer, the refractive index of which could be controlled by changing the composition of the hardener. Subsequently, a plane plate of PC was pressed against the substrate to remove excess epoxy polymer and to protect the waveguide from the top side. The whole sandwich was now treated 2 h at 70°C for hardening of the epoxy resulting in a "buried"

waveguide which was then endface polished by standard procedures.

3   Experimental results

The index step of our waveguides was chosen to be $6 \cdot 10^{-3}$ leading to monomode operation at $\lambda = 1.3\mu m$ wavelength. Fig. 5 shows a typical strip cross-section. The corresponding electric field pattern was obtained by a finite element calculation. Refractive index profile measurements were performed using the refracted near- field (RNF) technique.

Fig. 5 :

Schematic representation of a typical waveguide cross section and the corresponding field pattern calculated by the method of finite elements

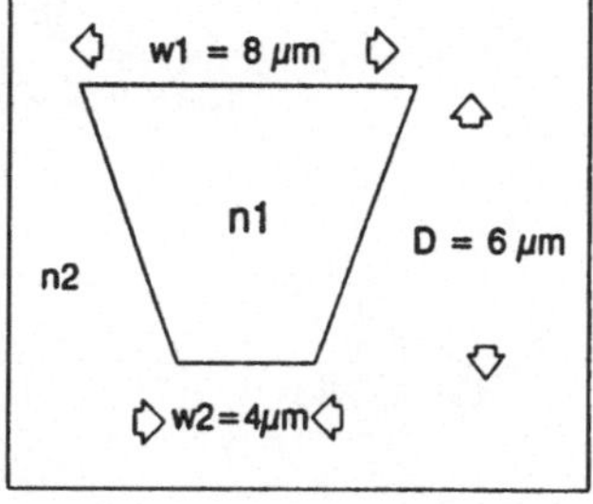

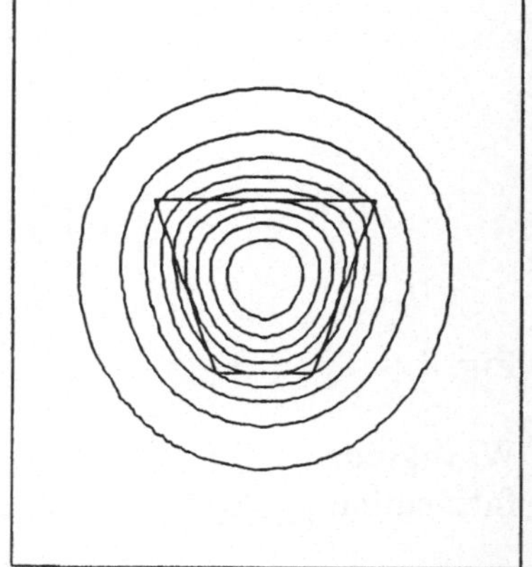

For loss measurements laser light of 1.32μm wavelength was coupled to the strip waveguides through an optical fiber (York HB 1200). The total insertion loss (fiber-waveguide-fiber) of 40mm long waveguides was measured to be 5.5dB. Taking into account the coupling losses by field mismatch and imperfect endfaces, this corresponds to a waveguide attenuation of about 1dB/cm.

We could identify two major contributions to the losses which are NIR-absorption and macroscopic defects acting as scattering centers. The scattering can be reduced by perfecting the original resist pattern, by using extremely purified epoxy polymer and by an optimized hardening procedure.

The NIR-absorption of a 1cm thick bulk sample of the epoxy polmer was measured as shown in Fig. 6. It turned out that ~0.6dB of the waveguide losses at 1.32µm were due to absorption, mainly due to harmonics of N-H vibrations.

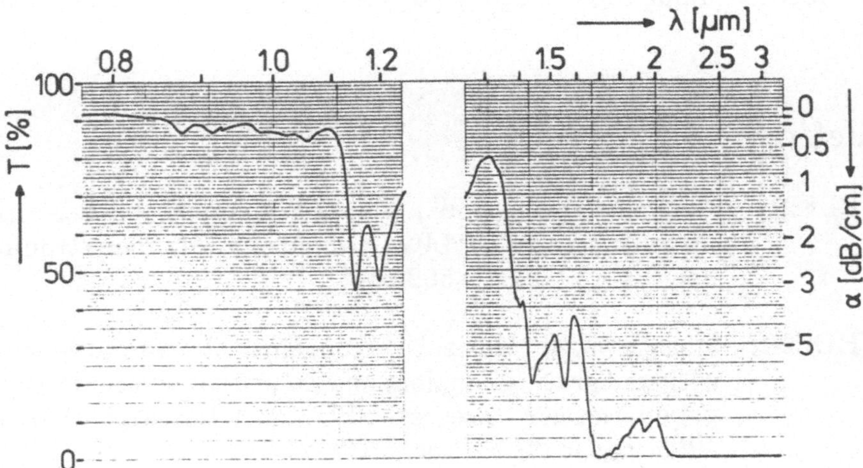

Fig. 6 : Spectral transmission of the epoxy polymer with 1.32µm laser line indicated. The right y-axis shows the pure extinction loss in dB/cm.

4   Resume

In summary, we have successfully applied replication techniques to monomode polymer waveguide fabrication. By means of a metal master we embossed waveguide grooves into PC, which were subsequently filled with an epoxy polymer. After covering the structure with a second PC plate and hardening of the epoxy, we got all-polymer "buried" channel waveguides with a cross section well adapted to fiber coupling. Our first samples showed losses of about 1dB/cm at λ=1.32µm, wich were primarily due to absorption of the core material. NIR-absorption arises mainly from harmonics of C-H and N-H vibrations of the organic molecules and is, therefore, a general problem for application of optical polymers at communications wavelengths. However, great improvements have been achieved recently [IMA 91] by using deuterated polymers.

At present, we are going to fabricate samples with a fluorinated polymer as the core material in order to reduce NIR absorption. Furthermore, investigations of the birefringence and of long-term phase drifts in our waveguides are on the way.

Acknowledgement:

The authors are gratful to Dr. M. Märtin from Jenoptik Carl Zeiss Jena GmbH for supplying the epoxy polymer, and for useful discussions.

References:

[BLE91]   Bley, P.; Blacher, W.; Menz, W.; Mohr, J.: Description of microstructures in LIGA-technology. Microelectronic Enge-neering 13 (1991) 509-512

[HOS90]   Hosokawa, H.; Horie, N.; Yamashita, T.: Simultaneous fabri-cation of grating couplers and an optical waveguide by photo-polymerisation. Integrated Photonics Research , Hilton Head 1990, paper MF6, Proc. pp.26-27

[IMA91]   Imamura, S.; Yoshimura, R.; Izawa, T.: Organic channel waveguides with low loss at 1.3µm, OFC 91, paper TuF6

[KRE90]   Krevelen, D. W. van: Properties of polymers. Elsevier Amster-dam-Oxford-New York-Tokyo 1990

Examples and Potential Applications of LIGA Components in Micro-Optics

J. Göttert, J. Mohr, C. Müller

Nuclear Research Center Karlsruhe
P.O. Box 3640
W-7500 Karlsruhe 1

Abstract:

The miniaturization of passive optical components such as lenses or prisms with dimensions of several micrometers up to a few millimeters calls more and more for novel manufacturing techniques for optical components. One possibility is offered by the LIGA technique which by combination of X-ray lithography, electroforming and molding allows such components to be manufactured in any two-dimensional shape, structural heights up to several 100 micrometers with very smooth and vertical side walls, in polymethyl methacrylate (PMMA), which is transparent in the visible and near IR-regions, as well as in metal. Such components are applied in 3D-microoptical assemblies combined with other hybrid-integrated optical elements. In addition, compact and efficient coupling elements have been built for use with multimode glass fibers. Besides, in a light guiding resist multilayer both simple light guiding components and more complex structures such as a planar grating spectrograph are manufactured with small attenuation losses. Multimode fibers are coupled in an optimum way via integrated fiber fixing grooves.

1.   Introduction

During recent years, quite a number of elements from so-called integrated optics have been introduced in microoptics [Vo88]. These are mostly waveguide structures for single-mode applications which are fabricated by optical lithography and diffusion processes into special substrates such as $LiNbO_3$ [Po90]. Light sources and detectors are coupled to these integrated optical chips (IOCs) via monomode glass fibers [Li88]. These IOCs can be manufactured to very stringent specifications. Also, extremely small and high-performance semiconductor light sources and detectors have been built [Wi83]. However, until this day no suitable microoptical components are available by which, in combination with these miniaturized electrooptical elements, appropriate simple microoptical setups have been built, as e.g. necessary for coupling the light into optical fibers with a high efficiency [Alt91]. The reason is that no suitable techniques are presently available to reduce in size microoptical components such as prisms, lenses or beam splitters in a freely selectable configuration and structural size.

By the LIGA (German acronym for Lithographie, Galvanoformung, Abformung) technique developed at the Karlsruhe Nuclear Research Center (KfK) a process technology has become available which allows microoptical components of any cross-sectional geometry to be manufactured with heights up to several hundred micrometers and minimum detail dimensions in the micrometer range as well as a structural accuracy in the submicrometer range [Be86]. The smoothness of the walls is in the nanometer range [Mün87]. The LIGA technique relies on X-ray lithography for primary pattern generation in polymers. Metallic microstructures are fabricated by electrodeposition in these plastic templates [Man89]. The subsequent molding steps permit mass fabrication and widening of the material variety [Hag89], especially in case of plastics, which is of interest above all in optics applications. Further development of the process as well as the fabrication of various microstructures capable of application in microsystems are done at the KfK. First products have been commercialized by MicroParts GmbH, Karlsruhe, a joint venture of different German companies.

The microoptical components generated by the LIGA technique may be arranged individually as well as adjusted with respect to each other as a microoptical bank. Besides, simultaneous patterning of mounting supports allows further hybrid microoptical components to be integrated to build up even complex microoptical beam paths. By the adjusted arrangement of microoptical components with respect to fiber fixing grooves compact fiber optical coupling elements can be manufactured in addition.

By modifying the irradiation process elements have been fabricated which allow the light to be reflected from one function level (LIGA level) in direction of the normal, and coupling it e.g. into a second function level lying above the first level. By this, very compact optical function levels positioned to each other can be combined into a highly integrated stacked assembly [Bre91].

Use of a light guiding three-layer resist in X-ray lithography permits to manufacture relatively long, low-loss optical elements such as a planar grating spectrograph [An88].

In this paper the LIGA process will be described with its modifications relevant to application in microoptics. The properties of the passive microoptical components demonstrated by a number of examples and their potential applications are discussed.

2.   LIGA Process

The manufacture of microoptical structures according to the LIGA technique is represented schematically in Fig. 1.

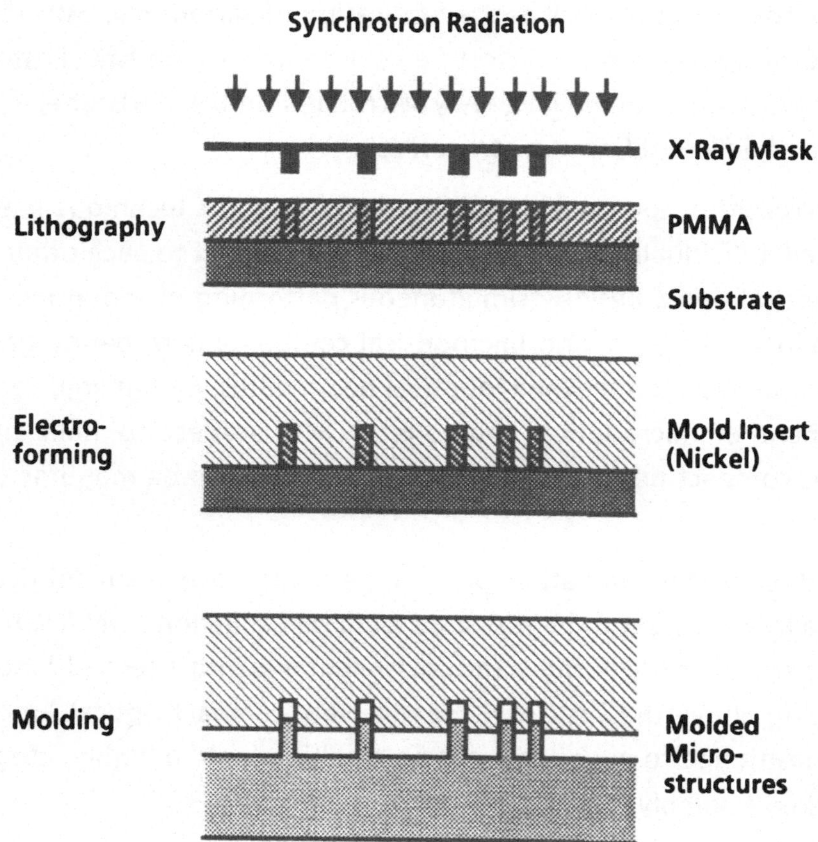

Fig. 1: Schematic of the process steps involved in the LIGA process.

By X-ray lithography, the first step of the LIGA process the absorber pattern on an X-ray mask is transferred into a resist layer of several hundred micrometers thickness by means of shadow projection with synchrotron radiation. The lithographic process is done at the ELSA-Storage Ring of the University in Bonn at an electron energy of 2.3 GeV resulting in a characteristic wavelength of 0.5 nm. The resist used is polymethyl methacrylate (PMMA) whose resolving power is excellent [Mo88]. The irradiated resist parts suffer from destruction of the polymer chains due to dose exposure and, hence, their molecular weight diminishes. In a subsquent development step these parts are dissolved selectively while the unirradiated parts remain unchanged. Thus, the resist covered during irradiation by the gold absorbers of the mask forms the primary microstructure. These plastic microstructures may already be the end product of manufacture; this is the case for most of the microoptical structures described in this paper. Further process steps follow in manufacture of metallic micropatterns and in mass production in which the microstructures produced by lithography serve as the primary template for electrodeposition of metal. To produce metallic secondary patterns the plastic mold is filled with metal to the upper limit only. Subsequently the plastic material is dissolved. On the other hand, to produce a mold insert the metal is grown up to 4 - 5 mm thickness above the height of the microstructures. Using the mold insert tertiary plastic structures are generated either by reaction or injection molding or also by means of relief printing. By this technique the mold insert is pressed into PMMA under vacuum conditions at temperatures slightly above the glass transition temperature and under defined conditions of pressure. After a phase of cooling the mold insert is withdrawn [Har92].

To compare the qualities achieved, a prism structure manufactured by X-ray lithography and one made by relief printing are represented in Fig. 2. In both cases, the structural heights are approx. 140 µm. No difference in the quality of the side walls, an important criterion of application in optics, can be observed on the two structures. This demonstrates that mass production of microoptical structures by molding is feasible without loss of quality.

The sequence of process steps described allows structures to be manufactured on a base plate with side walls running normal to it. On account of inclination of the mask/specimen unit relative to the X-ray

Fig. 2: SEM-picture of a microprism structured by X-ray lihtography (left) and by relief printing (right).

beam (Fig. 3) also side walls inclined relative to the base plate have been structured [Bl91].

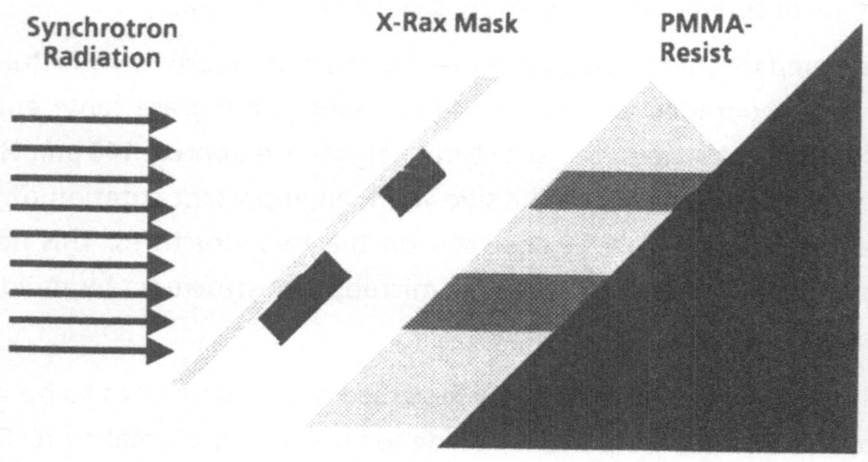

Fig. 3: Scheme of the lithographic step to structure side walls with arbitrary angle of inclination relative to the base plate.

The angle of inclination corresponds to the tilting angle of the mask and the specimen. This is attributable to the fact that in the resist the boundary between the irradiated and the unirradiated materials is very distinct [Mo88].

If different inclinations are to be achieved on one base plate, several irradiations are necessary, with the respective parts to be covered by a diaphragm. Figure 4 shows such an edge of an inclined side wall with an angle of inclination of 45°. In this example no change in quality compared with vertical side walls can be found by visual inspection. Also the optical function of total internal reflection at the boundary PMMA-air is guarenteed as can be seen in Fig. 4.

Fig. 4: SEM-picture of an edge of an inclined side wall with an angle of inclination of 45° (left).
Total internal reflection of an HeNe laser beam at the PMMA-air interface (right).

A three-layer resist has been developed for manufacture of light guiding optical elements [An90]. On a base plate made of epoxy phenolic resin a first layer consisting of a copolymer of PMMA (78 %) and tetrafluoropropyl methacrylate (TFPMA) (22 %) is polymerized. The PMMA/TFPMA copolymer was selected because it is also well suited for X-ray patterning. With this composition, combined with a PMMA core layer, a numerical aperture of 0.2 is obtained. However, also other apertures can be achieved by variation of the composition of the copolymer. Instead of the metal plates currently used in the LIGA process, plates made of epoxy phenolic resin have been used in this case because the thermal expansion coefficient of these plastic plates is better adapted to the resist material so that tensile cracking is reduced. To improve the adhesion of the copolymer layer to the base plate, the surface of the epoxy phenolic resin plate is chemically activated prior to polymerization of the first layer [Mül91].

The first layer is milled off to a thickness which corresponds to the thickness of the cladding of the optical fibers used and then a foil made of PMMA is welded onto it which is likewise adapted by milling to the thickness of the fiber core. A cover foil consisting of the same copolymer is provided on the PMMA core foil, likewise by welding; it constitutes the top cladding layer of the light guiding assembly (Fig. 5).

Welding is performed slightly above the glass transition temperature and at pressure. Due to welding a defined refractive index profile can be generated because the diffusion of the two different molecular chains is actually rather small, but still sufficient to guarantee good adhesion of the two layers. Moreover, by welding the danger of stress corrosion cracking is clearly diminished which would have an adverse effect on the transmission properties. As appears from measurements of the refractive index as a function of the height, the variation of the refractive index between two successive layers takes place within an interface layer of less than 10 µm thickness. Therefore in a first approximation an optical waveguide with a weak gradient index profile is achieved [An90].

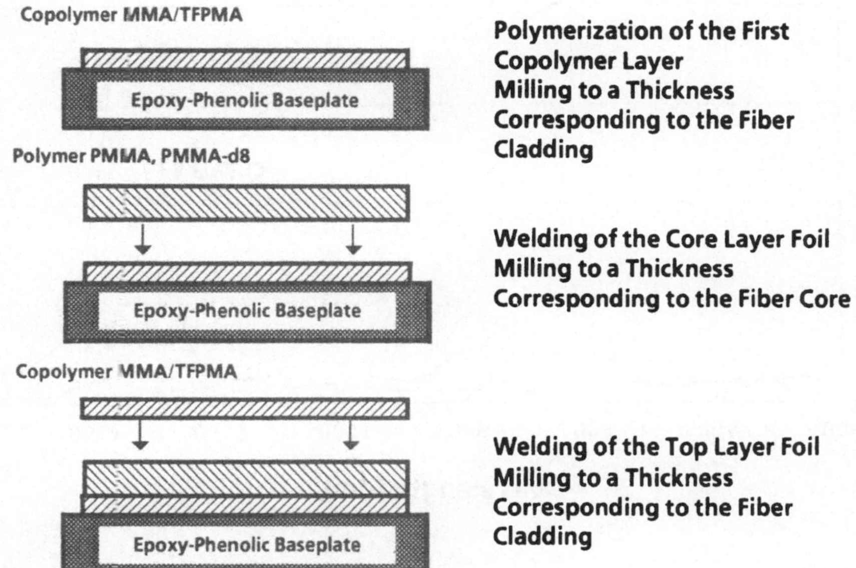

Fig. 5: Scheme of the fabrication steps of a light guiding multilayer resist.

3. Basic Optical Features

3.1 Material Attenuation

The high transparency of the PMMA resist material in the visible and near IR regions is a major prerequisite of manufacture of microoptical components using X-ray lithography. Measurements on cylinder shaped bulk specimens made of the resist material PMMA show a mean attenuation of approx. 0.2 dB/cm in the range from 600 to 1300 nm. At the absorption peaks at 900 nm and 1180 nm this value rises up to 0.4 dB/cm and 3 dB/cm, resp. To manufacture low-loss structures for the near IR-region, deuterated PMMA (PMMA-d8) is used as a core material. Attenuation of this material occurs in the spectral range from 600 to 1300 nm at about 0.1 dB/cm [Gö91a]. For the three-layer resist the at-

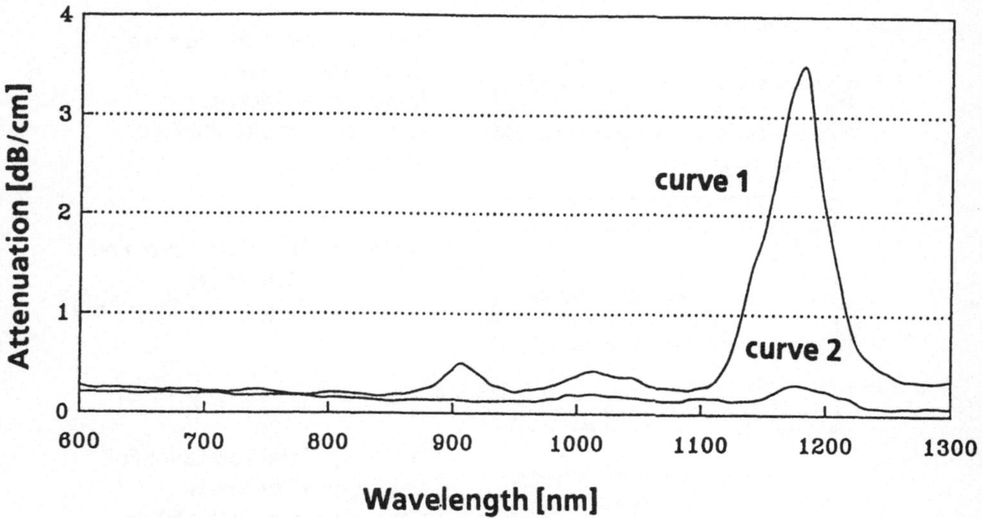

Fig. 6: Attenuation of strip polymer waveguides structured by X-ray lithography in a three-layer resist with a core of PMMA (curve 1) and PMMA-d8 (curve 2), resp..

tenuation is slightly larger which is caused by the relatively great roughness of the side walls of the strip waveguides and by defective points in the three-layer resist. These values may be improved by an advanced process technology.

3.2   Investigations of the Surface Quality

Besides a small attenuation above all the roughness of the optical interfaces and the homogeneity of the material are of crucial importance to optical applications. To be able to determine the surface roughness of the side walls manufactured by the LIGA process, phase shifting interferometry was applied in a collaborative effort with the Erlangen University [Schwi85]. At the Institute for Applied Optics of Erlangen University a modified Linnik interferometer was mounted which allows a specimen surface of approx. 560 x 770 µm to be examined with a resolution of 1.3 µm according to the Rayleigh criterium [Bre92].

In Fig. 7 a photograph of the interference pattern which results from superposition of the wave fronts reflected by a LIGA structure of 500 µm height and an ideal reference surface is shown. Over approx. 400 µm

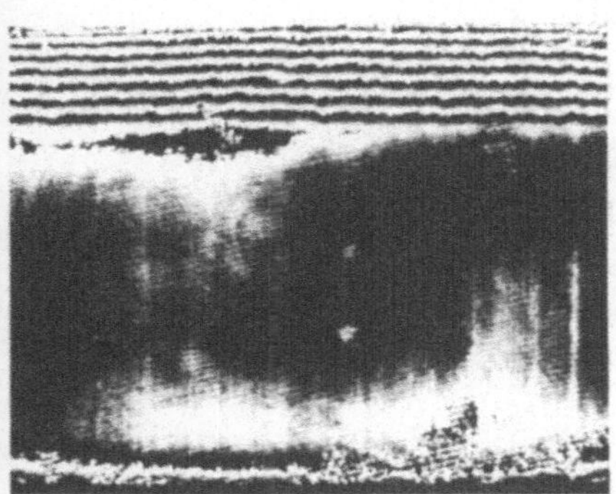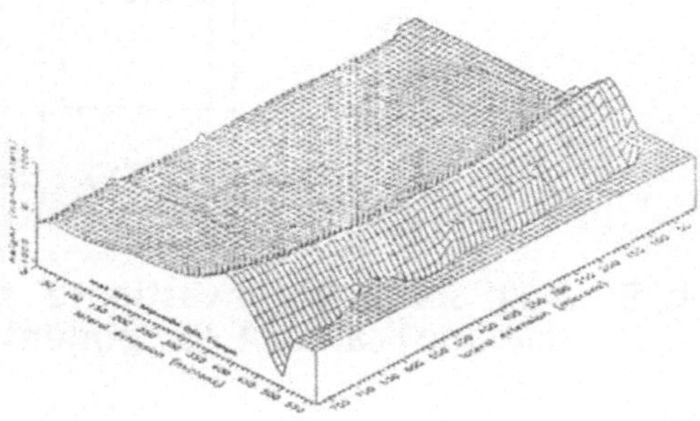

<u>Fig. 7:</u> Photograph of the interference pattern on a LIGA side wall (left) and calculated surface profile (right).

height only an interference strip appears which corresponds to a mean deviation from an ideal surface of $\lambda/2$ whereas the interference strips in the upper part of the structure which follow each other closely are indicative of distinct roundness. This roundness is attributable to a dose impact in the shadowed area in the top layers which can be avoided by use of an optimized mask. The calculated surface roughness of the side wall is represented on the right side of Fig. 7. The mean roughness is about 40 nm and is, consequently, of sufficient quality for most of the applications in optics.

The influence of material homogeneity and of the surface roughness on the image formation is studied by reference to the test setup represented in Fig.8. The mask pattern (letter F) illuminated with a white light source is projected by a microscope objective(10 x /0.2) onto the side wall of a LIGA structure. About 4 % of the incident light is reflected

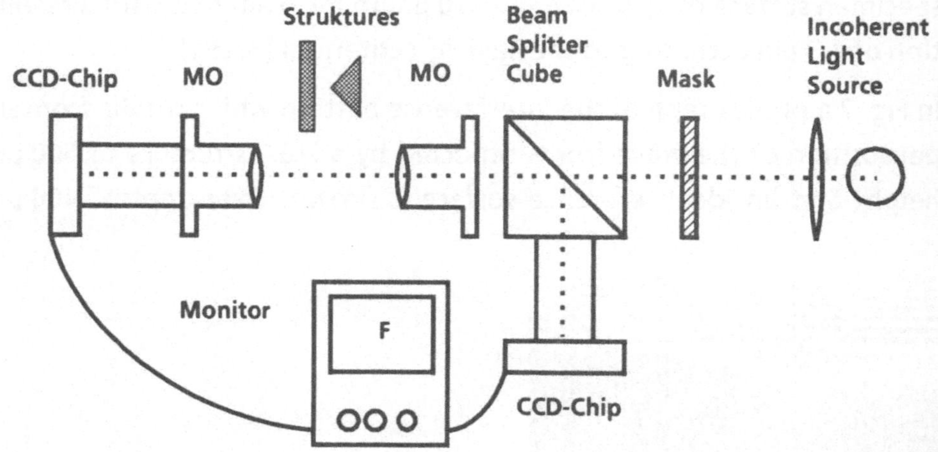

Fig. 8: Test setup to investigate the image formation quality of microoptical LIGA components.

backward and is projected onto a CCD-chip by the same objective via a beam splitter. The image formation of the mask pattern can be evaluated on a monitor. If the CCD-chip is positioned straight ahead and detects the image projected onto the rear side of a plane parallel plate, the image formation can be examined in the transmission mode after it has passed through two interfaces of a LIGA structure.

In Figs. 9a to 9c the relevant images have been represented by the example of the "letter F" which is composed of single points. Figure 9a shows for comparison the image of the letter reflected by an optical mirror. In this representation the single image points composing the letter are about 25 µm in lateral length. In Fig. 9b the image reflected by a LIGA side wall has been represented; Fig. 9c shows the image after it has passed through a plane parallel plate of 3 mm length manufactured in a 500 µm thick PMMA resist by means of X-ray lithography.

All three images can be well compared with each other and demonstrate that the homogeneity of the resist material as well as the quality

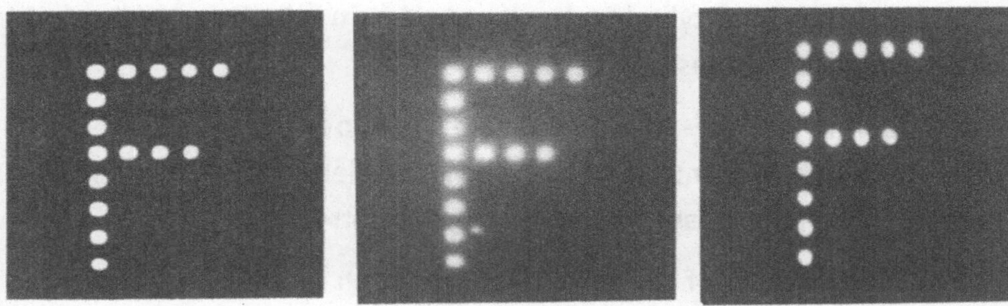

Fig. 9a - 9c: Image of the letter **F** formed by an optical mirror (left), by a LIGA side wall (center) and after transmission through a plane parallel LIGA plate (right).

of the side walls achieved by the LIGA process are best suited for optical applications involving incoherent light.

Investigations made with coherent light show that local variation in density as well as impurities cause the passing light beam to get widened by scatter and diffraction and result above all in a considerable deterioration of the image formation quality. For that case polymerized foils, which can be made out of high-purity monomer material and free from defects should be preferred over the resist layers produced by polymerization from the casting resin.

Using molding techniques also other classes of polymers can be structured, e.g. polycarbonates. For these materials the requirements of optics and also of other boundary conditions such as thermal stability can be better satisfied compared with the methacrylates.

4.   Microoptical Components

4.1   Examples of microoptical structures

The results described as well as the freedom in two-dimensional design offered by the LIGA process suggest to fabricate simple microoptical ele-

232

ments such as prisms, cylindrical lenses and beam splitters as well as light paths composed of these elements.

The following Figs. 10 - 12 will illustrate the above statements by examples. The functioning mode of the individual elements is explained by a parallel HeNe-laser beam passing through the structures.

In Fig. 10a the path of the laser beam through an inverting prism has been represented.

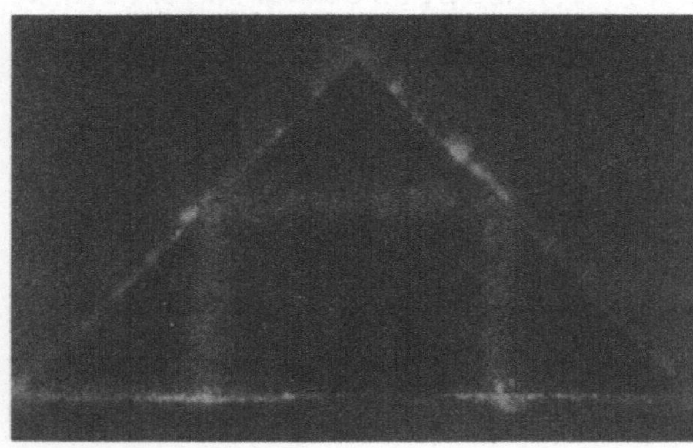

Fig. 10a: Light path through an inverting prism.

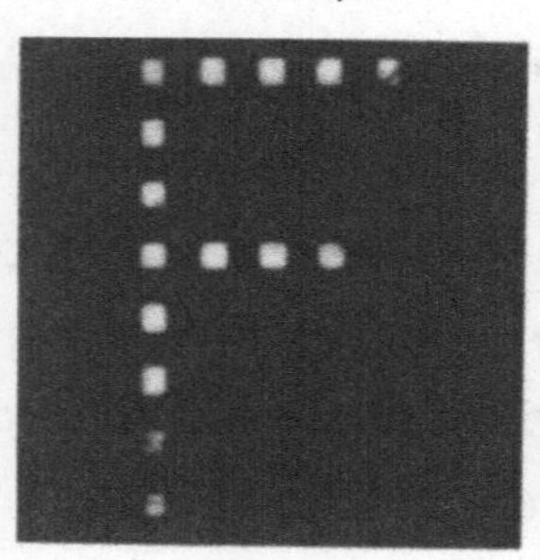
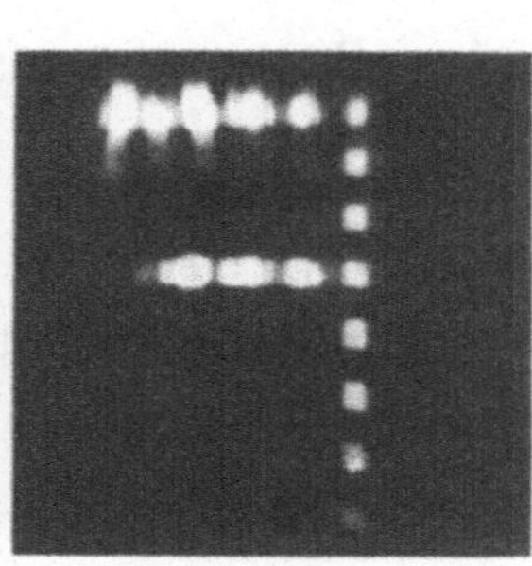

Fig. 10b: Letter F imaged on the entrance surface (left) and on the exit surface (right) of the inverting prism using the test setup described in Fig. 8.

The incident light beam is totally reflected two times at the prism hypotenuses arranged at ±45° with respect to the direction of incidence, and it leaves the prism in direction opposite to the direction of incidence. To investigate the quality of image formation the test setup represented in Fig. 8 is used. The mask pattern (letter F) is projected on the entrance surface of the inverting prism. The reflected intensity is detected by a

CCD-chip, the image quality is shown in Fig. 10b on the left hand side. On the right hand side of Fig. 10b the image projected on to the exit surface is represented. The image is inverted and also some of the pixels are distorted due to defects in the prism. In general, if the prism is perfect details of the image in the order of 5µm can be resolved [Bre92].

Figure 11a shows how the laser beam can be focussed by a cylindrical lens. The incident parallel beam is focussed to become a thin line which runs at a distance corresponding to the focal length of the lens. The theoretical focal length of the plano-convex lens is calculated from the radius of curvature (R = 1490 µm) to be f = 3 mm. This is confirmed by the measurement (Fig. 11b).

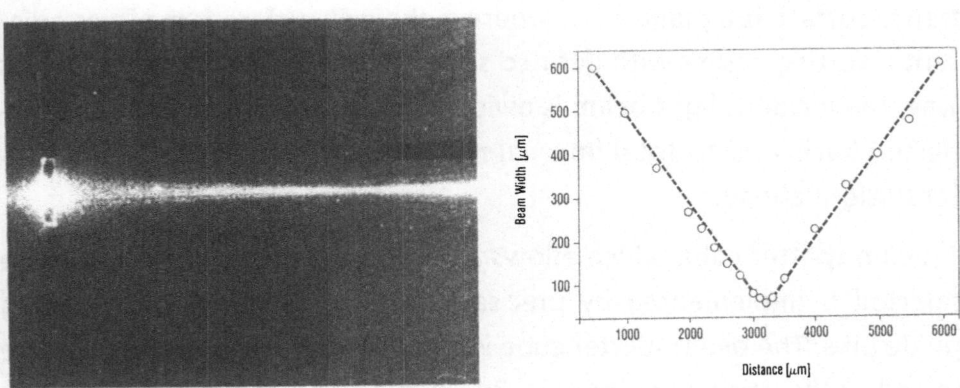

<u>Fig. 11:</u> Cylindrical Lens focussing a parallel laser beam (left). Measured 1/e² beam width as a function of the distance from the lens (right).

Figure 12 shows by the example of a Brewster telescope an optical setup consisting of two prisms positioned to each other. The laser beam which enters from the left hand side is shifted a small amount laterally and, in addition, is expanded in one direction.

Splitting up of a homogeneous, parallel light beam hitting a microoptical structure in normal direction into several small individual beams can be done by means of a geometric beam splitter [Gö91b]. Here, the en-

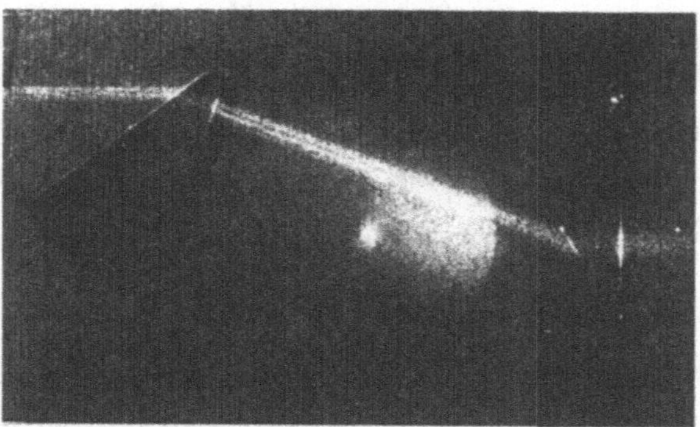

Fig. 12: Light path in a Brewster telescope.

trance surface is a plane area, whereas the exit surface is made as adjacent inverting prisms with defined spacings between each other. In this way, the incident light beam is divided into beams which are either reflected backward by total internal reflection or leaving the beam splitter straight ahead.

A beam splitter cube which allows any ratio of intensity division to be selected is implemented by precise fitting of two LIGA prisms using guide pins. The beam splitter cube is composed of two prisms which are joined at the prism hypotenuses. By covering one of the prism's hypotenuse with a vapor deposited metal layer in a defined manner any desired distribution of intensities can be achieved.

The described examples of microoptical structures illustrate the potentials inherent in the LIGA process for applications in microoptics. Especially by exact positioning of several structures on one substrate microoptical light paths can be implemented without additional adjustment into which further optical elements can be integrated with the help of mounting supports structured in the same process so as to "fit precisely and comply with the required function". In this way a complete, miniaturized optical bank can be built.

4.2   Application in Fiber Optics

An attractive potential application of those microoptical components is in the field of multimode fiber technology. Here, above all the capability of exact positioning of mechanical mounting supports is used advantageously in which the multimode fibers can be positioned very precisely with respect to the microoptical component without any additional adjusting operations. The principle of a 1 x 2-fiber coupling element is explained by the schematic representation of Fig. 13. The intensity emitted by fiber 1 is partly passed through the parallel PMMA structure into fiber 2, partly decoupled by total internal reflection at the PMMA-air interface of the prism hypotenuse in fiber 3.

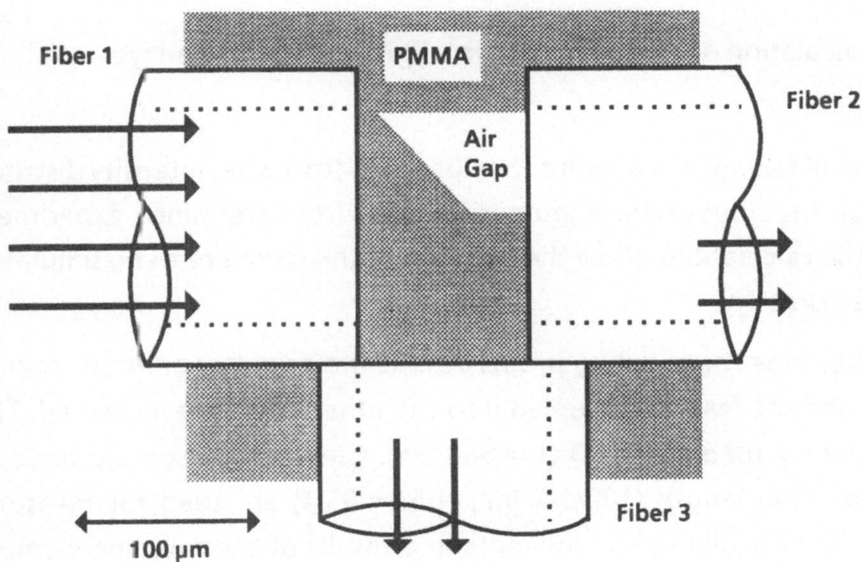

Fig. 13: Principle of a compact 1x2 -fiber coupling element.

In order to be able to describe the coupling principle of such structures and to optimize the design of the arrangement with a view to minimizing losses, a ray trace program was elaborated. The major features of the ray trace program have been compiled in Table 1.

<u>Table 1:</u> Features of the Simulation program.

- Step index and gradient index fibers can be used as light sources.

- Weighting the individual beams in conformity with the near-field intensity distribution measured.

- Input of any optical interface without defined sequence in the light path.

- Detection possible by means of position sensitive detectors or optical multimode fibers.

- Calculation of the beam path in a light guiding three-layer resist.

Above all taking into account the fiber data (core size, intensity distribution on the fiber endface, numerical aperture) determined experimentally the calculations allow the function of the structure to be simulated very accurately.

To determine the coupling properties the intensity I_0 is emitted from fiber 1 and the fractions coupled into the other fibers are measured. The light source used is an LED ($\lambda = 850$ nm). Gradient index multimode fibers of 10 m length (100/140 µm, N.A. $= 0.24$) are used for measurement. The calculated and the measured results of the coupling element explained in Fig. 13 have been compiled in Table 2.

The non-uniform intensity distribution among the two output fibers is largely due to variations in dimension at the coupling element. Consequently, less intensity is transmitted straight ahead (fiber 1 $=>$ fiber 2). These discrepancies can be compensated by an appropriate design cor-

<u>Table2:</u> Measured and simulated intensities coupled between the fibers connected by the coupling element explained in Fig. 13.

Connected Fibers	Coupling Element Made by X-ray Lithography	Coupling Element Made by Relief Printing	Ray Trace Simulation
Fiber 1 => Fiber 2	34%/1.67dB	32%/1.94dB	40%/0.97dB
Fiber 1 => Fiber 3	42%/0.76dB	42%/0.76dB	48%/0.18dB

Measurements made in cooperation with Robert Bosch GmbH, Stuttgart.

rection. Moreover, compared with the calculated coupling losses, about 0.6 dB higher losses per fiber have been measured. The underlying reasons are partly structural defects, partly deposition of dust.

Table 2 also lists the data of structures manufactured by lithography and molding, resp. Within the tolerance limits of the measurement accuracy no differences exist between the structures manufactured by the different methods which confirms that the geometry of the structures has been precisely adhered to in the course of the manufacturing steps.

The represented structure does not conform to the optimum arrangement of a 1 x 2-coupling element. Above all, not all beams are completely reflected because of the 45° prism angle so that they are lost in fiber 3. A design of a coupling element for use with multimode fibers (100/140 µm, N.A. = 0.28) optimized in terms of coupling losses is represented in Fig. 14. The coupling losses calculated from the simulation are in the order of 0.3 dB/fiber. If experience accumulated in previous experiments is taken into account, due to fabrication and material induced losses one can expect an overall attenuation of the coupling element of approx. 0.8 dB/fiber.

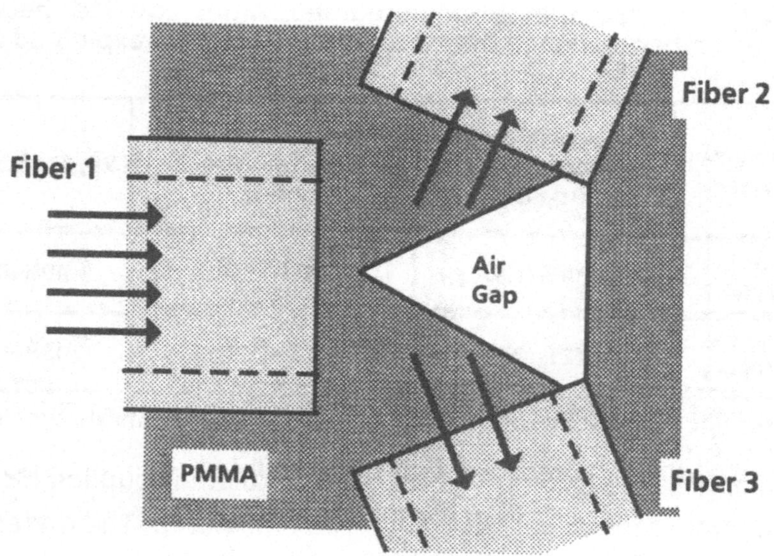

Fig. 14:   Optimized layout of a 1x2-fiber coupling element.

The freedom in patterning together with the flexibility in structural height allow such 1 x 2-coupling elements to be optimized for any multimode fibers. Thus, they offer an attractive alternative to the previously used Y-couplers [Gei86], and, using the molding technique, they can be made in a very accurate and reproducible manner.

4.3   Applications in a Light Guiding Resist Structure

Fiber forks with typical angles of aperture of 1 - 2° and a correspondingly great structural length are reasonably used to manufacture 1xn-coupling elements. With such structural lengths typical fiber spacings are several millimeters which makes a light guiding material advantageous. As an example the minimum structural length was calculated of a 1 x 3-fiber fork with which a uniform distribution of intensity among the three output channels is achieved (Fig. 15). Uniform distribution of the intensity among the three output fibers is achieved for 1.7 mm

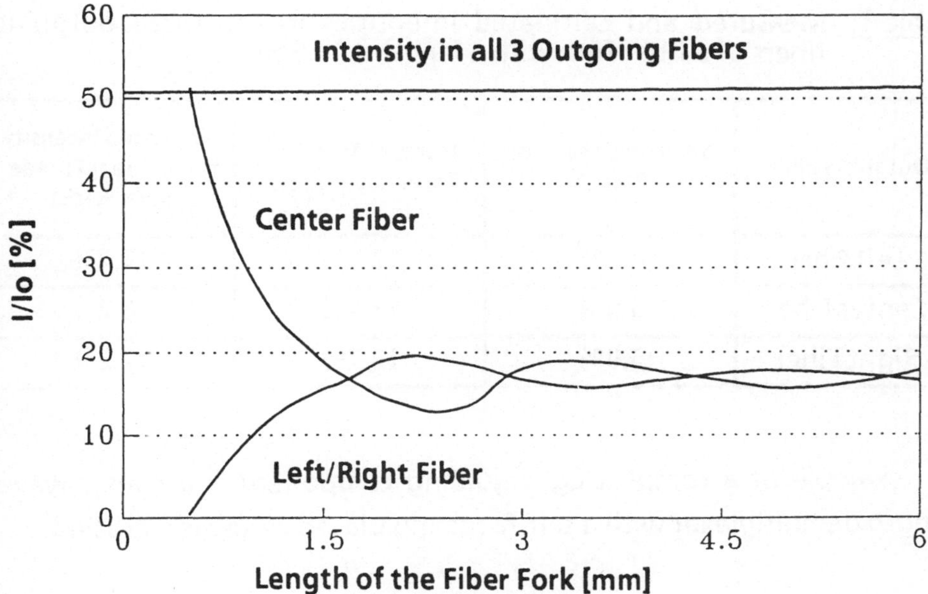

<u>Fig. 15:</u> Simulated intensity distribution among the outgoing fibers of a 1x3-fiber fork as a function of the fork length.

structural length at the earliest. It should be added in this context that due to the uniform illumination of the total decoupled area the total losses of the whole device will amount to approx. 3dB (1dB per fiber).

In the experiment a fork structure of 4.6 mm length (corresponding to an angle of aperture of 1.4°) was manufactured and measured using step index multimode fibers (100/140µm, N.A.=0.28). The measured values have been summarized in Table 3. It is evident from a comparison that the transmitted intensities from a pure PMMA structure are lower by a factor 2.5/fiber which underlines the advantage offered by the light guiding three-layer. The comparison with the simulated values shows that instead of approximately 17 %/fiber only 12.5% of the incident intensity was detected in each of the outgoing fibers. These losses are attributable mainly to defects in the structure.

<u>Table 3:</u> Measured and calculated intensities in the three outgoing fibers of a fiber fork with a length of 4.6mm.

Outgoing Fiber	Structure Made of PMMA	Structure Made of a Three-layer Resist	Simulated Intensity for an Ideal Threee-layer Resist
Left Fiber	4.9%	12.3%	17.2%
Center Fiber	5.0%	13.0%	16.2%
Right Fiber	5.0%	12.7%	17.2%

An example of a complex light guiding component is a planar wavelength demultiplexer with a self-focussing blazed reflection grating.

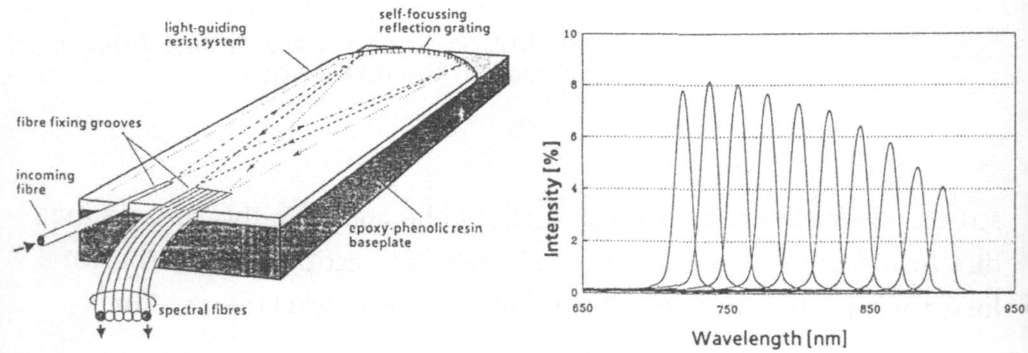

<u>Fig. 16:</u> General layout of the planar grating spectrograph (left). Measured intensities in the ten spectral fibers (right).

The light of a white light source coupled into the demultiplexer is distributed in the spectral range from 720 nm to 900 nm among ten spectral channels. Figure 16 shows the schematic layout of the demultiplexer together with the coupling fiber and the ten spectral fibers. The light is coupled into the element by means of a glass fiber (50/125 μm, N.A. = 0.2), diffracted at the grating and reflected onto the endfaces of the spectral fibers. The three-layer resist has been adapted to the geometry of the glass fiber. The light guiding core layer is 50 µm in thickness; each of the totally reflecting cladding layers is about 37.5 µm thick. The coupled light covers a path of 47 mm length in the element. The intensity coupled into the ten spectral fibers has also been represented in Fig. 16 versus the detected wavelength. It can be clearly seen that the element splits up the incident light and distributes it among the single spectral fibers which is in conformity with the calculations. The relatively low transmitted intensity of approx. 8 % at the maximum is attributable to the roundness of the teeth of the reflection grating. In this case the grating is no longer blazed effectively and portions of the incident intensity are diffracted into different orders and are no longer detected by the spectral fibers. This is evident from Fig. 17 where the result of

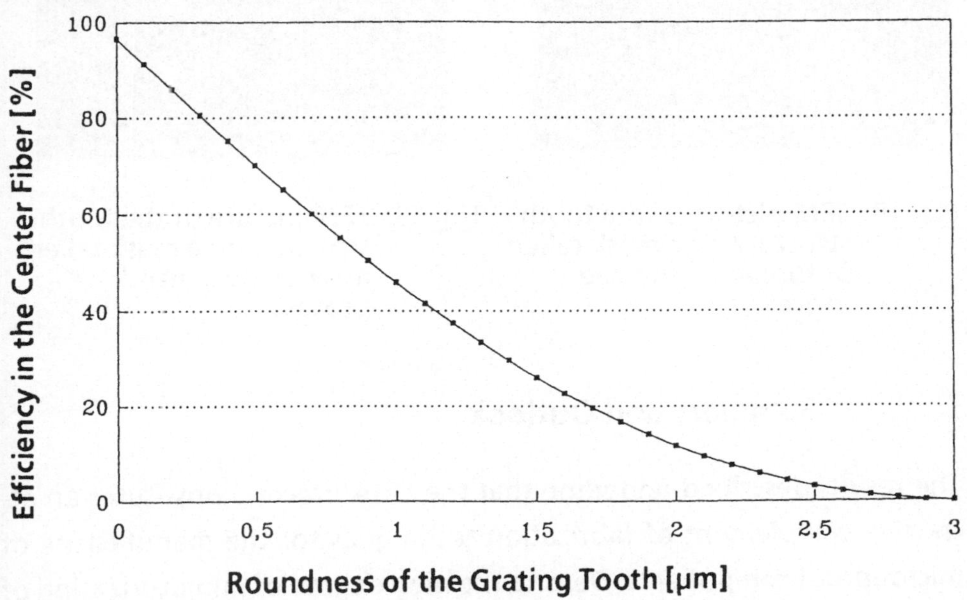

Fig. 17: Simulated intensity transmitted into the center spectral fiber as a function of the roundness of the grating pitches.

simulation of the maximum transmitted intensity of a reflection grating blazed into the second order has been plotted versus the corner roundness. The simulation shows that for 1 µm corner roundness the intensity detected in the central fiber decreases by 45 %. The tooth roundness of the grating measured so far is about 1.2 µm (see fig. 18) which gives a maximum theoretically detectable intensity of 40 %. A comparison of the SEM-picture of the tooth structures in Fig. 18 with those in Fig. 19 shows that following an improvement of process technology clearly smaller roundness radii can be achieved so that it seems possible to enhance the transmitted intensity of the demultiplexer to attain a value around 25 %.

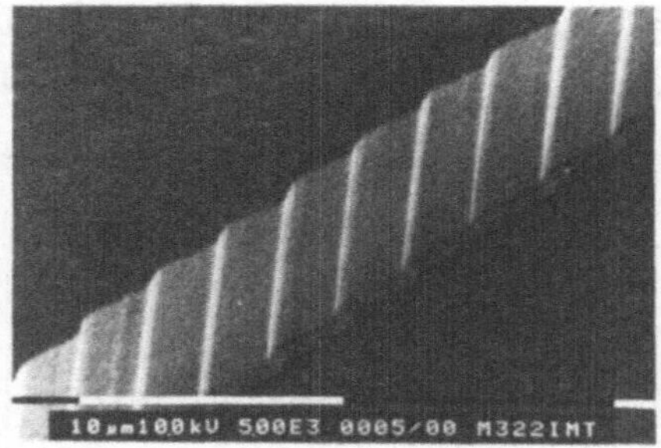

Fig. 18: SEM-picture of the tooth structure on a mask taken before an optimized process.

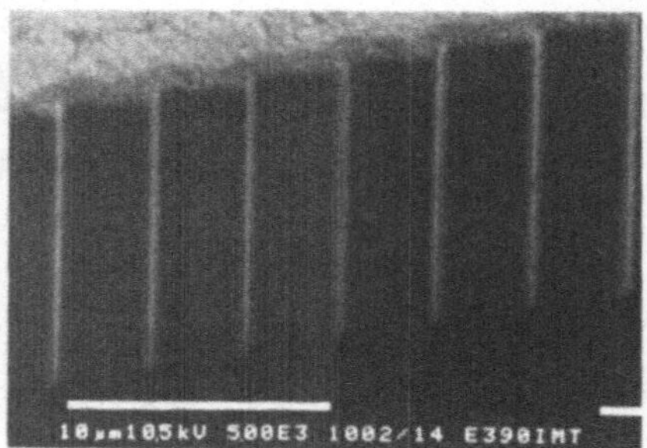

Fig. 19: SEM-picture of the tooth structure on a mask taken after an optimized process.

5. Summary and Outlook

The results described underline that the LIGA process constitutes an attractive complement of fabrication techniques for the manufacture of microoptical components. Despite the high degree of miniaturization of passive optical elements it has been possible to fabricate high performance and low loss elements in PMMA both by X-ray lithography and by molding. Due to the freedom in patterning it is no problem to compose, besides individual components, also light paths made of several components which have already been fully adjusted. By some simple examples such as microprisms, cylindrical lenses and geometric beam splitters the properties of the microoptical structures have been clearly demonstrated.

By combining miniaturized passive function elements with multimode fibers precisely positioned with respect to them using fiber fixing grooves compact and effective fiber coupling elements have been achieved. These elements are flexibly adaptable to any multimode fibers and constitute an alternative to the previously used Y-couplers. Due to the short light paths in the material, these effective coupling elements react rather insensitively to material attenuation and also divergence losses caused by the multimode fibers are not very much noticeable, too.

Since, moreover, the three-layer resist presented makes available a light guiding system capable of patterning by X-ray lithography the possible applications of the LIGA process are not restricted exclusively to the compact elements. Also for optical paths of several 10 millimeters length material attenuation is sufficiently low. This has been demonstrated by the example of the 1x3-fiber fork as well as by a grating spectrograph with self-focussing reflection grating.

The results described show that patterning and optimization of individual microoptical components or of assemblies combined with glass fibers has so well advanced that first microoptical product specimens with defined properties can be manufactured using the LIGA process. It has been shown, especially by the example of the fiber coupling elements, that the elements molded by relief printing do not deviate in terms of coupling losses from the structures manufactured by lithography. These results underline the potential of molding as a mass manufacturing process for microoptical components. Moreover, this offers the possibility to use also polymer materials characterized by higher attenuation values, but also e.g. by better thermal stability than is the case with PMMA. Thanks to electroplating such fiber coupling elements can be built up also in metal. This provides further potential applications of these elements, e.g. under conditions characterized by extremely stringent temperature requirements.

On the basis of results achieved so far besides the optimization of individual structures also consideration of the system as a whole is to be pursued consistently in more in-depth activities. For instance, profiting by the LIGA process and the achievable slope in patterning as well as by combination of microoptical LIGA structures with other three-dimensional elements capable of hybrid integration through mounting supports positioned in the light path, it is possible to manufacture complex microoptical assemblies.

Besides, compact micro- and fiber optical sensor systems can be manufactured by which, e.g. material data of a fluid passing through an integrated cuvette can be determined. For instance, in combination with the grating spectrograph, a spectral analysis of the fluid can be made in a spectral range to be defined in advance. The described fiber coupling elements can be applied in a fiber optical sensor system for which bidirectional data transfer is requested and substitute there the conventional, relatively large and expensive Y-couplers. By combination of the structures shown with moveable micromechanical components manufactured by the LIGA process [Bu91], too, further potential applications in optics are opened up.

In telecommunication networks single mode fibers and waveguide structures are used. The results described also offer first possible solutions in applying the LIGA technique in this field. Investigations will be carried out at the Institute for Microstructure Technology (IMT) at the Nuclear Research Center, Karlsruhe, to manufacture coupling elements for applications involving single-mode fibers.

Acknowledgements

The authors wish to thank the Physical Institute of the University Bonn for the supply of the X-ray source, the research group of Dr. K.-H. Brenner, Institute for Applied Optics, University of Erlangen-Nürnberg, especially Mr. S. Kufner and Mr. A. Müller, for making the interferometric measurements, as well as Dr. H. Sautter of the Robert Bosch GmbH, Zentrale Forschung, Stuttgart, for discussions and measurements related to the fiber coupling elements.

References

[Alt91]   Althaus, H.:
"Anwendungen von Mikromechanik, Mikrooptik und Mikroverbindungstechniken in der Halbleiterindustrie dargestellt am Beispiel von Optoelektronischen Bauelementen", OTTI 1. Symposium Mikrosystemtechnik, Regensburg, 1991.

[An88]   Anderer, B., Ehrfeld, W., Münchmeyer, D.:
"Development of a 10-Channel Wavelength Division Multiplexer / Demultiplexer Fabricated by an X-Ray Micromachninig Process", Int. Congress on Optical Science and Engineering, Proc. SPIE Vol. 1014, Micro-Optics I, Hamburg, 1988.

[An90]   Anderer, B., Ehrfeld, W., Mohr, J.:
"Grundlagen für die röntgenlithographische Herstellung eines planaren Wellenlängen-Demultiplexers mit selbstfokussierendem Reflexionsgitter", KfK-Bericht Nr. 4702, Kernforschungszentrum Karlsruhe, 1990.

[Be86]   Becker, E.W., Ehrfeld, W., Hagmann, P., Maner, A., Münchmeyer, D.:
"Fabrication of Microstructures with High Aspect Ratios and Great Structural Heights by Synchrotron Radiation Lithography, Galvanoforming and Plastic Moulding (LIGA Process)", Microelectronic Engineering, 4, 1986.

[Bl91]   Bley, P., Göttert, J., Harmening, M., Himmelhaus, M., Menz, W., Mohr, J., Müller, C., Wallrabe, U.:
"The LIGA Process for the Fabrication of Micromechanical and Microoptical Components", Micro System Technology, ICC Berlin, 1991.

[Bre91]   Brenner, K.-H.:
"3D-Integration of Optical Systems", ECO4, Proc. SPIE 1506, Micro-Optics II, Den Haag, 1991.

[Bre92]   Brenner, K.-H., Göttert, J., Kufner, M., Kufner, S., Mohr, J., Moisel, J., Müller, A., Sinziger, S., Testdorf, M.:
"3D-Integration of Microoptics with the LIGA Process", to be published.

246

[Bu91]   Burbaum, C., Mohr, J., Bley, P., Menz, W.:
            "Fabrication of Electrostatic Microdevices by the LIGA Technique", Sensors
            and Materials, Tokyo, 1991.
[Gei86]   Geisler, J., Beaven, G., Boutruche, J.P.:
            "Optical Fibres", EPO Applied Technology Series, Vol. 5, Oxford, New York,
            Toronto, Sydney, Frankfurt, 1986.
[Gö91a]   Göttert, J., Mohr, J., Müller, C.:
            "Mikrooptische Komponenten aus PMMA, hergestellt durch Röntgentie-
            fenlithographie", VDI-Tagung Werkstoffe der Mikrotechnik, Karlsruhe,
            1991.
[Gö91b]   Göttert, J., Mohr, J.:
            "Characterization of Micro-Optical Components Fabricated by Deep-Etch
            X-Ray Lithography", ECO4, Proc. SPIE 1506, Micro-Optics II, Den Haag,
            1991.
[Hag89]   Hagmann, P., Ehrfeld, W.:
            "Fabrication of Microstructures of Extreme Structural Heights by Reaction
            Injection Molding", The Journal of the Polymer Processing Society, 1988,
            Int. Polymer Processing IV (3), 1989.
[Har92]   Harmening, M., Bacher, W., Bley, P., El-Kholi, A., Kalb, H., Kowanz, B.,
            Menz, W., Michel, A., Mohr, J.:
            "Molding of Threedimensional Microstructures by the LIGA Process",
            MEMS, Travemünde, 1992.
[Li88]   Lilienhof, H.-J.:
            "Integriert-Optische Sensoren in Glas - Herstellung und Anwendung",
            AMA-Seminar Faser- und Integriert-Optische Sensoren, Heidelberg, 1988.
[Man89]   Maner, A., Ehrfeld, W.:
            "Electroforming Techniques in the LIGA Process for the Production of Mi-
            crodevices", Proc. INTERFINISH 88, 12th World Congress on Metal Finish-
            ing, Paris, 2, 1988. Materials & Manufacturing Processes, 4(4), 1989.
[Mo88]   Mohr, J., Ehrfeld, W., Münchmeyer, D.:
            "Analyse der Defektursache und Genauigkeit der Strukturübertragung bei
            der Röntgentiefenlithographie mit Synchrotronstrahlung", KfK-Bericht Nr.
            4414, Kernforschungszentrum Karlsruhe, 1988.
[Mül91]   Müller, C.:
            "Analyse des Leistungsvermögens eines mit der Röntgentiefenltihogra-
            phie hergestellten optischen Demultiplexers", Diplomarbeit an der Univer-
            siät Karlsruhe, 1991.
[Mün87]   Münchmeyer, D., Ehrfeld, W.:
            "Accuracy Limits and Potential Applications of the LIGA Technique in Inte-
            grated Optics", in: M. Weck, Editor, Micromachining of Elements with Op-
            tical and Other Submicrometer Dimensional and Surface Specifications,
            Proc. SPIE 803, 1987.
[Po90]   Pohlmann, T.:
            "Integrated-Optic Modulators and Switches for GHz-Applications", OPTO
            7, Nürnberg, 1990.

[Schwi85] Schwider, J., et. al.:
"Echtzeitinterferometrie für die Optik-Prüfung", Optica Applicata, Vol XV, 1985.
[Vo88] Voges, E.:
"Technologie der Integrierten Optik in der Sensorik", Technologie-Trends in der Sensorik, VDI/VDE-Technologiezentrum, Berlin, 1988.
[Wi83] Wilson, J., Hawkes, J.F.B.:
"Optoelectronics - An Introduction", Englewood Cliffs, New Jersey, London, New Dehli, Singapore, Sydney, Tokyo, Toronto, Rio de Janeiro, Wellington, 1983.

Photonic Microsystems from LIGA Technology

H.O. Moser, W. Ehrfeld, H.-D. Bauer, P. Kistenmacher, H. Schift

IMM Institut für Mikrotechnik GmbH
Ackermannweg 10
6500 Mainz 1

Abstract

Due to its almost complete freedom in lateral structuring, to the large choice of materials, and to the cheap mass production feature, LIGA technology offers new possibilities to realise photonic microsystems, either integrated or microoptical. In particular, advantages can be expected with respect to systems combining optically nonlinear with linear materials and concerning flux and flux density of the optical energy flow. It is found that the optical Kerr coefficient n_2 of an interesting nonlinear material must exceed a lower limit, i.e., $n_2 > (\lambda\Delta\phi/2\pi I_0)\alpha$, in order to be useful for photonic microsystems. Herein, λ is the wavelength, $\Delta\phi/2\pi$ the phase shift caused by the nonlinear part of the microsystem, I_0 the energy flux density of the incident light, and α the absorption coefficient of the nonlinear material. Bubeck and coworkers have given relationships between α and n_2 for a variety of organic materials. In case of a recently described phthalocyaninatoruthenium complex the optical Kerr coefficient and the 3rd order susceptibility should satisfy $n_2 > 4\cdot10^{-14}$ m^2/W and $\chi^{(3)} > 4\cdot10^{-8}$ esu, respectively, at $\lambda = 680$ nm, $\Delta\phi/2\pi = 0.01$, and $I_0 = 100$ MW/cm^2 . This material comes close to these requirements.

1. Introduction

Considering the potential of LIGA technology [BEC86, EHR91, MOH91, AND90] in microfabrication, in particular, the great flexibility in lateral structuring by means of lithography, the wide choice of materials including pure metals, metal alloys, plastics, and ceramics, and the possibility of cheap mass production by plastic injection molding and die stamping, it seems worthwhile to revisit the field of photonic microsystems [FRI88, JAE88, MIT90, MOL91, PRA91, DES92] under these aspects for new possibilities to create integrated or microoptical devices including optically nonlinear materials. Though the optical nonlinearity may be of second as well as of third order the work described below concentrates on all-optical microsystems using $\chi^{(3)}$ - materials.

In the following, we first discuss some general aspects of photonic microsystems based on LIGA technology, then describe specific devices, and derive an estimate for the magnitudes which both, Kerr coefficient and $\chi^{(3)}$, must attain in order to be useful. Finally, the question of applications is shortly addressed.

2. Relation between photonics and LIGA technology

LIGA technology, with LIGA being a German acronym combining lithography, electroforming, and plastic moulding, offers an utmost flexibility in lateral shaping, a large choice of materials, and the possibility of cheap mass production.

As the X-ray mask needed for the synchrotron radiation deep lithography step of the LIGA process is produced with an electron beam writer almost arbitrary lateral contours of the microstructures can be shaped. Besides the usually vertical surfaces obtained from irradiation at normal incidence, oblique or conical surfaces may be generated by tilting and rotating the target before or eventually

250

during irradiation [EHR91]. In this way, mirrors to couple or to extract radiation into or from the device may be realised. In addition, a second layer containing a different micropattern may be applied on top of the first, already processed one allowing for the fabrication of vertically piecewise variable microstructures [EHR91].

The classical materials for integrated optics and microoptics with the LIGA technology, such as polymethylmethacrylate (PMMA) as well as partially fluorinated methacrylates and copolymers thereof, keep their importance for photonic systems, too, at least for the optically linear part of a device. They are joined by numerous nonlinear organic polymers with an extended delocalised π-electron system such as polyphenylenevinylene (PPV), polydiacetylene (PDA), polythiophene (PT), polyphenylacetylene (PPA) [BUB90, ETE91, NEH90], and a recently described phthalocyaninatoruthenium complex [GRU92], further by organic chromophores including Rhodamin 6G and phthalocyanine derivatives [BUB91, KAL89], in particular in combination with PMMA, furthermore, by polymers with side chain chromophores [STA90], and "oligorylenes" which can be synthesized with a precisely controllable length [SCH91], and, finally, by organically modified ceramic polymers ("ormocers") [SCM90]. Moreover, a hybrid combination with anorganic materials, such as glass or $LiNbO_3$, is also conceivable. In this case, a small piece of that material could be glued to the substrate, and, then, the polymer material could be cast around and structured by lithography. Of course, other ways of microassembling such hybrid systems could be pursued.

For microdevices to be made predominantly from plastics, mass production methods such as micro injection molding [HAG89] or die stamping [KNO91] can be partly applied. In this way, plastic materials which are not suitable as X-ray resists, such as polycarbonate (PC), become also available.

The two main processes, direct synchrotron radiation deep lithography of a 3-layer-resist and stamping in combination with filling, are represented schematically in fig. 1 a and b.

In the first case, a 3-layer resist is prepared on a substrate made of epoxy phenol resin which is radiation insensitive. The core of the resist is normal PMMA, while the cladding consists of a copolymer of PMMA and tetrafluoropropyl

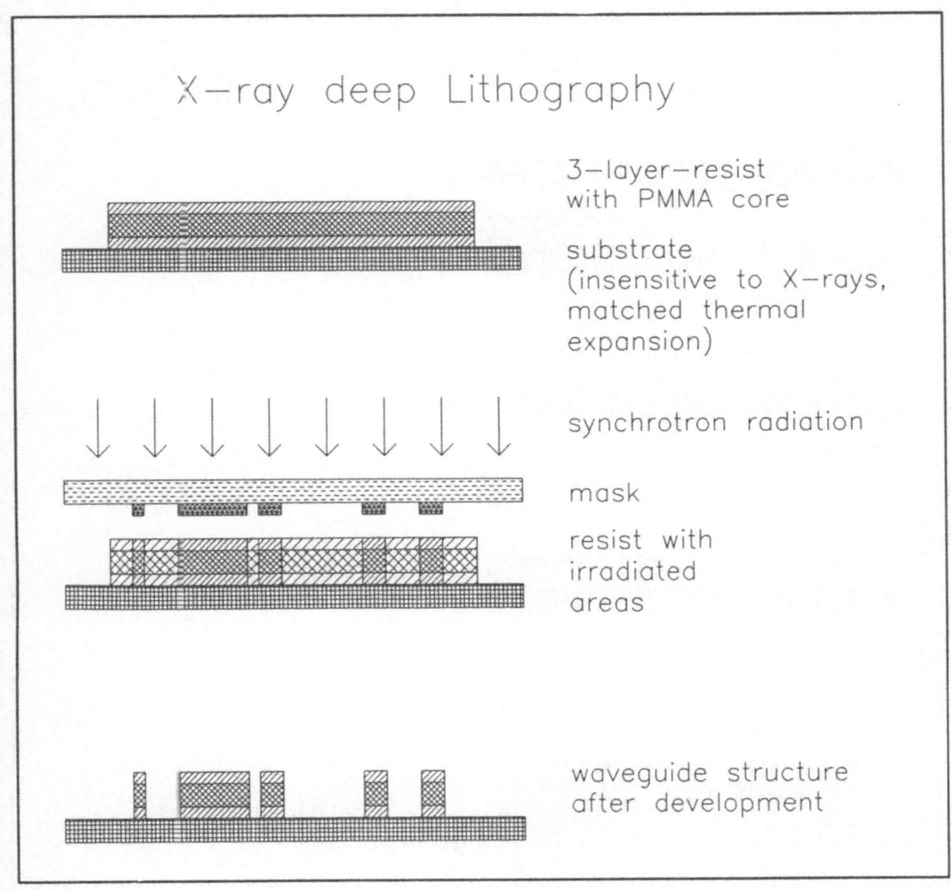

Fig. 1 a : Forming waveguides by direct lithography in a 3-layer resist.

methacrylate which allows to control the refractive index. This system is irradiated through an X-ray mask. On developing, the final waveguide structure is formed [AND90, MOH91]. In the second case, a die stamp is formed by LIGA and used to stamp a plastic cladding material. Then, core material is filled in and the whole is covered by a layer of cladding material.

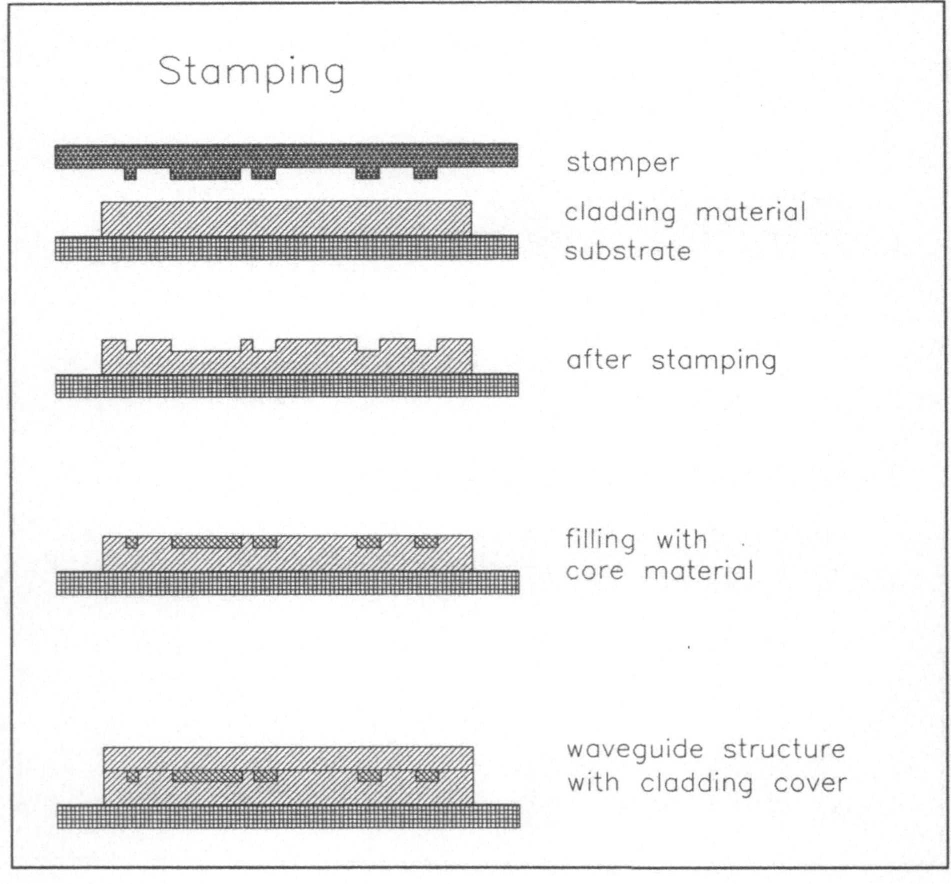

Fig. 1 b : Forming waveguides by stamping a slab of cladding material with a die produced by means of LIGA , filling the hollow structures with core material, and putting a cover of cladding material.

So far, the predominantly used production processes and materials in photonics have been those of semiconductor technology and fiber optics [EAT91], as can be seen from examples such as the fast amplitude-dependent switch based on a directional coupler [FRI88] or the soliton device for long-distance telecommunications [MOL91]. The interest in polymers for photonic applications is growing, mainly because of the numerous possibilities of cheaply patterning devices and of tailoring optical properties on a molecular level [KEI92, STR92].

3. Microsystems based on LIGA technology

One of the important features LIGA is offering for microsystems is the use of different polymers with different optical properties and functions within a single system. The basic idea is that the waveguide structures can be made with a linear low-damping material, e.g., PMMA, while the nonlinear material is concentrated on specific areas where a nonlinear effect is to be produced. In this way, the optical functions, linear and nonlinear, are separated. The lithographic patterning allows to maintain sharp boundaries between different areas, i.e., materials. In this way, microcuvettes can be formed in which the nonlinear material is poured or injected.

A further characteristic advantage of LIGA technology is the free choice of the waveguide cross section. Typically, height and width may vary from a few μm up to several 100 μm where the width could even be larger than these values. In this way, the number of possible modes in the waveguide can be controlled. In proportion to the cross sectional area, the energy flux can be made high to realize optical power devices. On the other hand, materials with lower damage thresholds may be used. Furthermore, the energy density may be increased to favour nonlinear optical effects by shaping the waveguides, e.g. through tapering, by adding focusing structures such as lenses or diffractive optical elements, and by using resonant structures.

254

Using polymer materials the refractive indices can be controlled within a wide range. In particular, the difference of refractive index between core and cladding can be chosen significantly larger than in integrated optics, so that larger numerical apertures can be set. Filling liquid polymeric material into already patterned structures may have the side effect that the index profile will not remain steplike, but is smoothed out by diffusion. Annealing should further help in realizing GRIN-like index profiles.

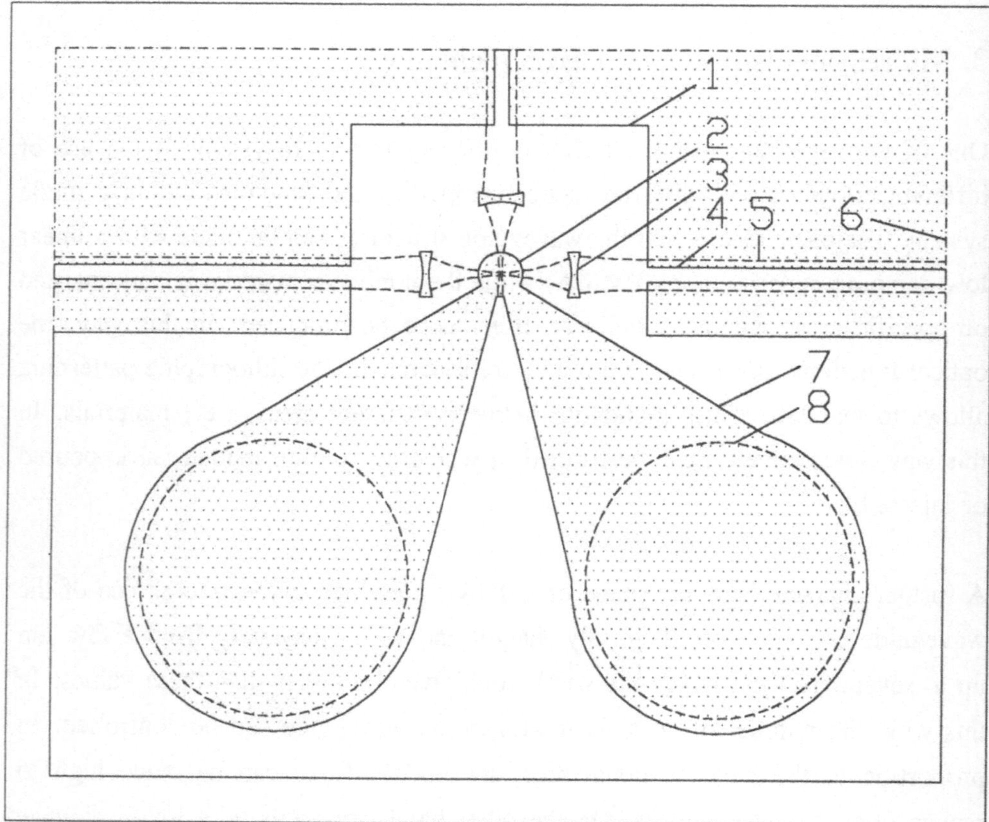

Fig. 2: Set-up to focus light from three monomode waveguides into a microcuvette. Patterned waveguide 1, microcuvette 2, cylinder lens 3, monomode waveguide 4, extension of fundamental mode into cladding material 5, cladding 6, filling channel 7, filling bores in cover plate 8.

Fig. 2 shows a schematic of a device which could be used as a monomode switch or as a phase conjugating mirror. Light from several monomode strip waveguides is focused through cylinder lenses into a microcuvette filled with a nonlinear material. To fill and vent the microcuvette two channels are provided, narrowing from the macroscopic, though small, access holes down to the dimensions of the microcuvette, fig. 3 showing a cross section of the filling and venting channel structure. In the small areas of the lenses and the microcuvette the wave is not guided in the 3-layer-resist case. This can be partly remedied by adding reflective metal layers on both, base and cover plate. In the stamp-and-fill case the wave may be guided in these areas, though with a different numeric aperture.

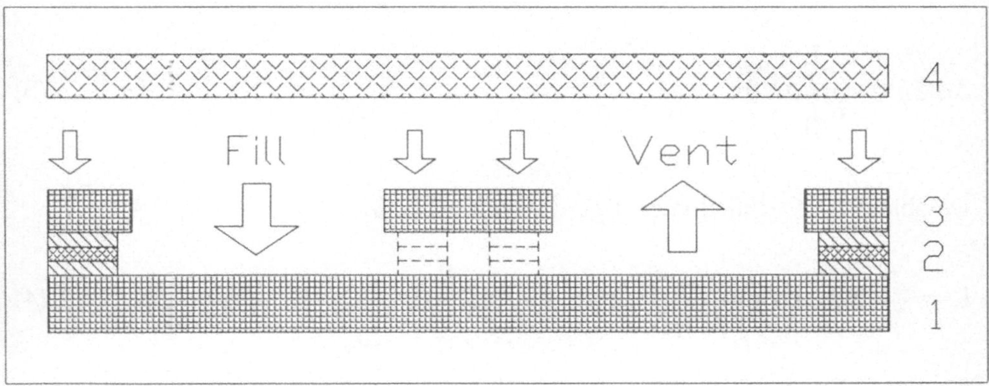

Fig. 3: Cross section of 3-layer waveguide with substrate 1, patterned waveguide 2, cover plate with filling bores 3, and sealing plate 4. Filling, venting, and sealing are indicated schematically by arrows.

4. Magnitude of nonlinear effect

The usefulness of nonlinear optical effects and of devices exploiting them is determined by the magnitude of the effect. We derive an estimate by asking for the phase shift $\Delta\phi$ of a wave of intensity I_0 induced by a region with length L filled with nonlinear material characterized by a nonlinear refractive index n including a Kerr coefficient n_2, the latter corresponding to a 3rd order susceptibility $\chi^{(3)}$, and an absorption coefficient α at a wavelength λ. SI units are used throughout. Starting from Beer's law

$$I = I_0 \exp(-\alpha x), \tag{1}$$

including the optical Kerr effect

$$n = n_0 + n_2 I, \tag{2}$$

and the phase shift

$$\Delta\phi(L) = \frac{2\pi}{\lambda} \int_0^L \Delta n \, dx \tag{3}$$

we obtain L as a function of n_2, α, n_0, λ, $\Delta\phi$, I_0

$$L = -\frac{1}{\alpha} \ln \left(1 - \frac{\lambda \Delta\phi \, \alpha}{2\pi \, I_0 n_2} \right) \tag{4}$$

This expression limits the second term in the argument to <1 leading to a lower limit for the Kerr coefficient

$$n_2 > \frac{\lambda \Delta\phi}{2\pi \, I_0} \alpha \quad . \tag{5}$$

Inserting typical values, e.g., $\lambda = 633$ nm, $\Delta\phi / 2\pi = 0.01$, $I_0 = 100$ MW/cm^2, and $\alpha = 10^4$ cm^{-1}, we obtain $n_2 > 6.3 \cdot 10^{-15}$ m^2/W. To the extent to which the second term approaches 1, e.g., by growing absorption, dropping n_2, or decreasing intensity I_0, the interaction length L needed to produce a wanted phase shift $\Delta\phi$ grows very strongly exhibiting a kind of cut-off towards smaller values of n_2. This dependence can be used, in turn, to adjust the interaction length since the $1/\alpha$ - term tends to give very small values for L whereas a useful device needs a certain length, or volume, in particular, when involving crossed beams.

We now discuss three special cases. For vanishing α or large n_2, eq. (4) becomes

$$L = \frac{\lambda\Delta\phi}{2\pi\,I_0 n_2} \qquad . \tag{6}$$

Empirical relationships between n_2 and α have been found [BUB91, GRU92] which can be written as

$$n_2 = p_k\,\alpha^k\,, \tag{7}$$

with $k = 1, 2$, $p_1 = 2.63\cdot10^{-21}/n_0^2$ m^3/W , and $p_2 = 2.63\cdot10^{-27}/n_0^2$ m^4/W. Inserting (7) into (4), we obtain

$$L = -\left[\frac{p_k}{n_2}\right]^{1/k} \ln\left(1 - \frac{\lambda\Delta\phi}{2\pi I_0}\,p_k^{-1/k}\cdot n_2^{\,1/k\,-\,1}\right) \tag{8}$$

and condition (5) reads

$$n_2^{\,1\,-\,1/k} > \frac{\lambda\Delta\phi}{2\pi\,I_0 p_k^{1/k}}. \tag{9}$$

For a linear relationship between n_2 and α, i.e., $k=1$, which holds for a class of π-conjugated polymers including poly (3-decylthiophene) PT, poly (phenylacetylene) PPA, poly (p-phenylene vinylene) PPV, etc. [BUB91], the argument of the logarithm becomes independent of n_2 leaving L again proportional to $1/n_2$

$$L = -\frac{2.63\cdot10^{-21}/(m^3/W)}{n_0^2} \ln\left[\,1 - 3.8\cdot10^{20}/(W/m^3)\,\frac{\lambda\Delta\phi n_0^2}{2\pi I_0}\right]\frac{1}{n_2}. \tag{10}$$

Condition (5) does no longer involve n_2, but simply requires

$$(3.8 \cdot 10^{20}/(\text{W}/\text{m}^3)) \frac{\lambda \Delta \phi n_0^2}{2\pi I_0} < 1 \ ,$$

which must be satisfied, essentially, by a sufficiently large light intensity I_0 . The required minimum value, assuming $\lambda = 633$ nm, $\Delta \phi /2\pi = 0.01$, and $n_0 = 1.6$, is $I_0 = 620$ MW/cm^2.

For the second class of materials, represented, e.g., by a recently described phthalocyaninatoruthenium complex [GRU92] or rhodamin 6G dye [BUB91], and characterised by n_2 varying as α^2 , i.e., k =2, we find

$$L = -\sqrt{\frac{2.63 \cdot 10^{-27}/(\text{m}^4/\text{W})}{n_0^2 n_2}} \ln \left(1 - \frac{\lambda \Delta \phi}{2\pi I_0} \sqrt{\frac{n_0^2}{2.63 \cdot 10^{-27}/(\text{m}^4/\text{W}) \ n_2}} \right), \quad (11)$$

and condition (5) leading to

$$n_2 > \frac{1}{P_2} \left[\frac{\lambda \Delta \phi}{2\pi I_0} \right]^2 \ .$$

This is a real lower limit for n_2 which, in the case of the phthalocyaninatoruthenium complex with k = 2, assuming $\lambda = 633$ nm, $\Delta \phi /2\pi = 0.01$, $I_0 = 100$ MW/cm^2 and $n_0 = 1.6$ results in $n_2 > 3.9 \cdot 10^{-14}$ m^2/W. Interestingly, the light intensity enters squared. For $\lambda = 680$ nm, with $n_0 = 2.3$, $n_2 > 4 \cdot 10^{-14}$ m^2/W. Note that the n_2 values used in the foregoing discussion are related to the common $\chi^{(3)}$ through [BUT90, GIB85]

$$n_2/(\text{m}^2/\text{W}) = \frac{5.26 \cdot 10^{-6} \ (\chi^{(3)}/\text{esu})}{n_0^2} .$$

Thus, the Kerr coefficient estimated above translates into $\chi^{(3)} > 2 \cdot 10^{-8}$ esu and $\chi^{(3)} > 4 \cdot 10^{-8}$ esu , respectively.

In fig. 4, the length of the interaction zone L is plotted versus the Kerr coefficient n_2 for the cases (6), (10), and (11) .

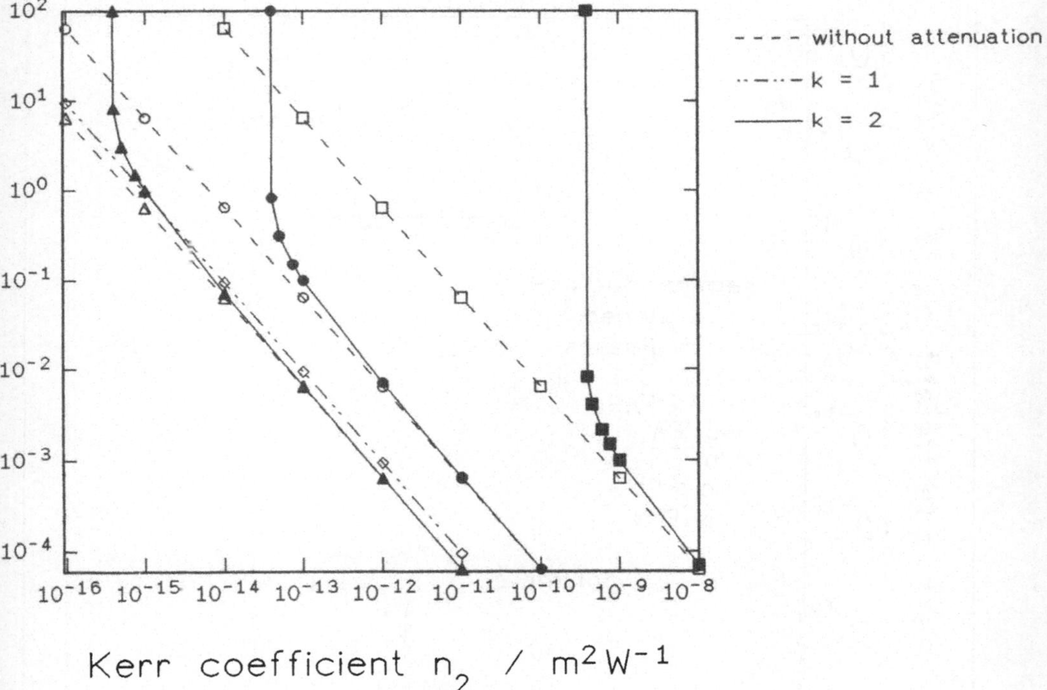

Fig. 4: Length of space filled with nonlinear material and producing a phase shift of $2\pi/100$ versus Kerr coefficient. Solid lines and filled symbols: k=2 materials, dash-dotted line and open diamond: k=1 materials, dashed lines and open symbols: vanishing absorption, squares, circles, and triangles stand for 1, 100, and 1000 MW/cm^2, respectively.

Fig. 5 shows an overview of the Kerr coefficient n_2 of various materials versus switching time. It can be seen, that the upper border of the polymer area is close to or fulfills the requirements. The material with k=2 shows a certain advantage compared to those with k=1. The length of the interaction zone is within the range useful for microsystems and can be tailored by adjusting the magnitude of the logarithm via I_0 .

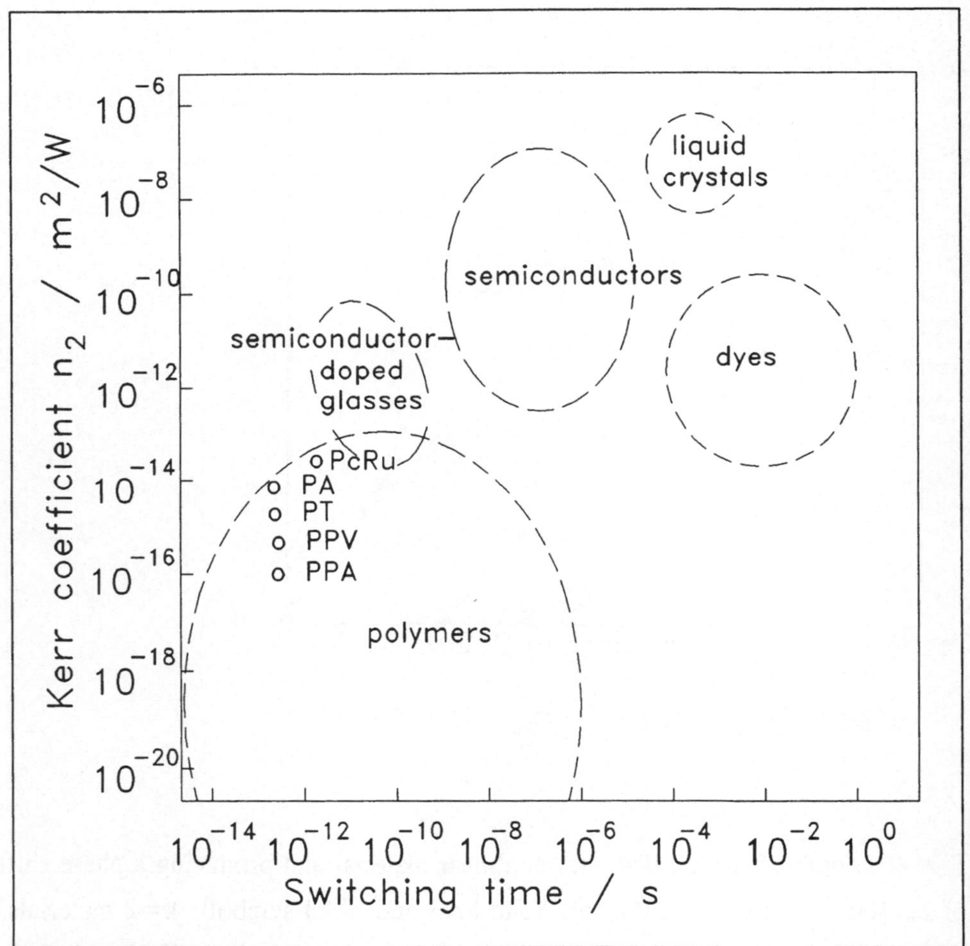

Fig. 5: Optical Kerr coefficient versus switching time for various materials (adapted and modified from [KAL89]).

According to [GRU92], $\chi^{(3)}$ can be made larger for k=2 materials at the expense of longer switching times. This would facilitate the task for slower devices as they might be useful for photonic safety systems in hazardous environments or for avoiding electromagnetic interference.

5. Potential Applications

The field of potential applications of polymer based LIGA-structured all-optical photonic microsystems encompasses that of more conventional photonics, including, namely, optical computing, and optical signal and image processing [PRA91]. In addition, LIGA photonic microsystems promise to be fabricated cheaply and in large quantities, so leading to broader ranges of application. However, as pointed out in [PRA91], the development of this field comprises high risk, and high payoff, and will need time.

Whereas no specific applications are imminent, we feel that, to the extent to which materials and patterning techniques become available, cheap devices for switching and signal processing might emerge performing comparator or trigger functions, restoring signals, or acting as protection against light levels too high for sensors.

Acknowledgment

This work is carried out in collaboration with Max Planck Institute for Polymer Research and is funded by the Volkswagen foundation.

References

AND 90   B. Anderer, KfK 4702, Kernforschungszentrum Karlsruhe, 1990.

BEC 86   E.W. Becker, W. Ehrfeld, P. Hagmann, A. Maner, D. Münchmeyer, Microelectronic Engineering 4(1986)35.

BUB 90   C. Bubeck et al., Makromol. Chem., Makromol. Symp. 37(1990)239.

BUB 91   C. Bubeck et al., Chem. Phys. 154(1991)343.

BUT 90   P.N. Butcher, D. Cotter, The Elements of Nonlinear Optics, Cambridge University Press, 1990, pp. 306.

DES 92   E. Desurvire, Scientific American, Jan. 1992, pp. 96.

EAT 91   D.F. Eaton, Science 253(1991)281.

EHR 91   W. Ehrfeld, D. Münchmeyer, Nucl. Inst. and Meth. A303(1991)523

ETE 91   S. Etemad et al., in "Organic Molecules for Nonlinear Optics and Photonics", pp. 489, J. Messier et al., eds., Kluwer 1991.

FRI 88   S.R. Friberg et al., Opt. Lett. 13(1988)904.

GIB 85   H.M. Gibbs, Optical Bistability: Controlling Light with Light, Academic Press, Orlando, 1985, pp. 375.

GRU 92   A. Grund et al., J. Phys. Chem., 1992, in press.

HAG 89   P. Hagmann, W. Ehrfeld, Intern. Polymer Processing IV(1989)3, pp. 188.

JAE 88   compare, e.g., D. Jäger, Laser und Optoelektronik 1(1988)46; H.-J. Eichler et al., ibid. pp. 59.

KAL 89   A. Kaltbeitzel et al., in "Electronic Properties of Conjugated Polymers III", Springer Ser. in Sol. St. Sc. Vol. 91, pp. 220, H. Kuzmany et al., eds., Springer 1989.

KEI 92   N. Keil, these proceedings.

KNO 91   K. Knop, Phys. Bl. 47(1991)901.

MIT 90   F. Mitschke, Phys. Blätter 46(1990)463.

MOH 91   J. Mohr, J. Göttert, C. Müller, P. Bley, KfK-Nachrichten 23(2-
3)(1991)93.

MOL 91   L.F. Mollenauer, J.P. Gordon, S.G. Evangelides, Laser Focus
World 27(1991)159.

NEH 90   D. Neher et al., Synthetic Metals 37(1990)249.

PRA 91   P.N. Prasad, D.J. Williams, Introduction to Nonlinear Optical Effects
in Molecules and Polymers, John Wiley & Sons, Inc., New York,
1991

SCH 91   S. Schrader et al., Synth. Met. 41-43(1991)3223.

SCM 90   H. Schmidt, M. Popall, SPIE Vol. 1328(1990)249.

STA 90   J. Stamatoff et al., Die Angew. Makromol. Chemie
183(1990)151.

STR 92   B. Strebel, these proceedings.

Force Transducers on the Basis of Piezoelectric Copolymers

B. Halstrup, M. Crede, R. Kassing
Institute of Technical Physics
University of Kassel

Abstract

This report presents a novel method for the production of force transducers
for scanning microscopy using piezoelectric polymers as sensitive materials.

Introduction

During recent years scanning tunneling microscopy (STM) has become a well established means for the characterization of surfaces. This method can even yield atomic resolution by measuring the tunneling current between tip and sample.

Nevertheless STM is restricted to the characterization of electrically conducting surfaces. Therefore scanning force microscopy was developed for studying insulating samples. A scanning force microscope records the interaction force between a measuring tip and the surface under investigation. The tip is mounted on the free end of an elastic silicon bar (cantilever), which is deflected by the force. Hitherto the deflection is measured by different optical methods, which require sophisticated equipment (e. g. interferometer).

Therefore direct conversion of the deflection into an electrical signal — without any optical detours — is desirable.

Force measurement by piezoelectric effect

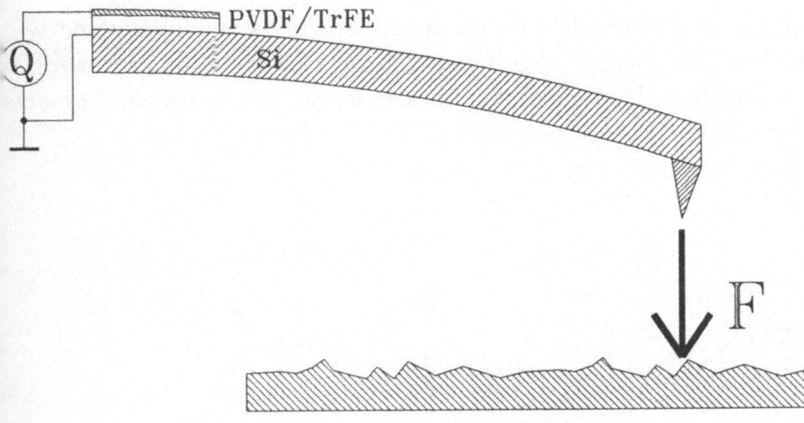

Figure 1: Interaction between cantilever and sample

As mentioned earlier the investigation of surfaces is done by micro–machined silicon cantilevers. One has to distinguish between static and dynamic modes of measurement. Static measurements are performed by drawing the cantilever across the sample with the tip pressed against the surface. Thus an image of surface topography is gained. If a dynamical measurement is carried out the cantilever is excited by a piezoelectric actuator at an excitation frequency almost equal to its resonance frequency. By approaching the sample the interaction between tip and investigated surface leads to a shift of the the resonance frequency. From this frequency shift the desired information can be drawn.

The mechanical properties of a cantilever — rigidity, resonance frequency, etc. — are determined by density (2.33 g/cm^3) and Young's modulus ($E_{Si} \approx 170$ GPa) of silicon as well as by length l, width b and thickness d of the cantilever. If such a cantilever is bent by a force F acting on its free end, a maximum strain η_{max} is induced at the clamped end:

$$\eta_{max} = \frac{6F_0 l}{E_{Si} b d^2}.$$

Applying Hooke's law yields the maximum stress $\sigma_{max} = E_{Si} \cdot \eta_{max}$.

The forces to be detected cover the range between 10^{-9} and 10^{-12} N /1/. The conversion of these forces can be achieved by using the piezoelectric effect.

A material is said to be piezoelectric, if it reacts to the application of mechanical strain by generating an electrical polarization. The corresponding electrical field is proportional to the strain; the designation of the proportionality factor depends on the mutual orientation of applied strain and generated electrical field. If they are collinear the factor is named d_{33}, if they are perpendicular, it's called d_{31}.

Silicon itself shows no piezoelectric effect. Therefore it has to be covered by a piezoelectric layer. For this purpose copolymers of polyvinylidene fluoride and trifluorethylene are well suited. The monomers possess large electric dipole moments. If poled by an electric field the copolymers become piezoelectric.

The following picture shows both of the monomeres.

Figure 2: Vinylidene fluoride   Trifluorethylene

Production of silicon cantilevers

Technological methods, which were once developed for semiconductor industry nowadays find new applications, e. g. the production of micromechanical elements. The force transducers for scanning microscopy are prepared by using appropriate etching , lithography and deposition methods. Especially by plasma etching force transducers of different geometries can be produced. The measuring tip at the free end of the cantilever is produced by an intricate etching process, which combines both dry and wet etching techniques /3/, /4/.

Preparation and characterization of piezoelectric polymer layers

Copolymers of three different compositions were investigated. The ratios of PVDF to TrFE were 50/50, 65/35 and 75/25. The first step was to solve the copolymers in an appropriate organic solvent. The results are summarized in the table below.

Then small amounts (some ml's) of the solution were spun on a rotating silicon wafer. This had been covered before with thermal oxide and metallization. Afterwards the copolymer layer was metallized and poled at a temperature of 90 °C. For

Solubility of copolymers			
	50/50	65/35	75/25
Ethylmethylketon	+	+	-
Dimethylformamid	+/-	+/-	+/-
Propylencarbonat	+/-	-	-

Table 1: Solubility of copolymers in different solvents

two hours a field of 1 MV/cm was applied. With electrical fields of some MV/cm an electrical breakdown can occur in the polymer layer. Therefore precise knowledge of layer thickness was necessary. By keeping a fixed ratio of solvent to solute and carefully controlling the speed of rotation every desired layer thickness between 1 and 15 μm could be achieved with an accuracy of half a micron.

After the poling procedure single cantilevers were separated mechanically. They served as samples for determining the piezoelectric coefficients. The determination of d_{31} and d_{33} was carried out in different ways. To measure d_{33} a perpendicular load was applied to a silicon wafer bearing a piezoelectric layer. The coefficient d_{31} was determined by loading a cantilever at its free end and detecting the generated charge.

The copolymer layer, metallized on both top and bottom, can be regarded as a plate capacitor. The application of a perpendicular force F on area A leads to a mechanical stress $\sigma = F/A$ and a corresponding strain $\eta = F/(E_{Co} \cdot A)$. The generated charge Q at the condenser plates yields an electric field of $Q/(\epsilon\epsilon_0 A)$. Thus follows:

$$d_{33} = \frac{E_{Co} \cdot Q}{\epsilon\epsilon_0 F_0}.$$

The constants E_{Co} and ϵ are Young's modulus and relative dielectric constant of the copolymer under investigation.

If a cantilever is loaded at its free end by a force F, a surface charge Q is generated on the piezoelectric layer. Let the layer cover an area A_c, then d_{31} is given by:

$$d_{31} = \frac{E_{Si} bd^2 Q}{6\epsilon\epsilon_0 A_c l F}.$$

The results of both measurements are presented in the picture beneath.

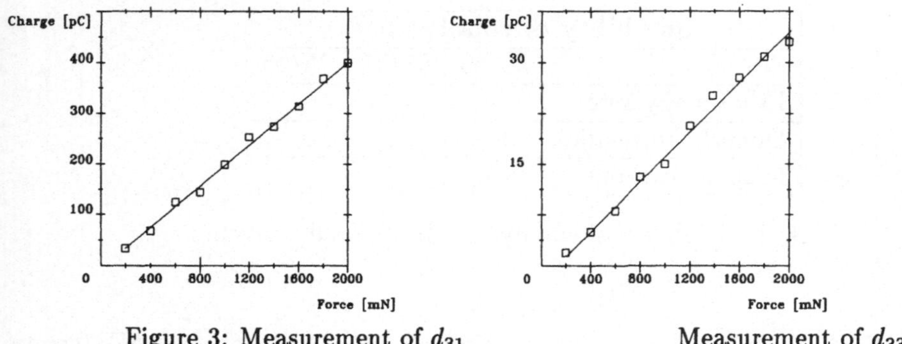

<table>
<tr><td align="center">Figure 3: Measurement of d_{31}</td><td align="center">Measurement of d_{33}</td></tr>
</table>

Summary

By joining established semiconductor technologies with methods hitherto applied in engineering sciences only microstructure technology offers numerous possibilities for the construction of novel sensors and actuators. This report chose as an example the integration of a piezoelectric polymer layer into a silicon force transducer. In the future the application of such layers could perhaps substitute the sophisticated optical apparatus which is in use nowadays. Such a system promises reduced costs due to batch production as well as easier handling. Therefore further investigations to develop the presented methods seem worthwhile.

References

/1/ Nonnenmacher, M. , Rastermikroskopie mit Mikrospitzen, Thesis, Kassel, 1990

/2/ Nonnenmacher, M. , J. Greschner, O. Wolter, R. Kassing, Scanning force microscopy with micromachined silicon sensors, J. Vac. Sci. Technol. B, 2 Mar/Apr1991

/3/ Wolter, 0. , Th. Bayer, J. Greschner, Micromachined silicon sensors for scanning force microscopy, J. Vac. Sci. Technol. B, 2 Mar/Apr1991

/4/ Rangelow, I. W. , R. Kassing, Trockenaetzprozesse in der Mikromechanik, 'Mikroperipherik', VDI Verlag, Okt. 1988

Applications of Integrated Optics in the Field of Measuring, Technology and Sensorics

Dr.E.Ruske, Jenoptik GmbH, O-6900 Jena,

Carl-Zeiss-Straße 1

Applications of integrated optics in the measurement technology mainly refer to the area of interferometry, and in this connection first priority is attached to the measurement of optical path. In this case the use of electrooptical materials such as $LiNbO_3$ is especially attractive, as the high frequency modulation of the laser ligth in one of the interferometer arms makes it possible to employ to advantage the very effective heterodyne evaluation equipment and to achieve a resolving power within the range of 10nm.

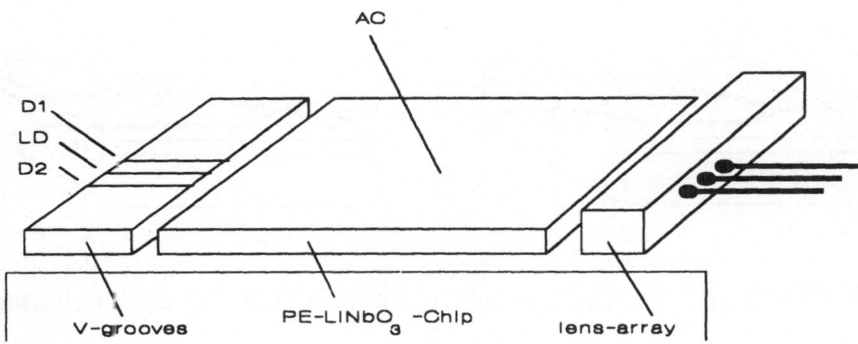

Fig. 1 Basic design of an optical measuring head

The basic design of such an optical measuring head is shown in figure 1. The layout contains a $LiNbO_3$ chip on which the interferometer strucutre was generated by means of proton exchange.
The radiation of a laser diode (LD) is coupled into the chip via a fibre and then distributed to a measuring and two reference channels. The exiting

bundles of rays are collimated by a lens array and retroreflected to the chip by three reflectors. Two receivers (D_1, D_2) analyse the interference signals. The modulator is also integrated into the chip. The modulation frequency varies within the MHz-range, the chip size is indicated as about 3cm x 1cm [Bau 92].

On the other hand the integrated optics provides new possibilities of realizing a multitude of chemical, medical and biological sensors.

Two basic structures could be stated as an example in order to explain the principle based on the theory of guided light in planar waveguides, especially the efficient utilization of the evanescent filed in connection with the total internal reflection. A sensor detecting the absorption of light is depicted in figure 2 whereas figure 3 shows a sensor responding to phase changes.

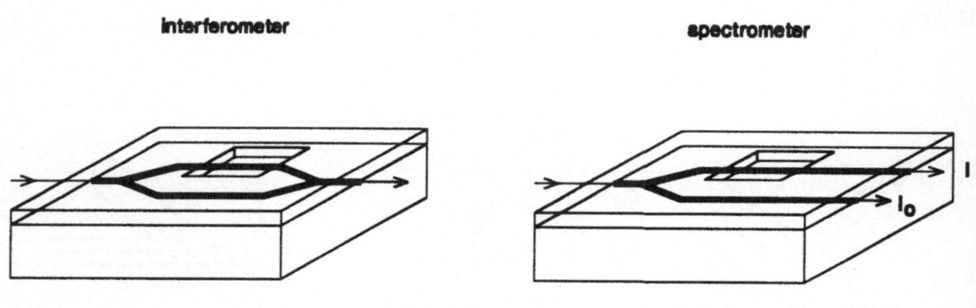

Fig. 2 Optical chip as interferometer   Fig.3 Optical chip as spectrometer

As we are especially occupied with the realization of two medical sensors based on the detection of aqueous mediums I would like to give you a summary of our examinations concerning the response (sensivity) of these sensors because, especially for the detection of water owing to its low influence on the guided light it is necessary to optimize the waveguide.

For the first time Harris and Polky [Pol 72] reflected on the sensitivity, furthermore the absorption spectroscopy with guided light was investigated by Posner [Pcs. 85] whereas Tiefenthaler and Lukosz [Tie89] concerned themselves with the detection of refractive index changes.

The waveguide system consists of a substrate (n_s), a waveguiding film (n_f) and the coating (cover)(n_c).

The effective refractive index N which represents the measured value according to Tiefenthaler as a function of the refractive index of the coating (cover) n_c can be expressed as

$$\frac{\partial N}{\partial n_c} = \frac{n_c}{N} \cdot \frac{P_c}{P} \cdot \frac{\left[2(N)^2 - 1\right]^\rho}{(n_c)^2} \tag{1}$$

$\rho=0$ for TE mode

$\rho=1$ for TM mode

$$P = P_s + P_f + P_c \tag{2}$$

P is the power of the guided modes.

P_s, P_f, P_c are the shares of the power in the substrate, the film and the coating respectively.

In the case of having water as the coating (cover) and $n_c<n_s$ the following can be stated:

When the value $\delta N / \delta n_c$ is added to the waveguide thinkness d_f then one can observe that the system is most sensitive to the basic mode TM_0.

The sensivity is high for $n_c \rightarrow n_s$, that is the substrat and the coating have the same refractive index. In the case of having water with $n_c = 1,33$ this can not be realized for the substrat.

The sensivity will also increase if the refractive rate of the waveguiding film is high provided that the refractive indexes of the coating (cover) and the subtrate are constant.

Assuming that glass is used as a substrat then the sensitivy achieved with the ion-exchanged glasses is low.

272

That is the reason why Tiefenthaler and Lukosz employed sol-gel layers (coatings) as waveguides on glass substrats. The latter have the advantages that one can achieve an optimum refractive index.

Similar results were obtained by Posner assuming an absorbing coating.

Consequently in the case of an aqueous coating (cover) it is favourable to work with monomode waveguides the refractive index of which is high in comparison with the glass substrate.

Seen from this point of view polymer waveguides seem to be also very prospective,especially if one takes into accoount the following aspects:

- a more cost-effective fabrication is conceivable
- it is possible to produce structures by means or embossing (e.g. coupling elements)
- in case of aqueous solutions the polymer-water interface seems to be easier to master then the glass-water interface
- the immobilization of a chemistry within the polymer waveguide may be thinkable thus the interaction cross-section will be determined by the waveguide itself and not only by the evanescent field
- the scope of possible applications will be extended by utilizing the electro- optical effects.

References

[Bau 92] J. Bauer, E. Dammann, E. Fritsch Secound Symposium of Microsystem Technology Februar 1992, Regensburg

[Pol 72] J. Polky, J. Harris J. opt. Soc. Amer. 62 (1972) 1081

[Pos 85] T. Posner, W. Karthe, R. Müller, Exp. Technic of Physics 33 (1985) 3, 241

[Tie 89] K. Tiefenthaler, W. Lukosz, JOSA D 62 (1989) 209

Realization of IO-Polymer-Components and Present State in Polymer Technology at HHI Berlin

N. Keil

Heinrich-Hertz-Institut für Nachrichtentechnik Berlin GmbH
Department Optical Signal Processing
Einsteirufer 37, D-1000 Berlin 10, FRG, Tel. ++ 49 30 310 02 590

Abstract

Polymer technology shows the potential to fabricate and integrate the basic components of optical crossconnect elements which are important processing units in a future transparent optical network.
The fabrication and characterisation of integrated optical polymer components and the present state in polymer technolgy at the Heinrich-Hertz-Institut in Berlin is presented.

1 Introduction

A future integrated broadband communication network (IBCN) is expected to make use of optical signal processing techniques such as the coherent multi-carrier (CMC) technique to increase the system capacity for transmission and switching of optical broadband signals [STR 89].

The most promising candidate for such an optical communication scheme is a transparent optical network wherein the switching takes place in the space-frequency domain [KEI 91]. Important processing units in a transparent optical network are optical crossconnect elements, which provide the following main signal processing functions:

- power splitting and combining,

- selective and broadband space routing,

- optical channel separation and insertion and

- all optical frequency conversion.

For an integrated optical crossconnect element, the following basic signal processing devices are needed:

- broadband power splitters and combiners,

- optical space switches,

- optical filters and

- optic-optical frequency converters.

Fig. 1.1 gives the impression that polymer technology is the ideal technology for the integration of crossconnect elements because the material system shows linear as well as nonlinear optical behaviour, which means that optical components like directional couplers, passive stars and wavelength multiplexers, thermo-optic or electro-optic tunable waveguide switches and filters as well as parametric components like optical frequency converters and optical amplifiers using second order nonlinear waveguides can be fabricated.

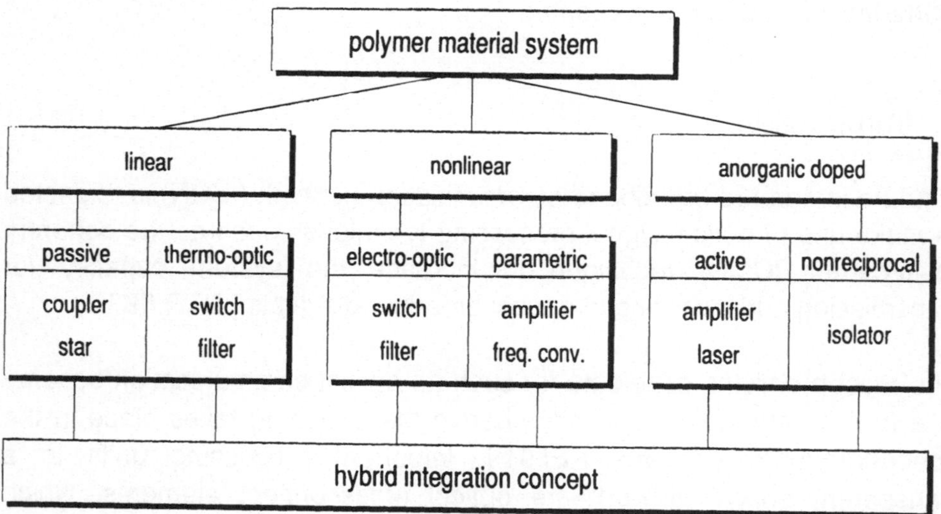

Fig 1.1 Polymer technology at HHI.

Because the organics can be doped with anorganics like rare earth elements, it seems to be realistic to expect amplifiers and lasers with rare earth doped polymer waveguides. Even integrated optical isolators using waveguides doped with magneto-optic elements are conceivable.

From the system point of view, polymer technology shows general structural advantages because parametric effects are broadband and nonreciprocal and the integration technology is conceptional hybrid, which means great flexibility at low cost.

2 Linear Components

The waveguide material for linear components used in the Heinrich-Hertz-Institut is PMMA and a commercially available photoinitiator. The technology to realize monomode waveguide structures is very simple: spin coating followed by a photolithographic waveguide structuring [FRA 86].

All basic components like strips, y-branches, directional couplers, Mach-Zehnder interferometers, etc. are disigned for the 1,55 μm wavelength region, because our main goal is to replace all the fiber devices which are currently used in the coherent system experiment of our group.

Fig. 2.1 illustrates the two important processing steps: exposure and annealing treatment.

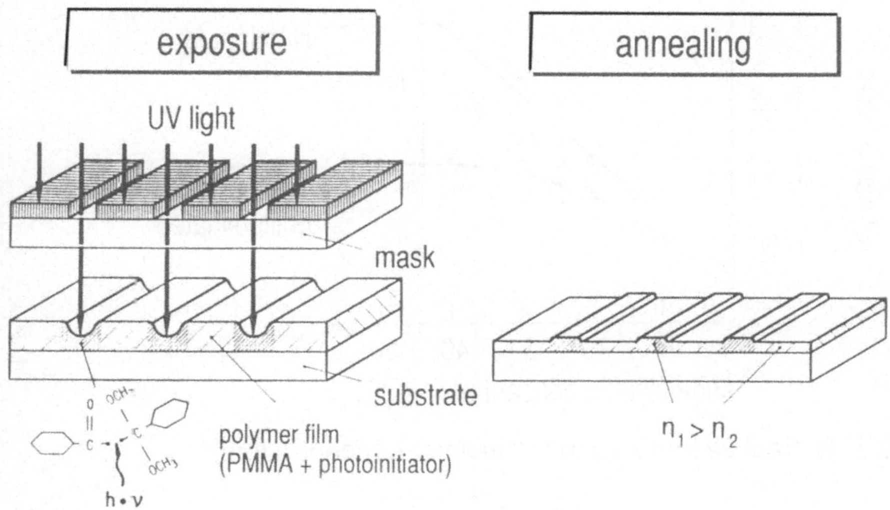

Fig 2.1 Fabrication process.

The fabrication process starts with a mixture of PMMA and photoinitiator deposited on a substrate by spin coating. We use different substrate materials like optical glasses and silicon wafers with SiO_2 buffer layers. Film thickness is in the order of 3 to 5 µm.

The film on the substrate is exposed by UV light through a mask having the desired layout. During UV exposure, a photochemical reaction takes place in the illuminated regions as indicated by the broken molecule, the refractive index is increased here and the layout of the photomask is transfered to the polymer film.

In order to develop and fix this refractive index change, an annealing treatment in an oven is followed. During this step, the nonreacted photoinitiator molecules in the nonilluminated areas will evaporate out of the film while the PMMA molecules remain causing a decrease in refractive index in that regions.

Because of the material loss, a final rib structure can be observed at the end of the process.

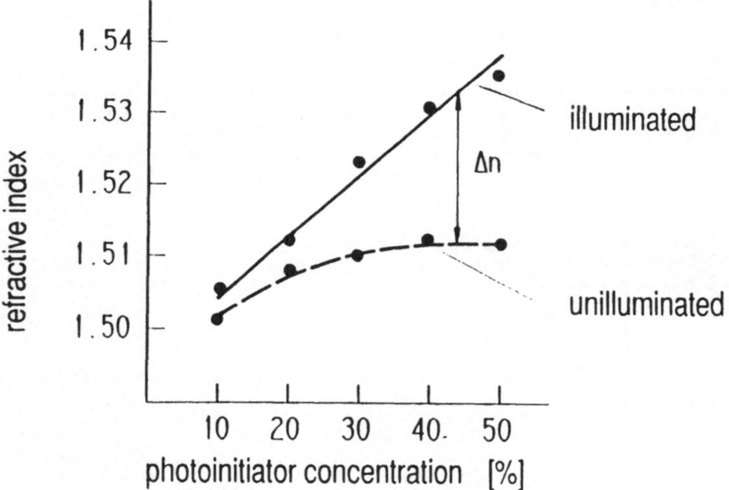

Fig 2.2 Refractive index vs. photoinitiator concentration.

Fig. 2.2 shows the measured relationship between the photo-induced refractive index change and the photoinitiator concentration. The photoinitiator used here is benzildimethyketal.

The films were illuminated 1 hour and then baked for more than 24 hours at 60 to 65 °C. The values were measured by m-line spectroscopy.

It is obviously that one can cover a range from approximately n = 1,50 to n = 1,54 by a simple choise of the right concentration of photoinitiator.
The obtainable Δn value is between 0 and 3% depending on the photoinitiator concentration. This Δn value is tunable between its maximum value and zero, by simple exposure the whole film again, at this time without the mask, before the annealing treatment is followed. During this step, the former nonilluminated areas can be increased in the refractive index to the desired value. We call that process *post exposure*.

In Fig. 2.3 different parts of our test structures like polymer y-branches (Fig. 2.3a) and directional couplers (Fig. 2.3b) fabricated on Si/SiO$_2$ substrates are shown.
These waveguides are 5 μm wide and have a rib structure which is typical for our fabrication process.

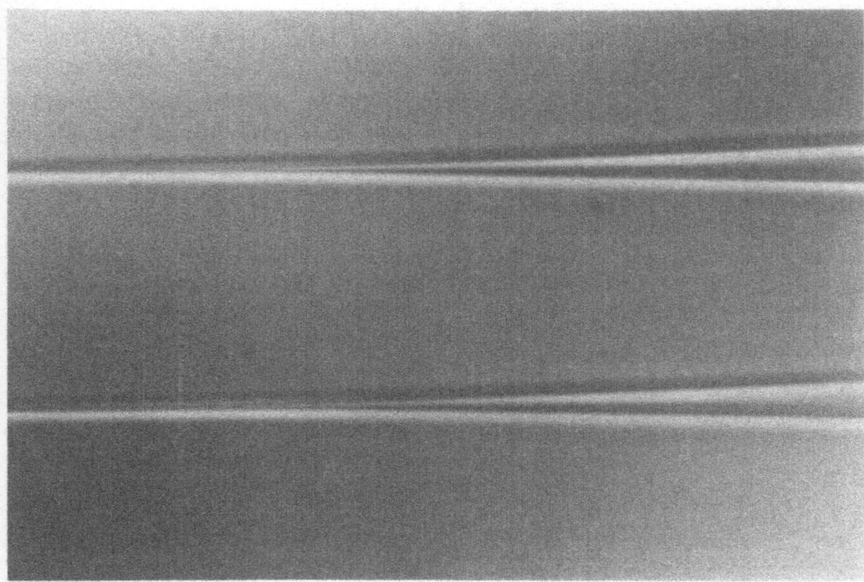

Fig 2.3a Polymer y-branches.

Fig 2.3b Polymer directional couplers.

In order to measure the near field pattern of our polymer waveguide structures, light from a 1,55 µm DFB laser diode was coupled in by a fiber pigtail. The output signal was monitored by a microscope objective connected to an IR camera.

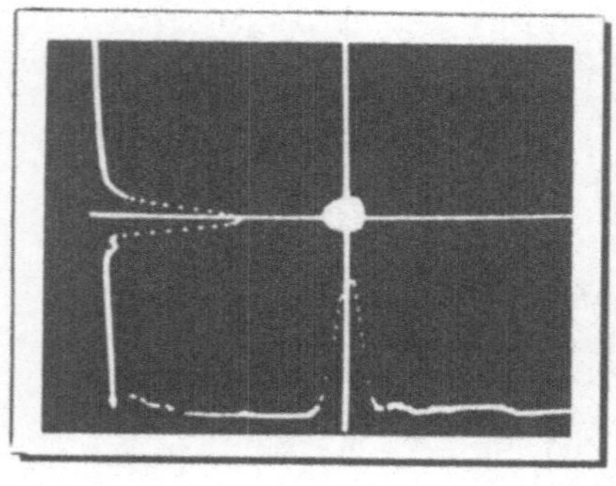

strip waveguide

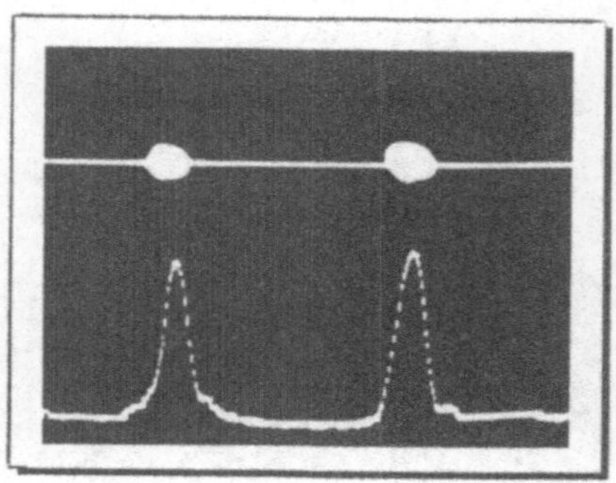

3 dB coupler

Fig 2.4 Near field pattern.

Fig 2.4 shows the result. As one can see from the field distribution, the strip is monomode in both axis (Fig. 2.4a) and the directional coupler is almost a 3 dB one (Fig. 2.4b).

A very important step in the fabrication of integrated optical devices is the fiber-chip coupling. Because of their low refractive index, polymer waveguides have a natural advantage over anorganic materials like $LiNbO_3$ or InP.

Fig. 2.5 shows as an example a y-branch with one input and two output fibers.

The optical endfaces where made by cutting the substrate on its backside and then cleaving it on a knife edge. The polymer film will rip along the cutting edge. This gives good optical endfaces without any additional polishing steps.

The optical fibers are fixed with a low refractive UV glue. With that procedure, an input loss of typically 2 dB/endface can be obtained.

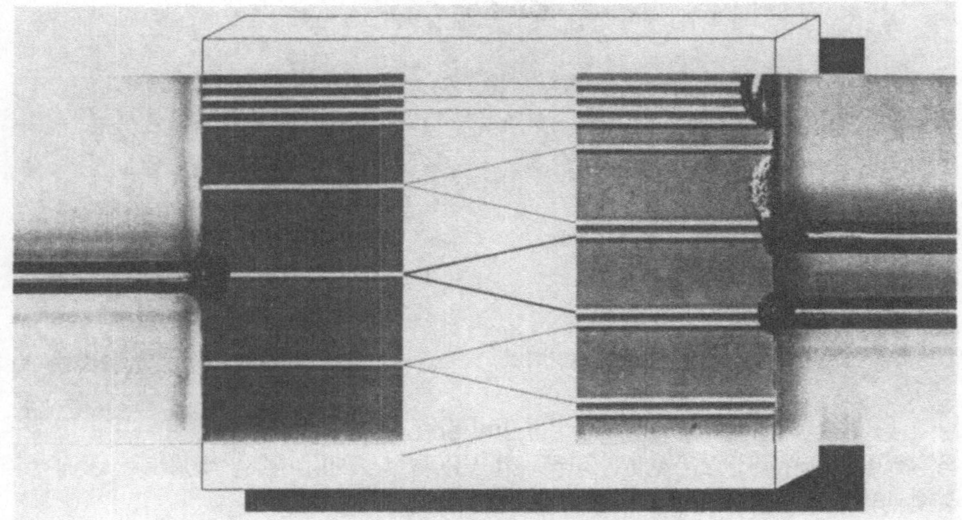

Fig 2.5 Fiber-chip coupling.

Fig. 2.6 shows the packaged device. There are in total 4 strip waveguides, 4 y-branches, 16 directional couplers and 8 s-bends on a 3 cm long chip. One y-branch is connected to the optical fibers.

An excess loss of 7 dB could be obtained, corresponding to a waveguide attenuation of approximately 1dB/cm at 1,55 µm wavelength, with respect to the input, output and splitting losses.

This polymer component is polarization independent and stable in a wide temperature range and is used as a power combiner in a coherent switching experiment.

Fig 2.6 Polymer optical power splitter.

Our next step is to increase the number of planar devices used in the system experiment. We will start with power splitters, directional couplers and passive stars.

The fabrication of thermo-optic tunable devices like polymer (2x2) switches and polymer optical filters is now under investigation.

3 Nonlinear Structures

The material systems for optical nonlinear structures in our laboratory are in general guest-host systems like azo doped PMMA or side-chain polymers like the DHANS molecule which was originally synthesized by D. Dorsch from Merck.

The technology used to form practical devices is in principal the same as reported for passive structures before. But there are some additional processing steps necessary, which can be explained by the multi-layer structure shown in Fig. 3.1, which is typical for electro-optic devices.

First, the electro-optic active layer is sandwiched between the upper and lower optical buffer layer. These buffer layers are made from low refractive index materials. They will isolate the nonlinear waveguide from substrate and electrodes, respectively to avoid additional attenuation.

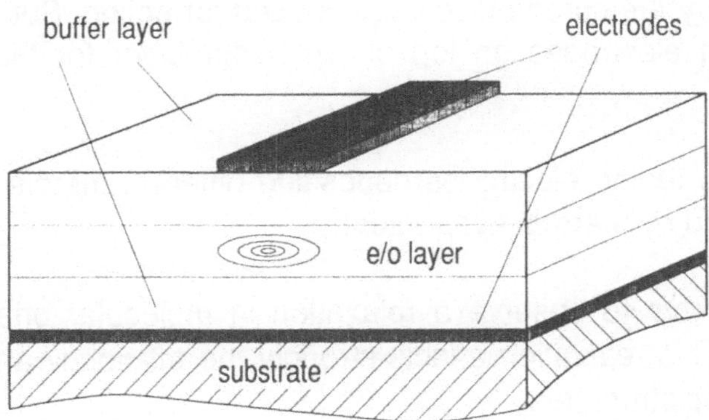

Fig 3.1 Multi-layer structure.

We have used different materials like epoxy, UV glue and SiO_2 as buffer layers and silicon and optical glasses as substrates to form such multi-layer structures.

We prefer the polymeric materials, because of their low dielectrical constants. Besides that they are easy to process by spin coating, which means cost effective fabrication. The thickness of the buffer layers is in the region of 3 to 5 µm.

Another very important processing step is the poling of the nonlinear dipol molecules which are embedded in the electro-optic film.
Because of their natural randomly orientation, these films will show no 2nd order nonlinear behaviour, that means also no electro-optic effect is detectable. By aligning the molecular dipols using an external electric field, one can overcome this.

The film is heated to that temperature, at which the molecular dipols are free to rotate. The application of an external electric field will force the molecules to aligne in the same direction. After this, by maintaining the electric field, the film is cooled to the lower temperature at which the molecular dipols are fixed. Now the field can be switched off.

In our laboratory the corona discharge is used for poling. But one can also use the tuning electrodes on top of the buffer layer for that orientation process.

We form these tuning electrodes depositing different metals like chrome, copper and gold by thermal evaporation.

However we have to observe a relaxation in molecular orientation after several time. This relaxation effects strongly the efficiency and stability of our electro-optic structures.

The most promising approach to overcome this relaxation problem is a chemical cross-linking of the dipolar molecules during poling, a process which was originally investigated by IBM [EIC 89].
This chemical cross-linking can be induced either by temperature or by UV light [MAN 91].

We will try both methods to improve the long term stability of our electro-optic systems.

A very interesting and important question is how to fabricate monomode waveguides in such optical nonlinear structures?

Fig. 3.2 shows a comparison of the multi-layer structures fabricated by the three different methods: photobleaching, photolocking and etching.
We have tried the two methods photobleaching and photolocking at the moment.

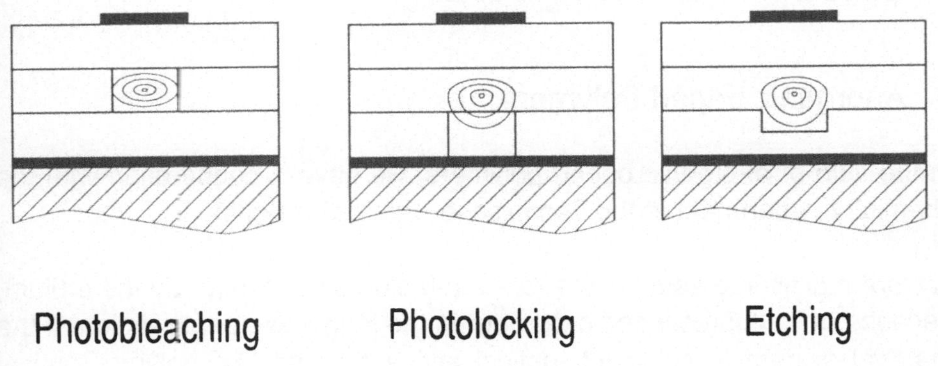

Fig 3.2 Comparison of multi-layer structures.

Photobleaching means that the waveguide is written by UV light as in the case of passive structures, but with one important difference. Here, the non-waveguide region is exposured, *bleached*, and the refractive index is decreased in that area.
With that process, we can obtain only relatively small Δn values and are almost fixed with the waveguide refractive index.
But this Δn value depends strongly on the dipol molecule used and so this is no disadvantage in general [DIE 90].

But more promising concerning flexibility is the fabrication process called *photolocking*.
Here a classical passive waveguide is formed in the lower buffer region by UV exposure and then coated with the higher refractive index electro-optic layer. By choosing different material combinations one can cover a wide range in waveguide geometry and effective index for monomode structures.

In the third process, a channel is etched, either by wet or dry etching, in the lower buffer layer and is then filled up with nonlinear waveguide material forming an inversed rib waveguide.
This process can be used with almost every material system, especially when the nonlinear molecule is insensitive to UV light.

We will start to investigate the etching process also, because we would like to get some own experience in advantages and disadvantages compared to other fabrication methodes.

4 Anorganic doped Polymers

In the field of anorganic doped polymers, we have a cooperation with the chemistry department at the Technische Universität Berlin.

We are mainly interested in the rare earth elements, especially the erbium, because of its fluorescence behaviour in the long wavelength region. If the material system is sufficient, optical active devices like polymer optical amplifiers and polymer lasers can be expected.

From our preliminary study, much higher concentrations of rare earth elements in polymers than in glass can be achieved. That will give much shorter devices, compared to the fiber amplifiers and fiber lasers reported at moment.

Fig. 4.1 shows the measured absorption spectra for an erbium-polymer compound.
Although the absorption peaks are all at the right position, we can not observe a real fluorescence at moment.

This could have several reasons. The most simple one is that the pumping wavelength is not perfectly matched to the absorption peak.
We started with the green light from a frequency doubled Nd:YAG laser. At moment a 1480 nm semiconductor laser is used for pumping, but more promising is the 980 nm wavelength region.

We will try to stimulate the material with that wavelength soon.

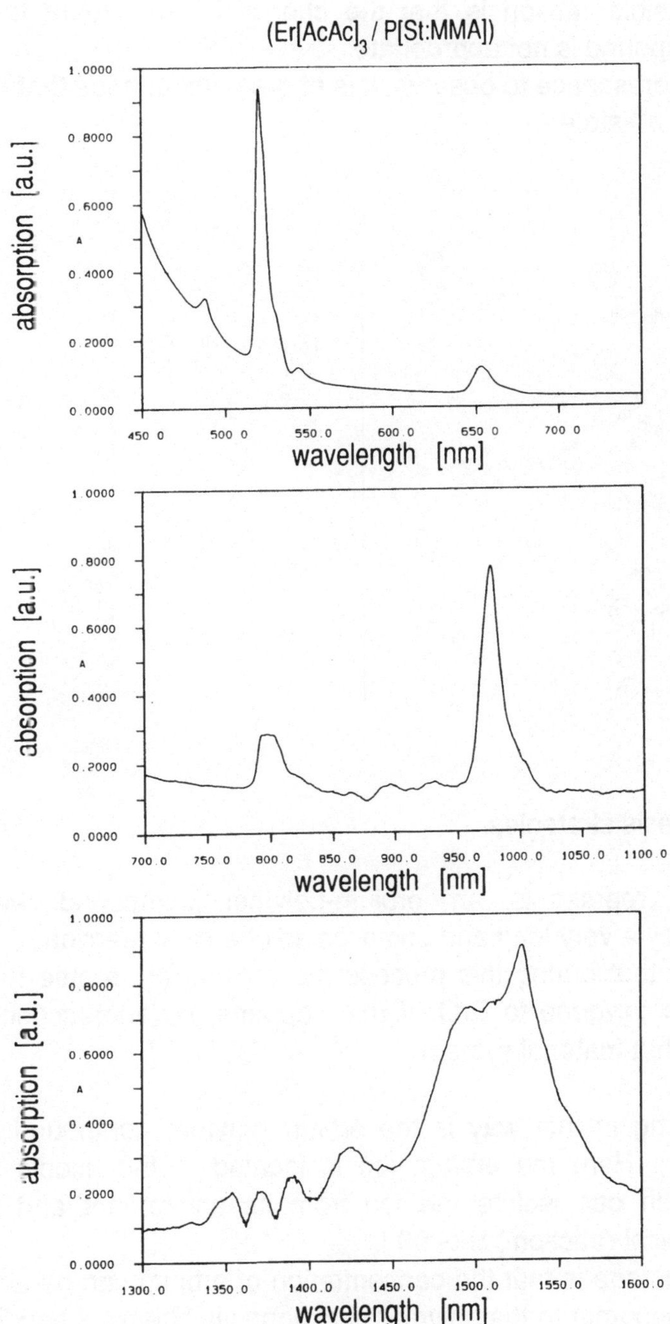

Fig 4.1 Absorption spectra of an erbium-polymer compound.

286

Another possible reason is that the chemical structure of the erbium-polymer compound is not appropriate.
For a real fluorescence to observe, it is of great importance that the erbium is in an ionic 3^+-state.

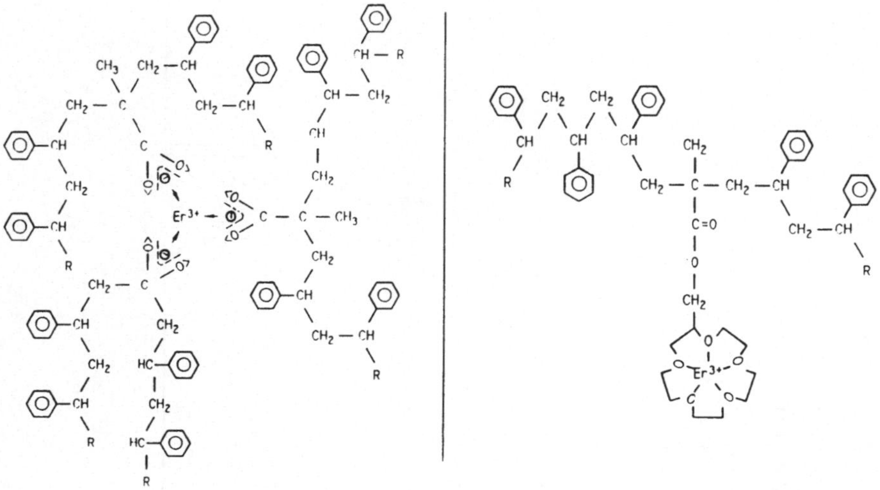

Fig 4.2 Sythesis strategies.

Fig. 4.2 (left) represents an erbium-polymer compound which was synthesized by a very long and complicated chemical reaction.
It is possible that during this process the erbium ion is able to react for example with oxygene to ErO. If this happens, no fluorescence can be observed in that material system.

More promising in that way is the erbium-polymer compound shown in Fig. 4.2 (right). Here the erbium ion is located in the middle of a ring complex which can isolate the ion from its enviroment and protect it against chemical reaction [HIG 90].
Another advantage is that the concentration of erbium can be adjusted by doping this monomer to the polymer backbone via chemical bondings.

The sythesis of this material system is now under investigation.

Nonreciprocal devices like optical isolators representing the second interesting group of optical components which may be fabricated from anorganic doped polymers.

Although no practical results are reported for organics at moment, there is one very promising approach for integrated optical isolators in an anorganic material system [SHI 91].

One should try now to combine the reported anorganic material system with an organic polymer as we do it similar with erbium.

5 Hybrid Integration Concept

With the future development of the optical signal processing technology towards the photonic key technology, the large scale integration of single optical components to functional devices on one single substrate is necessary.

At this time neither semiconductor technology nor polymer technology is able to dispose a totally integrated optical crossconnect element. These opto-electronic integrated circuits (OEIC) will require substrates in dimensions which are not practicable for conventional semiconductor technology today.
Because of the materials high refractive index, there are some additional problems in fiber-chip coupling. Furthermore, the technology itself is quite expensive.

On the other hand, polymer technology lacks off maturity and some of the key elements like the optic-optical frequency converter and the optical parametric amplifier are still not demonstrated even they could have functional advantages.
However the polymer technology is conceptional hybrid because polymer circuits can be fabricated on all kinds of substrate materials. That means that for example the optic-optical frequency converter and the optical amplifier can be firstly realized as a semiconductor element.
Fig. 5.1 shows our hybrid integration concept for an optical crossconnect element.

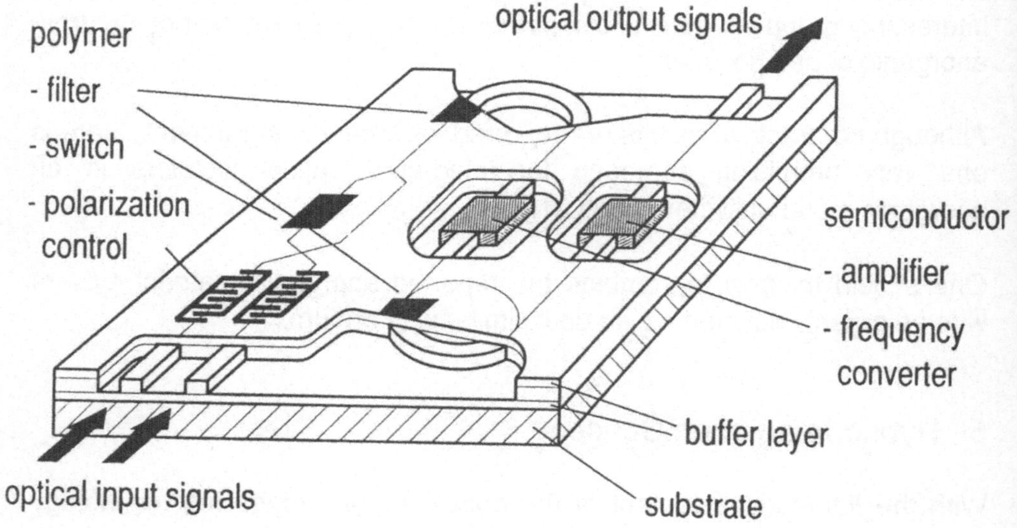

Fig 5.1 Hybrid integration concept.

On a large substrate made of glass, ceramics or polymer, two semiconductor islands are located working either as an optical frequency converter or as an optical amplifier.

The optical network is made of polymer waveguides and polymer components working as well as broadband power splitters and combiners, optical space switches and channel branching filters or as polarization control units.

If the material parameters (nonlinear coefficient, stability, etc.) are sufficient, the semiconductor elements can be exchanged by polymer components, also.

We will start with the polymer semiconductor interface by trying to integrate a semiconductor laser diode with a polymer monomode waveguide.

6 Conclusions

Fig. 6.1 shows a comparison between the conventional monolithic semiconductor integration technology and the proposed hybrid polymer integration technology, containing physical, technological and functional arguments.

argument	semiconductor	polymer
linear	+	+
nonlinear	+	+
active	+	anorg. doped
nonreciprocal	?	anorg. doped
substrate	InP, GaAs	glass, Si, InP, ...
fabrication process	epitaxy	lithography
attenuation (IR)	0,5 dB/cm	0,5 dB/cm
fiber/chip coupling	difficult	easy
costs	high	low
integration	monolithic	hybrid

Fig 6.1 Comparison of integration technologies.

Besides the functional advantages of polymer components like unidirectional frequency conversion and parametric amplification, the excellent physical properties of the material system in combination with a cost effective technology will lead to high efficient but low cost integrated optical polymer circuits (IOPC) in the future.

Acknowledgement

The author would like to thank Roman Weimann for preparing the erbium polymer compounds, Huihai Yao for the excellent cooperation during the last years and the Deutsche Bundespost TELEKOM for funding this work.

References

[DIE 90] Diemeer,M; Suyten,F; Trommel,E; McDonach,A; Copeland,J; Jenneskens,L; Horsthuis,W: Photoinduced channel waveguide formation in nonlinear optical polymers, Elec. Lett., Vol. 26, No. 6, 1990, pp. 379 - 380.

[EIC 89] Eich,M.; Reck,B.; Do,Y.; Wilson,C.; Bjorklund,G.: Novel second-order nonlinear optical polymers via chemical cross-linking-induced vitrification under electric field, J. Appl. Phys. 66 (7), 1989, pp. 3241 - 3247.

[FRA 86] Franke,H; Heuer,W: Photo-induced self-condensation, a technique for fabricating organic lightguide structures, SPIE Vol. 651, 1986, pp. 120 - 125.

[HIG 90] Higashiyama,N; Adachi,G: Effects of zinc (II) on fluorescence properties of divalent europium-poly(methacrylate containing 15-crown-5-structure) complex, Chem. Lett., The Chem. Soc. of Japan, 1990, pp. 2029 - 2032.

[KEI 91] Keil,N; Strebel,B; Yao,H; Krauser,J: Application of optical polymer waveguide devices on future optical communication and signal processing, SPIE 1991, San Diego, Proc. Vol. 1559, pp. 278 - 287.

[MAN 91] Mandal,B; Chen,Y; Lee,J; Kumar,J; Tripathy,S: Cross-linked stable second-order nonlinear optical polymer by photo-chemical reaction, Appl. Phys. Lett. 58, (22), 1991, pp. 2459 - 2460.

[SHI 92] Shintaku,T; Uno,T: Radiation-mode-converted optical waveguide isolator, ECOC'91/IOOC'91, Paris, Regular Papers, pp. 193 - 196.

[STR 89] Strebel,B; Bachus,E; Vathke,J: Switching in coherent multi-carrier systems, Globecom'89, Dallas, Conference Record Vol.1, 1989, pp. 32 - 36.

Applications of Integrated Optical Polymer Components on Coherent OFDM-Systems

B.Strebel

Heinrich-Hertz-Institut für Nachrichtentechnik Berlin GmbH

Einsteinufer 37, D-1000 Berlin 10, FRG

Abstract

Coherent optical carrier frequency techniques allow the superposition of distribution and dialogue services within a future IBN (Integrated Broadband Network). In laboratory experiments some subscriber connections were demonstrated until now using fiber laser technology [BAC 92].

The introduction in a more extended network needs an integration technology which can be applied to many identical circuits. Optical polymers offer an ideal technology for this purpose because this material can be treated with nonlinear, electrooptical, fluorescing and magneto-optical dope. Therefore all components of an optical transparent signal transport network may be realized. In a preliminary stage of development a combination of passive and electrooptic polymer waveguides with semiconductor laser components could be developed.

1. OFDM-Technique

The coherent optical frequency division multiplex (OFDM) technique or optical carrier frequency technique is motivated by making use of the large bandwidth of the optical fiber. Fig. 1 shows the attenuation of a monomode fiber as function of wavelength. In its low loss part 5,000 coherent optical carriers could be allocated in a 10 GHz spacing. The optical carrier frequency technique is similar to the electrical carrier frequency technique but offers much higher channel bandwidth and permits a digital angular modulation. Optical heterodyne receivers are employed.

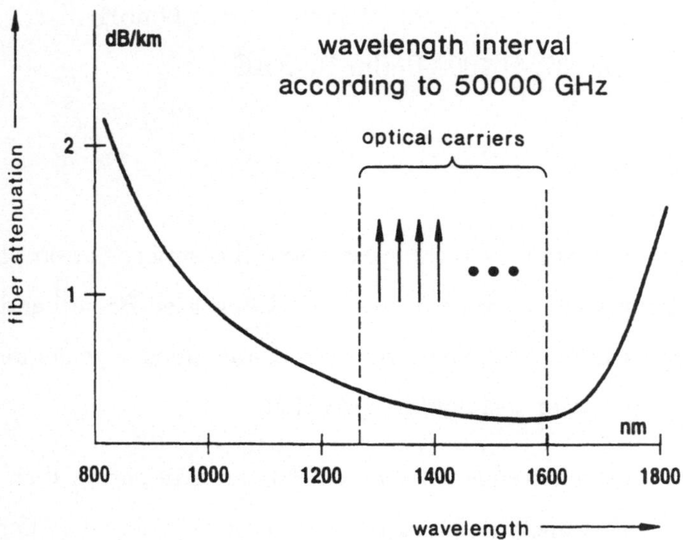

Fig. 1: Attenuation of monomode fiber

The future application fields of this coherent OFDM-technique are long distance transmission on heavy trunks. Further TV-distribution for highest capacity and flexibility as well as switching of broadband modulated optical carriers and carrier bundles are possible scopes. The long time goal of coherent OFDM-technique is a transparent optical integrated broadband network. It permits the superposition of distribution, dialogue, conferencing and remote control services.

2. Transparent optical network

The transparent optical network includes optical carrier frequency channels for the transport of broadband signals. The channels are defined as a limited window on the optical frequency axis. Within a channel the signals are transmitted transparently on optical carriers. Transparency means, that all signals remain on optical carriers between transmitters and receivers. Further the bit rate is free within the given channel bandwidth and the modulation spectrum will remain inchanged. Optical transparency includes also a change of the optical waveguide and optical frequency by means of waveguide switches and optical frequency converters. Fig. 2 shows the switching of information in the space and frequency domain.

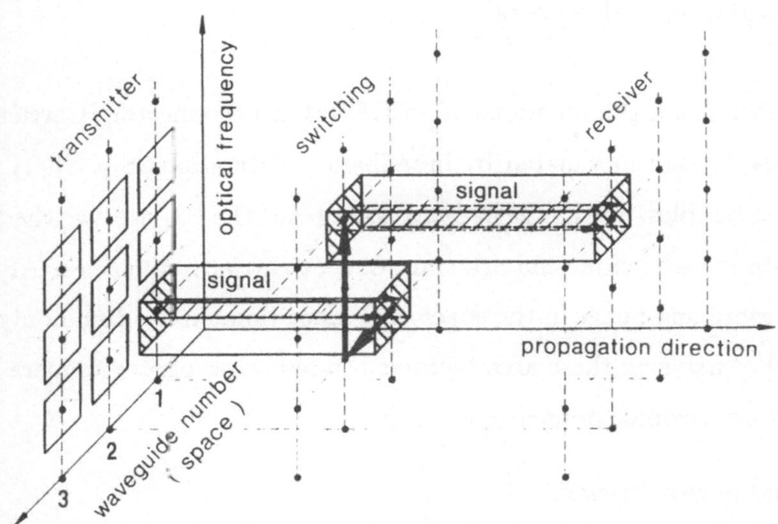

Fig. 2: Switching of information in space/frequency domain

Fig. 3 illustrates the function of a transparent optical network [KEI 91]. It is fed by optical transmitters sending informations on coherent optical carriers into a combination of frequency multiplexers and optical crossconnectors. The signal transport level is controlled by the network management. Monitoring signals describing the spectral content of local distributed network points are delivered to the management.

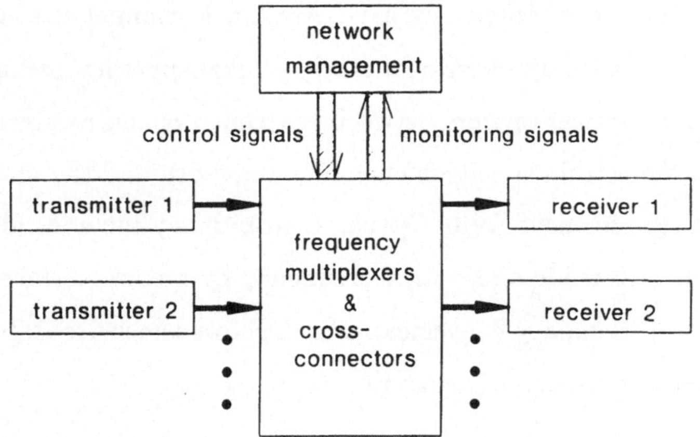

Fig. 3: Transparent optical network

Fig. 4 decribes again the functions of an OFDM-crossconnector. Carrier frequencies f_1, f_2, ... are modulated by broadband informations S_{11}, S_{12} ... and allocated on bundles. In the OFDM-crossconnector the $S_{\mu\nu}$ are switched to arbitrary output waveguides and arbitrary output carrier frequencies. Architectures for switching nodes in the space/frequency domain are commonly known [NIS 90]. Analysing these architectures the following photonic operations of OFDM-crossconnectors are requested:

- Broadband power division,

- channel branching,

- selective and broadband waveguide switching,

- switching of information between different optical carriers,

- channel filtering,

- amplitude levelling,

- polarization adjustment.

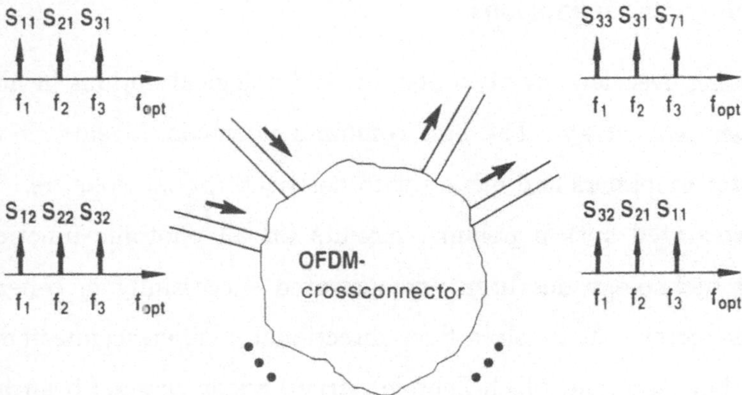

Fig. 4: OFDM-crossconnector

Fig. 5 shows some simple examples of optical waveguide devices for the operations in the space/frequency domain. Additionally monitoring and stabilization circuits are needed. Stabilization of filters and carrier frequencies entail great expenditure.

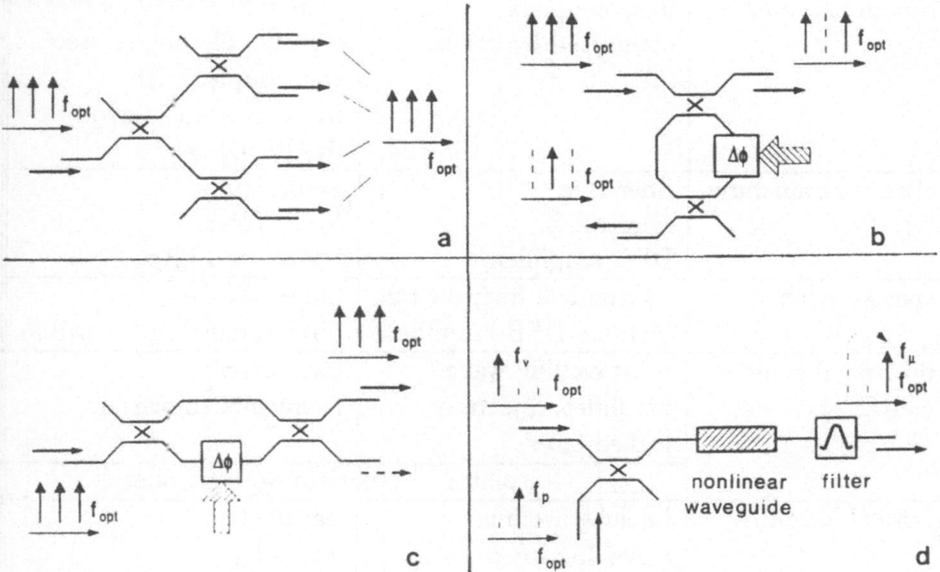

Fig. 5: Waveguide devices for space/frequency domain switching
a power divider, b channel branching filter
c waveguide switch, d optical frequency converter

3. Technological realizations

Fig. 6 gives an overview over two possible technological options in fiber-laser or polymer technology. The first combines fiber components as well as semiconductor amplifiers and mixers with compact optical isolators. The second uses integrated optical polymer circuits for all photonic functions. The fiber-laser technology has been demonstrated successfully in coherent optical crossconnectors. It involves fiber directional couplers connected by fused splices, fiber rings and Mach-Zehnder structures as channel branching filters and fiber switches or bistable DFB laser amplifiers. Carrier frequency conversion is performed by nonlinear travelling wave amplifiers and injection locked lasers. Additionally, compact isolators are inserted, and the overall performance exhibits strong polarization dependence.

devices	fiber/laser technology	opt. polymer technology
broadband splitter	fiber couplers connected by splices	planar Y directional coupler tree star coupler with free-space slab region [KAW 91]
channel branching filter	fiber ring fiber M.-Z. DFB-amplifier	planar ring M.-Z. filter transversal filter
space switch	mechanical fiber switch, bistable DFB-amplifier	planar M.-Z., directional coupler switch
optical frequency converter	n.l.travelling wave amplifier, injection locked laser	parametric frequency converter
	+ isolator + polarization controller	
optical amplifier	linear travelling wave amplifier	parametric amplifier, Er^{3+} doped amplifier
	+ isolator + polarization controller	

Fig. 6: Technological options

The new integrated optical polymer technology should provide an equivalent high functionality. In particular it should offer the integration of many identical switching circuits. The power deviders can be realized as planar Y, directional coupler tree or star with planar area. The most promising field for new devices are the optical channel filters. More sophisticated structures than the planar ring and the Mach-Zehnder configuration should be possible. The optical carrier frequency technique needs filters with steep slope characteristics or filters with tunable bandwidth and center frequency.

The switching structures can be derived from those of $LiNbO_3$-technology, known as directional coupler switch or Mach-Zehnder switch. Introducing nonlinear molecules to the circuits parametric frequency conversion and amplification should be possible. Principally also amplification should be considered by Er^{3+}-doping or isolation by filling magneto-optic material into the waveguides.

If the integrated optical polymer circuit technology could be realized, it would provide a package of system progresses compared to fiber-laser technology:

- The parametric amplification is low noise and nonreciprocal,

- the parametric frequency conversion is expected to be extremly broadband.

- the waveguide switches can operate very fast,

- channel filters of high complexity are possible because of the waveguide phase stability,

- isolators and polarization controllers may be constructed,

- optical detectors are not present in the primary circuit of the signal transport level and

- the integration of many identical switching circuits is the basic condition for a more extended public network.

4. Hybrid technology

Comparing the present state of fiber-laser technology with the integrated optical polymer circuit technology for a realization of switching circuits within the next years, than a hybrid technology should be considered. Fibers are used only for pig-tails with connectors. Power dividers, connecting waveguides, channel filters and switches would be fabricated in polymer technology. Highly functional components like optical frequency converters and linear amplifiers should be included as semiconductor components. Some research should be investigated for the introduction of polymer isolators and polarization control devices. The reason for a hybrid technology ist, that nonlinear optical materials are not sufficiently developed to fabricate parametric devices.

Fig. 7 demonstrates an example of a hybrid OFDM-crossconnect element. At the input all optical carriers are converted to a TM-polarization. The Mach-Zehnder switch belongs to the space switching stage. It selects the input waveguide and may be switched electro-optically or thermally. No high switching speed is required. For the selection of one optical carrier a tunable ring filter is used. The carrier frequency converter is inserted as semiconductor laser, locked to a master wave. The isolator may operate with a transversal magnetic field [SHI 91]. A lot of input and output waveguides are needed for spectral monitoring and filter stabilization. Therefore some output waveguides are connected to a multichannel heterodyne spectrometer. This

spectrometer is also a typical example for the application of hybrid technology
to subsystems of OFDM technique.

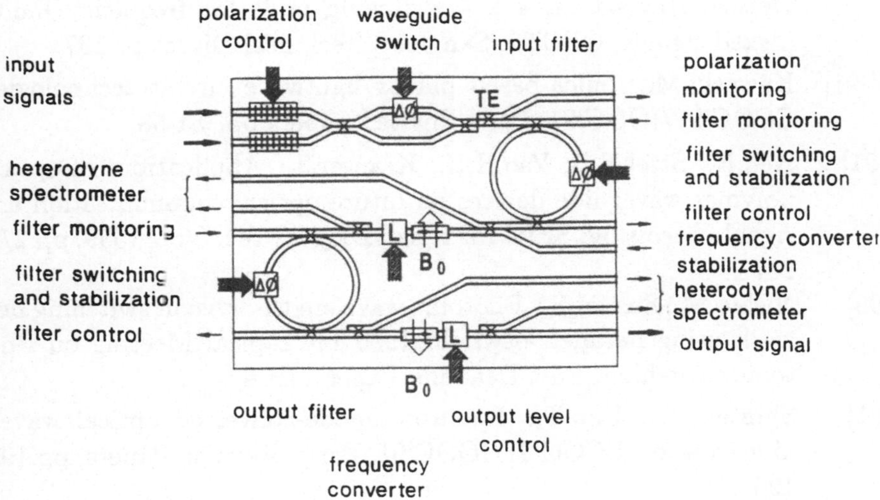

Fig. 7: Hybrid OFDM-crossconnect element

5. Summary

The transparent optical carrier frequency channel is a candidate for a
future integrated broadband communication network. Optical polymer circuits represent an excellent technology for transparent optical transmission
and switching via coherent optical carriers. As a first step, crossconnects,
frequency multiplexers and subsystem circuits could be realized in a hybrid
integration of polymer and semiconductor devices.

6. Acknowledgement

The author would like to thank the Deutsche Bundespost Telekom for
funding this work.

300

7. References

[BAC 92]   Bachus,E.-J.; Braun,R.-P.; Caspar,C.; Foisel,H.-M.; Strebel,B.: Optical transparent 4 x 4 switching node for frequency multiplexed signals, OFC'92, San Jose, Technical Digest p. 137

[KAW 91]   Kawachi,M.: Silica-based planar lightwave circuit technologies, ECOC'91/IOOC'91, Paris, Invited papers pp. 51-58

[KEI 91]   Keil,N.; Strebel,B; Yao,H.H.; Krauser,J.: Applications of optical polymer waveguide devices on future optical communication and signal processing, SPIE 1991, San Diego, Proc. Vol. 1559, pp.278-287

[NIS 90]   Nishio,M.; Suzuki,S.: Photonic wavelength-division switching network using parallel λ-switch, 1990 Int.Topical Meeting on Photonic Switching, Post Deadline Paper 14B-9

[SHI 91]   Shintaku,T.; Uno,T.: Radiation-mode-converted optical waveguide isolator, ECOC'91/IOOC'91, Paris, Regular Papers, pp.193-196

Linear Mode Beating and Nonlinear Mode Coupling in Resonant Optical Waveguides

F. Lederer, L.Leine, M. Mann[*], T.Peschel,
R.Muschall, U.Trutschel, Ch.Wächter,
C.Carigan[+], M.A.Duguay[+], and F.Ouellette[+]

Friedrich-Schiller-Universität Jena
Physikalisch-Astronomische Fakultät
Institut für Festkörpertheorie und Theoretische Optik
Max-Wien-Platz 1
0-6900 Jena
Germany

[*] Universität Osnabrück, FB 4 Physik,
W-4500 Osnabrück,
Germany

[+] Laval University,
Department cf Electrical Engineering
Sainte-Foy, Quebec
Canada G1k 7P4

1 INTRODUCTION

Conventional waveguiding which relies on total-internal reflection (TIR) has been well established for several decades. A great variety of devices as e.g. single waveguides, directional couplers, Y- and X-beam splitters, Mach-Zehnder interferometers have been implemented. Although many applications, based on the conventional scheme, work very successfully there are certain drawbacks inherent in this concept.

In particular, optical information transmission and processing applications need very effective fibre-waveguide interconnects. Due to the mismatch in the cross-section of fibres and waveguides lossy and technologically involved tapers are needed. Furthermore, the evanescent nature of mode coupling in directional couplers restricts the permitted guide separation to a few micrometers. Hence, involved and lossy bend sections have to be incorporated.

The conventional waveguiding scheme restricts the variety of materials used for manufacturing guides because the guiding film needs to have the largest refractive index. Additionally, the evanescent nature of the fields outside the core entails a rather deep penetration into the substrate material. Using a highly absorbing substrate thick buffer layers have to separate the core from the substrate to avoid high losses.

In order to overcome these drawbacks, even in the linear regime, one is encouraged to revisit the very fundamentals of the waveguiding concept.

Furthermore, waveguide configurations are promising candidates for all-optical switching and modulation due to the combination of diffractionless propagation of the fields (large interaction length) and the strong power confinement. Both peculiarities enhance the efficiency of the nonlinear processes that produce the needed induced index change.

The phenomena on which nonlinear guided wave devices are based can be categorized as weakly and strongly nonlinear. When the nonlinear contribution to the refractive index is much smaller than any variations in the linear refractive index that define the waveguide the guided field profile is assumed to be unchanged and perturbation methods can be used to describe the nonlinear mode coupling [STE 90]. The most studied weak nonlinear device to date has been the nonlinear directional coupler where the intensity of the input signal determines the routing of this signal, thus representing an all-optical switch.

In the case of large nonlinearities where the induced refractive index changes are comparable with the index discontinuities the very guiding mechanism is affected. The large induced index changes result in the arising of nonlinear guided waves (NGW) which exhibit power-dependent field profiles and propagation constants and can be unstable in certain domains of the nonlinear dispersion curve [BOA 91]. If one attempts to excite an unstable NGW (e.g. via end fire coupling) it cannot be captured by the guide and the power is expelled e.g.via a spatial soliton leaving no guided power at the waveguide output. Hence, there is a large difference in the transmission for either low or high input powers. This device may be used as an all-optical modulator.

The desirable material property, required for nonlinear guided wave applications, is an off-resonant purely dispersive, and thus ultrafast, Kerr nonlinearity (local dependence of the induced index change on the intensity as $\Delta n_{NL} = n_2 I$ [STE 90] because resonant nonlinearties are accompanied by absorption which can provide serious limitations to all- optical devices. Even the high-power excitation of the material below the band gap leads to serious absorption effects due to two-photon processes. Hence, it turned out that it is very favourable to use wavelength' which are situated below half the band gap of semiconductors. Nonlinear coefficients of $n_2 = 2.5 \cdot 10^{-17}$ m^2/W could be identified for a typical AlGaAs - material [AIT 91, SHE 91]. An ultrafast nonlinear directional coupler was shown to switch almost completely [AIT 91]. Off-resonant nonlinearities in polymers exhibit similar nonlinear coefficients [BUB 91, BAR 91]. Because these dispersive nonlinearities are relatively weak one has to look for waveguide schemes that react very sensitively to field-induced changes of the optical parameters. It turned out that the conventional guiding mechanism basing on total internal reflection as well as the related evanescent coupling between adjacent waveguides do not meet these requirement optimally. The guiding mechanism is relatively stable against induced index changes and the evanescent coupling is weak and restricts the separation of the coupler elements to few micrometers.

This is an additional motivation to investigate novel guiding schemes which rely on resonance effects such as one finds in Fabry-Perot resonators. We intend to show that these schemes meet better both the linear and all-optical requirements. The paper is structured as follows: In section 2 we introduce the concept of resonant waveguiding (ARROW's). Several linear applications as remotely controlled ARROW's and remote ARROW-couplers are studied in the succeeding sections. Eventually, we show that remote switching can be achieved in nonlinear ARROW-couplers.

2 FUNDAMENTALS OF RESONANT WAVEGUIDING
- THE ARROW CONCEPT -

An ARROW (<u>A</u>nti-<u>R</u>esonant <u>R</u>eflecting <u>O</u>ptical <u>W</u>aveguide) consits of a low-index film surrounded by a few (typical two or three) cladding and substrate films of high-index material (see Fig.2.1). This concept was introduced by Duguay et al. [DUG 86]. The guided waves are leaky waves with radiative losses of about 0.4 dB/cm for a properly designed ARROW [BAB 88].

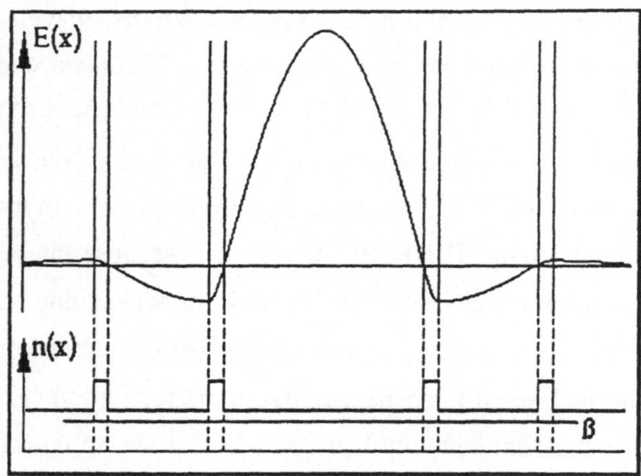

Fig.2.1 Refractive index profile, electric field and effective index β'(real part) of an ARROW

From Fig.2.1 one can recognize that the ARROW has two fundamental peculiarities, namely the real part of the effective index is less than all refractive indices of the configuration and, hence, the field is oscillating within all regions involved. Because the guiding effect relies on Fabry-Perot reflection large differences of the guiding properties between TE- and TM-polarized fields occur. ARROW's may exhibit radiation losses of less than 0.5 dB/cm for TE-polarized fields and more than 10 dB/cm for TM-polarized ones, thus representing perfect polarizers [BAB 88].

For these reasons we restrict ourselves to stationary ($\sim e^{-i\omega t}$) TE-polarized fields that propagate into the z-direction

$$E(x,z) = E(x)\, e^{ik\beta z} + c.c. \tag{2.1}$$

where $E(x)$ is the complex field profile and $\beta = \beta' + i\,\beta''$ the complex effective index. Introducing the real part of the transverse component of the wave vector as

$$k_{xi}' = k\sqrt{\varepsilon_i - \beta'^2}, \tag{2.2}$$

where the subscript "i" labels the film number, there are suprisingly well working, very simple formulas which permit the design of an ARROW. Firstly, the guiding film (core) should be resonant with respect to a fundamental mode (no node) which leads to

$$k_{x1}'d_1 = \pi \qquad \text{or} \qquad \beta'^2 = \varepsilon_1 - \frac{\lambda^2}{4d_1^2} \tag{2.3}$$

Hence, even for large thicknesses d_1 the ARROW remains a single-mode guide. Increasing d_1 β'^2 approaches the dielectric coefficient ε_1. This, in turn, improves the finesse of the Fabry-Perot cavity and leads to smaller radiative losses. Having found β'^2 the thicknesses of the reflecting films (d_2 ,d_3) can be calculated using the antiresonance conditions

$$k_{x2}'d_2 = (2m+1)\pi/2 \tag{2.4}$$

and $\qquad k_{x3}'d_3 = (2n+1)\pi/2 \qquad$ m,n - integers $\tag{2.5}$

If $\varepsilon_1 = \varepsilon_3$ holds, as in Fig.2.1, $d_3 = (l+1)d_1/2$ (l - integer) results consequently. Eqs.(2.3-2.5) permit the design of an ARROW, but provide no information about the radiative losses. The exact dispersion relations can be derived using the transfer matrix approach [LED 91] for the multifilm configuration under investigation as

$$(m_{11} + m_{22}) + i\,k_{x2}m_{12} - i\,m_{21}/k_{x2} = 0 \tag{2.6}$$

where the m_{ij} are the elements of the complete transfer matrix. Note that the resulting effective indices β as well as k_{x2} are complex-valued expressing the leaking nature of the guided waves. The imaginary part β'' is a measure for the radiation losses and is sketched in Fig.2.2 for varying thickness of the reflecting film '2'.

It is worth appreciating that these curves are typical for a high-finesse Fabry-Perot cavity (wide antiresonance "valleys" and narrow resonance peaks) although the refractive index difference is rather small. Furthermore, due to the wide antiresonance domains the fabrication tolerances are not very strict. Having in mind nonlinear applications one should slightly detune one of the mirrors from antiresonance starting with a reflector thickness near a resonance peak. Then, it can be anticipated that a nonlinearily induced change of the refractive index (Kerr effect) drives the system towards the resonance condition (high leakage rate). We have shown in a previous paper [MAN2 91] that the power in a nonlinear ARROW-cut-off modulator may be reduced by at least one order of magnitude with respect to conventional cut-off modulators.

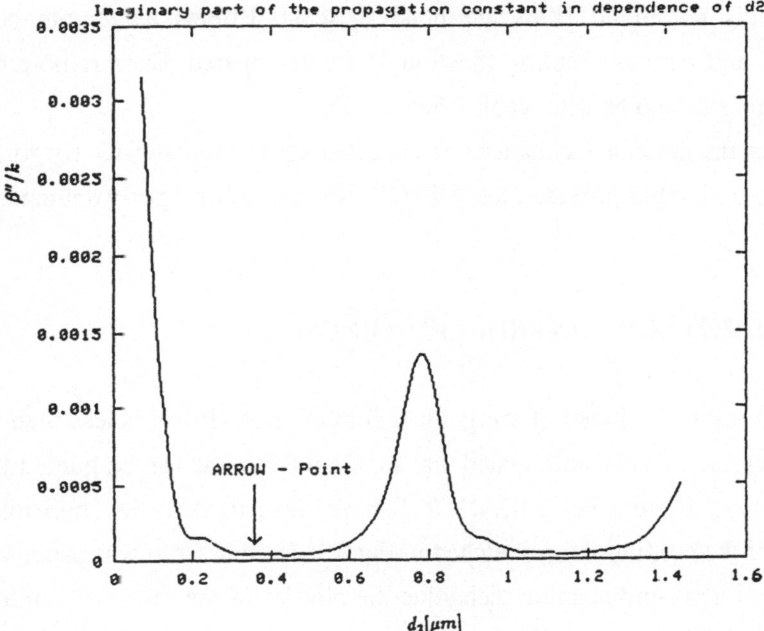

Fig.2.2 Imaginary part of the effective index β'' as function of the thickness of the reflecting film '2'; parameters: $\lambda = 1.06$ μm, $d_1 = 3$ μm, $d_3 = 1.5$ μm, $n_1 = n_3 = 1.53$, $n_2 = 1.69$

In summary, ARROW's exhibit the following unique features and advantages compared to conventional waveguides:

a) Even thick guiding films (up to 10 μm) are single-mode guides (fundamental mode) permitting an optimum matching between the ARROW-field and that of a wide fibre, hence, tapers can be avoided.

b) A greater variety of materials can be used because the core does not need to have the highest refractive index.

c) The optical confinement is very strong, which means, that thin buffer layers separate the guided field from an absorbing substrate.

d) ARROW's are optimum polarizers.

e) The fields within all films are nonevanescent. Effects as remote control (Section 3) and remote coupling (Section 4) are anticipated. Even remote nonlinear switching should be achievable (Section 5).

f) The resonant guiding mechanism is expected to respond sensitively to field-induced index changes provided the ARROW was initially properly detuned.

3 THE REMOTELY CONTROLLED ARROW

Because the refractive index of the guiding film of an ARROW is less than those of the reflectors, it can be anticipated that an ARROW-mode can be guided in air. Very recently, Cantin et al.[CAN 91] have investigated the transmission behaviour of the configuration sketched in Fig.3.1. In contrast to that paper where we have used a ray-propagation technique the physics of the remotely controlled ARROW is explained here in terms of a mode analyses and mode beating.
The air domain (d_c = 32 μm) between two pellicles ($n_1 = n_2 = 1.51$, $d_1 = 1.6$ μm, $d_2 = 1.43$ μm) forms the guiding core. If the thickness of the air-core is properly chosen the radiation losses are less than 0.25 dB/cm without substrate ($n = 1.5$).

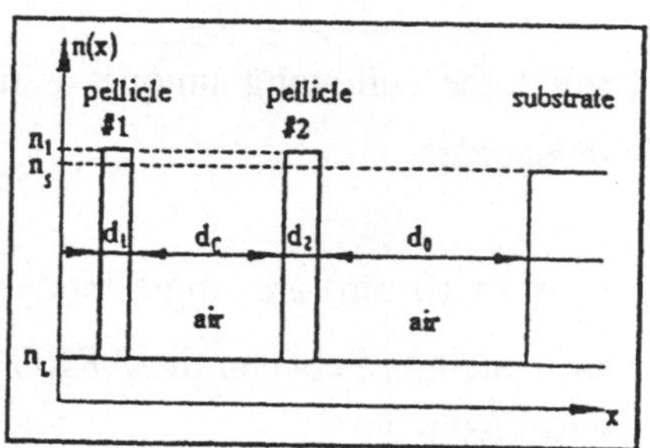

Fig.3.1 Remotely controlled ARROW with air-core

Because the fields are oscillating, even beyond the pellicles, the guiding properties should be controllable by the variation of the distance d_0 between pellicle #2 and the substrate.

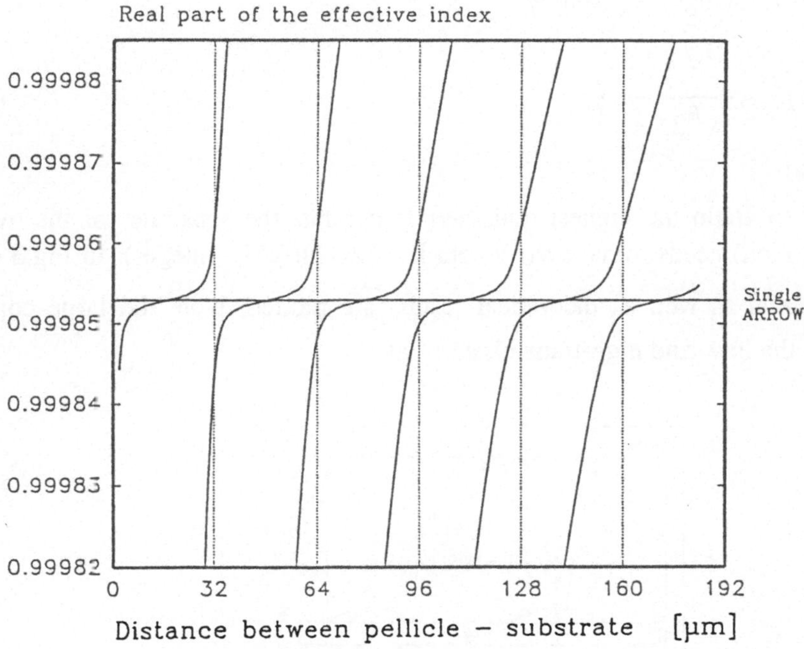

Fig.3.2   Real part of the propagation constant as a function of the distance d_0 for both the air-core ARROW without and with substrate

In Fig.3.2 the real part of the propagation constant is depicted both for the ARROW without and with substrate as a function of the distance d_0. It can clearly be recognized that for spacing distances $d_0 \approx (2n+1)d_C/2$ (n - integer) only one mode exists which crosses the straight single-ARROW curve (see Fig.3.3a). The transmission should approach unity for this distance because the presence of the substrate affects the radiative losses only slightly.

If $d_0 \approx md_C$ (m - integer) two modes with different propagation constants β_1' and β_2' coexist (see Fig.3.3b). These modes are excited with equal powers by a Gaussian input and beat each other in the course of propagation. If they are out of phase by π they cancel within the core and the transmission should drop to zero.

The half-beat length is given by

$$L_c = \frac{\lambda}{2(\beta_2' - \beta_1')}.$$

(3.1)

In order to attain the highest transmission contrast the substrate (or the overall configuration) needs to have the length $L = (2l+1)L_c$ (l - integer). In Fig.3.4 the experimental as well as theoretical results are plotted. Note the large contrast between the low- and high-transmission state.

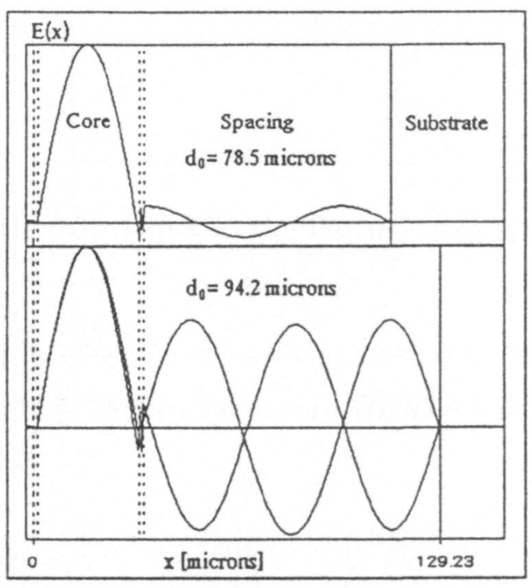

Fig.3.3 Mode profiles of an air-core ARROW with substrate; λ = 1.06 μm
a) d_0 = 78.5 μm, b) d_0 = 94.2 μm

The evolution of the intensity along the ARROW is shown both for the high- (d_0 = 78.5 μm) and the low-transmission state (d_0 = 94.2 μm) in Figs.3.5 and 3.6, respectively. The mode beating can clearly be identified in Fig.3.6.

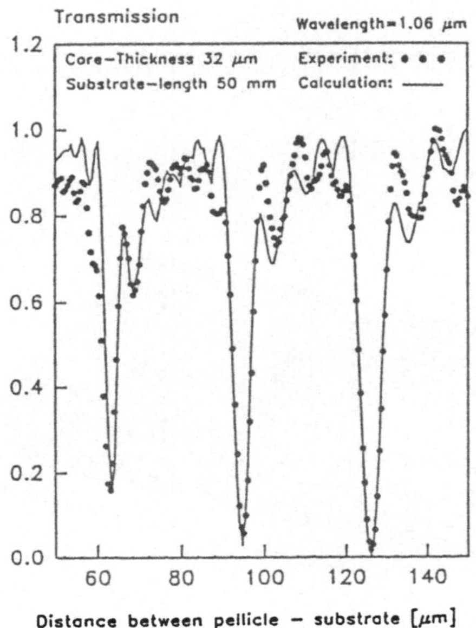

Fig.3.4 Transmission of the air-core ARROW as a function of the distance
pellicle #2 - substrate

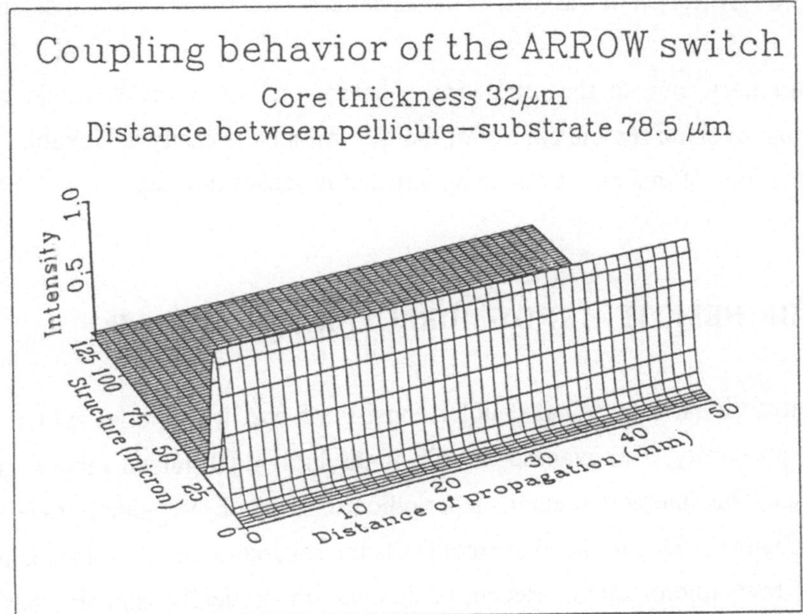

Fig.3.5 Evolution of the field in the air-core ARROW (high-transmission state)

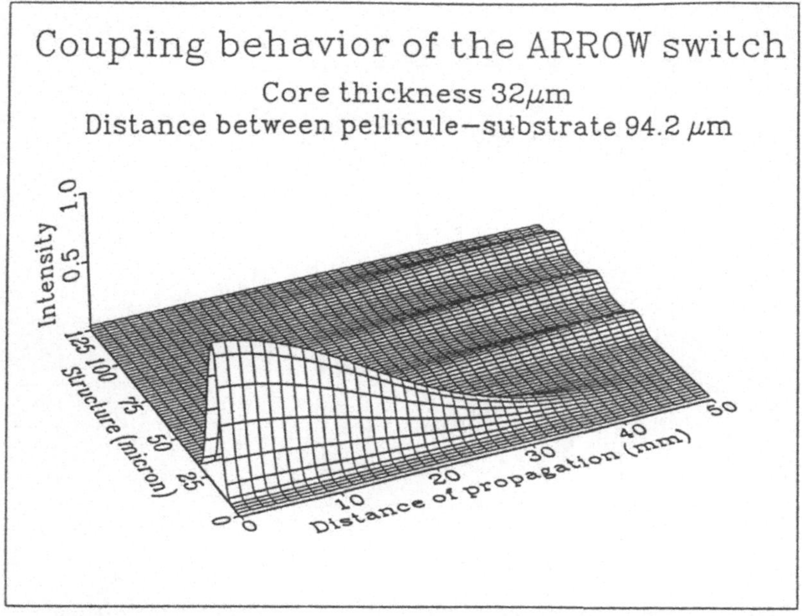

Fig.3.6 Evolution of the field in the air-core ARROW (low-transmission state)

In summary, due to the oscillating behaviour of an ARROW-mode within all regions involved remote control of the transmission is easily achievable. Potential applications of this effect can be anticipated in sensor devices.

4 THE REMOTE ARROW DIRECTIONAL COUPLER

A conventional directional coupler consists of two waveguides which are in a close proximity. The evanescent tails of the individual guided modes may then interact. This interaction entails a periodic exchange of the guided power between both channels. Due to the evanescent tails the separation of the guides is restricted to a few micrometers. Recently, it was theoretically shown that remote (nonevanescent) coupling over a distance of some ten micrometers can be achieved using a spatial soliton coupler [HEA 88]. This soliton coupler needs to exploit extremely large nonlinearities, not available till now in any material, and the experimental verification is very uncertain.

We employ a peculiarity of an ARROW (the oscillating behaviour of the fields in all regions) to propose a novel, linear remote coupler [MAN1 91]. The underlying physics is similar to that of the remotely controlled ARROW (see Section 3). Instead of a substrate we situate a second ARROW, which can capture the field from the first one, at a certain spacing distance d_0. The configuration is plotted in Fig.4.1.

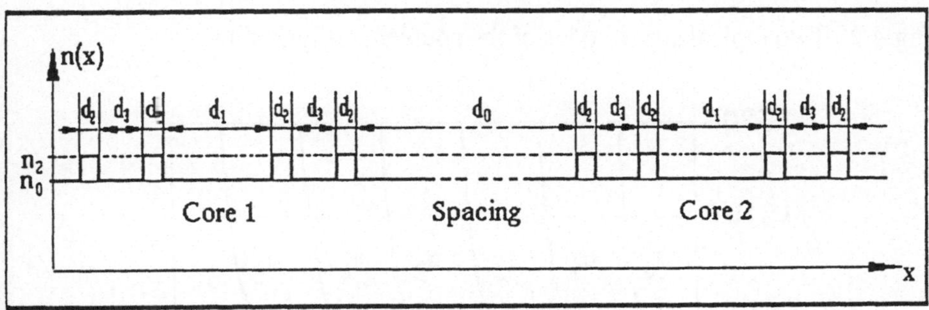

Fig.4.1   Refractive index profile of a remote ARROW- coupler;
parameters: $d_1 = 3$ μm, $d_2 = 0.36$ μm, $d_3 = 1.5$ μm, d_0 varying,
$n_0 = n_1 = n_3 = 1.53$, $n_2 = 1.69$, $\lambda = 1.06$ μm

If one solves the dispersion relation (2.6) for this multifilm configuration one can identify the supermodes (symmetric and antisymmetric ones due to the symmetry of the configuration) the field maxima of which are situated within the core regions (see Fig.4.2).

The dispersion curves are plotted in Fig.4.3 as a function of the spacing distance.

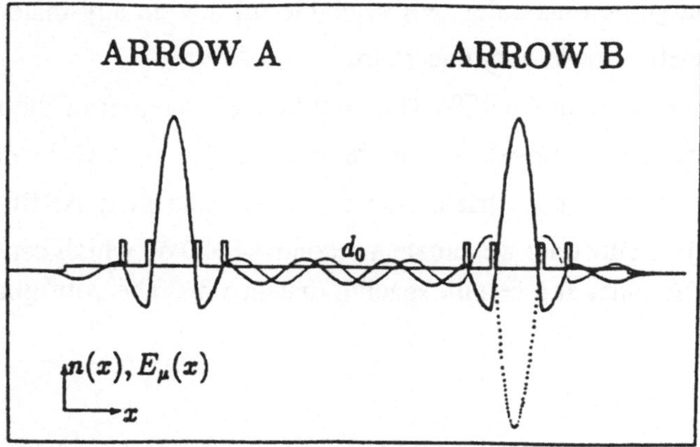

Fig.4.2 Two typical supermodes of the coupler configuration

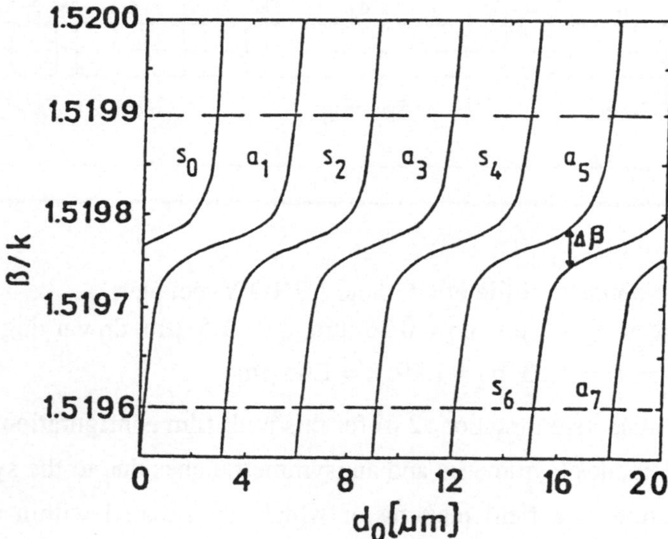

Fig.4.3 Real part of the effective index β' of the supermodes as function of the
spacing distance

The inspection of these curves discloses the peculiarities of an ARROW-coupler:

a) The dispersion curves are periodic with d_0, hence, in contrast to the conventional coupler the coupling behaviour reproduces periodically, lifting, at least theoretically, all limitations for the spacing distance.

b) There are domains where only two modes are involved $(d_0 \approx (2m+1)d_1/2)$ and the respective real parts of the effective indices differ only slightly.
c) Certain domains can be identified $(d_0 \approx nd_1)$ where three modes are involved.

Firstly, we study the case where only two modes take part in the process. As familiar, in the framework of supermodes there is no linear coupling, but a mere beating effect. The field can be written as

$$E(x,z) = E_s(x)\, e^{ik\beta_s z} + E_a(x)\, e^{ik\beta_a z} \tag{4.1}$$

s,a - symmetric, antisymmetric

For ARROW-configurations both the field amplitudes and the effective indices become complex-valued. But, due to the specific configuration chosen here the imaginary parts are very small (for details, see [MAN1 91]) and can be neglected for the further discussion. From Fig.4.2 it is evident that both modes are equally excited by a Gaussian beam incident on ARROW A. Due to the different effective indices the mode beating effect leads to the transfer of the power to ARROW B after the half-beat length

$$L_c = \frac{\lambda}{2|\beta_s{'} - \beta_a{'}|} \tag{4.2}$$

Fig.4.4 shows that the the half-beat length L_c has the same periodic dependence as the effective indices β/k. The power transfer is most complete if one chooses a maximum half-beat length. Because this length is somewhat large ($L_c = 14.7$ mm) it seems that one has to pay a price for the opportunity of remote coupling. Furthermore, the sensitive dependence of the half-beat length on the spacing distance could cause serious limitations for the application of this coupling concept. But, in the course of this section we are going to show how these drawbacks can be overcome.

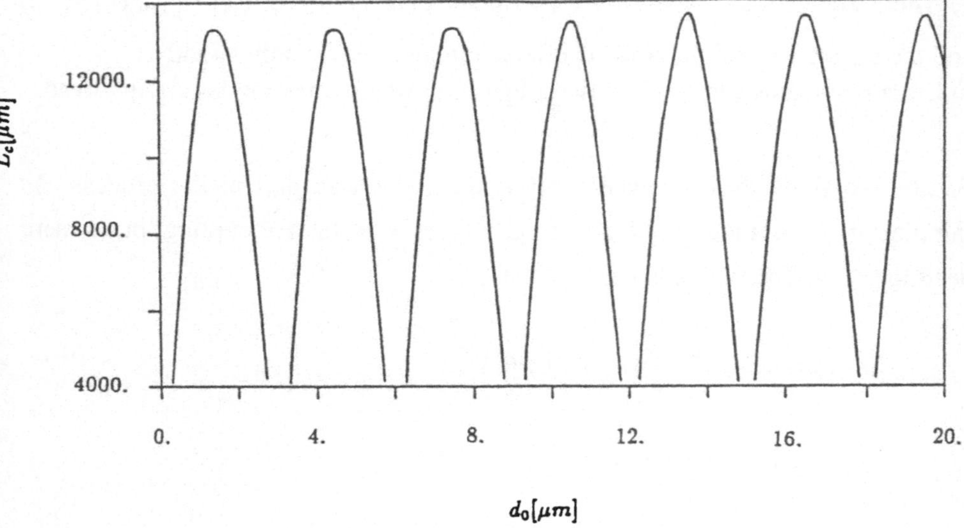

Fig.4.4 Half-beat length of the ARROW-coupler as a function of the spacing distance

In Fig.4.5 a BPM-simulation of the coupling process is displayed when the ARROW A was initially excited by a Gaussian beam, that could represent a typical output of a fibre, and the spacing distance was 16.5 μm.

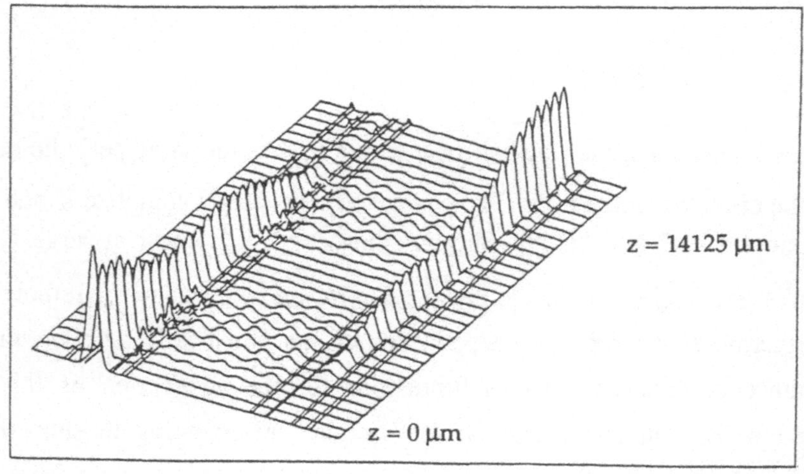

Fig.4.5 BPM-simulation of the power transfer in a remote ARROW-coupler

A further application of remote coupling, which may have some potential in communication schemes, is shown in Fig.4.6. A wide central ARROW 0 is excited by a field-matched fibre output. After some length the signal is equally distributed to the two outermost channels. Cutting the central channel after this length there is no further mode beating because the field profile represents one supermode of the configuration. Then, the power transfer to two fibres can be easily performed. Note, that contrary to conventional configurations no tapers and bends are needed.

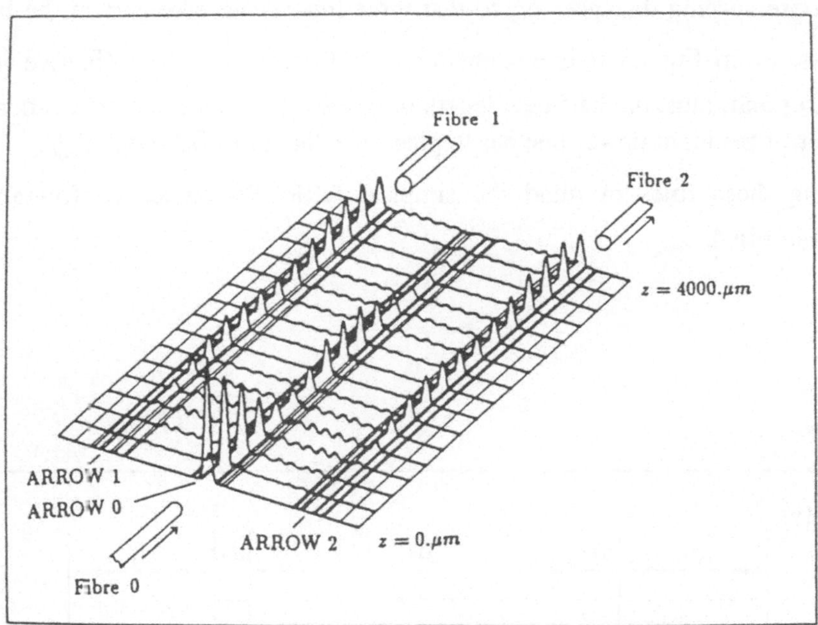

Fig.4.6 Equal power splitting in a remote three-core ARROW-coupler

At this stage, there are some potential objections against the ARROW-coupler configuration:

a) Too many films are involved in the configuration.

b) The tolerances with respect to d_0 are too strict.

c) The coupling length appears to be too large.

318

A more appropriate design may lift all these limitations. In the course of a detailed investigation of this structure it turned out that the following design rules have to be applied:

a) Shield the coupler as well as possible from the "outside world". This can be achieved by reducing the refractive index of both the substrate and the cladding media below the effective indices of the supermodes.

b) Increase the field within the spacing region by a slight detuning of the antiresonant condition for the inner reflectors. This leads automatically to a reduction of the number of films involved.

c) Choose spacing distances d_0 so that three modes can take part in the beating process. From Fig.4.3 it is evident that the difference of the effective indices increases and, thus, the half-beat length decreases. In order to achieve a complete transfer of the input power one has to guarantee that $\beta_{s1}' - \beta_a' \approx \beta_a' - \beta_{s2}'$.

Keeping these rules in mind the simplest ARROW-coupler configuration is shown in Fig.4.7.

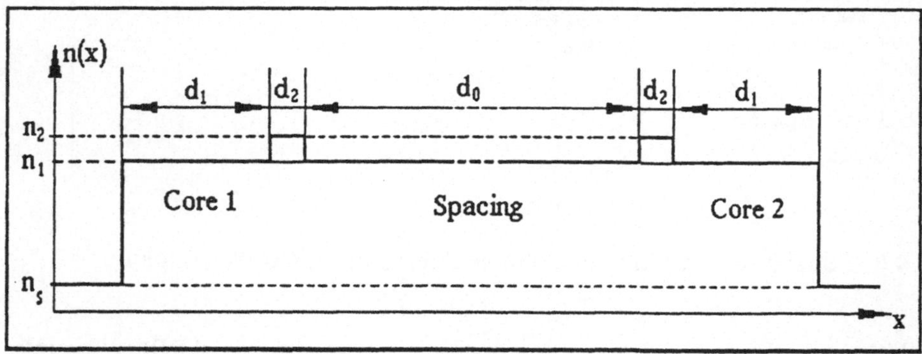

Fig.4.7 An optimized ARROW-coupler configuration with $n_c = n_c = 1.4$, $n_1 = n_0 = 1.53$, $n_2 = 1.57$, $d_1 = 5\ \mu m$, $d_2 = 0.8\ \mu m$, d_0 varying

The three pertinent modes beating each other are plotted in Fig.4.8. Because both symmetric modes are out of phase in the spacing region they cancel each other almost completely at the input and at the half-beat length.

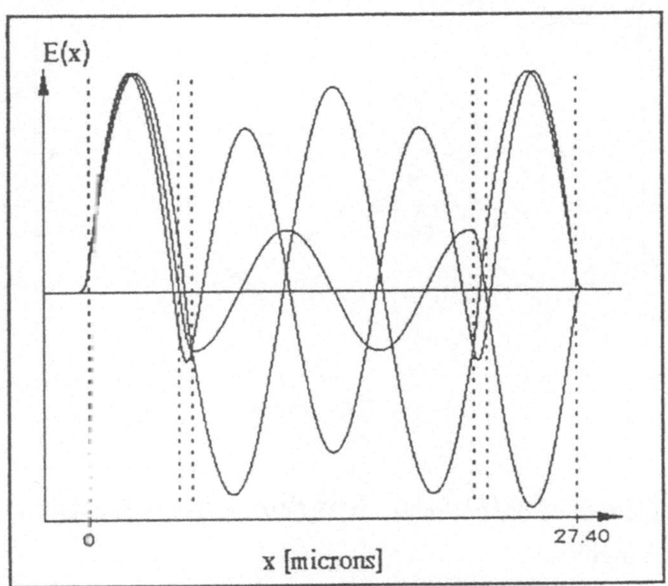

Fig.4.8 Profiles of the beating modes for $d_0 = 15.8$ μm

For a certain spacing domain the dependence of the half-beat length is shown in Fig.4.9 both for the two-mode and the three-mode beating process.

Firstly, it can be recognized that even for the two- mode case the half-beat length can be reduced by an appropriate detuning by a factor of 8. A further reduction occurs if one works in the three-mode domain. In summary, an overall reduction of L_c by a factor of 14 can be achieved with respect to the configuration shown in Fig.4.1. This is accompanied by an appreciable reduction of the number of films. The power transfer for the three-mode case was simulated by a BPM-calculation and is sketched in Fig.4.10 where the left-hand channel was excited by a Gaussian beam. Minor parts of the power are left within the left-hand reflector because an ordinary guided mode was excited there by the tail of the Gaussian input.

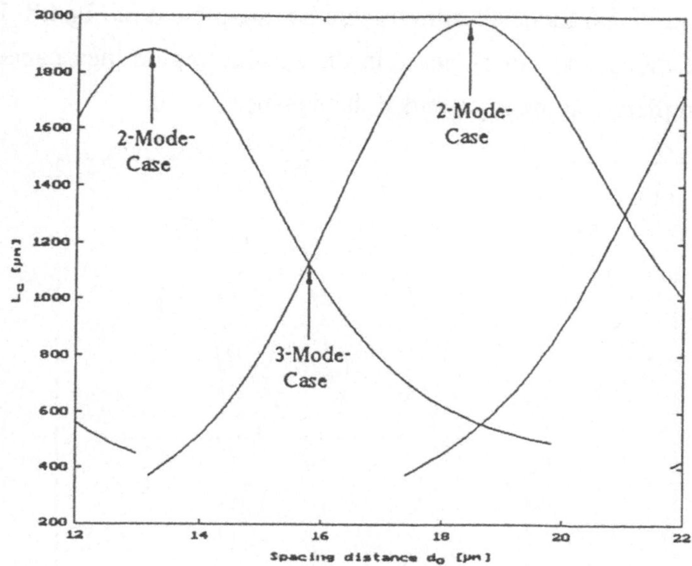

Fig.4.9 Half-beat length of the optimized ARROW - coupler if two or three modes are involved

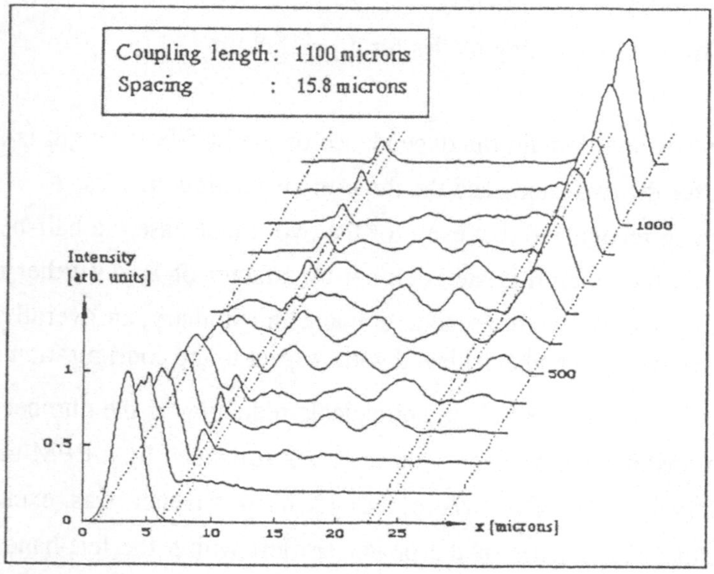

Fig.4.10 Evolution of a Gaussian input along the coupler configuration of Fig.4.7

In summary, we have shown that, even in the linear limit, an almost complete power transfer between remote channels in an ARROW directional coupler is achievable. It turned out that the configuration can be optimized with respect to the coupling length and the number of films needed.

Eventually, we have proposed a configuration that needs only two additional films with respect to a conventional coupler. This device is expected to have some potential in communication networks where fibre-waveguide links are present because the incorporation of lossy tapers and bends can be avoided.

5 THE NONLINEAR REMOTE ARROW SWITCH

The conventional nonlinear directional coupler is well investigated in the nonlinear guided wave literature (see for a summary of references [STE 90]. If, at least, one of the films that form the coupler is endowed with a third-order nonlinear material (e.g. a Kerr medium) the signal input in one channel may be routed between the two output channels varying the input power. If one fixes the coupler length to the half-beat length the signal is output at the cross-channel in the linear regime. Increasing the power above a critical value (critical power) the coupler gets nonlinearily detuned and the output appears at the initially excited channel. Thus, the nonlinear directional coupler represents an all-optical switch. The question arises whether this concept can also be applied for the remote ARROW-coupler introduced in Section 4. Recently, we have presented a rather comprehensive study of this device [TRU 91]. The configuration under investigation is the same as shown in Fig.4.1 except that we assume a homogeneous Kerr nonlinearity as

$$P_{NL}(x,z) = \varepsilon_0 \; \alpha \; |E(x,z)|^2 \; E(x,z), \tag{5.1}$$

$$\alpha = \varepsilon_0 \; c \; n^2 \; n_2$$

and the refractive index of the two outermost media was decreased below the effective indices of the supermodes. The latter modification simplifies the theoretical analysis appreciably. The nonlinear wave equation reads as

$$\Delta E(x,z) + k^2 n^2(x) E(x,z) = - \mu_0 \omega^2 P_{NL}(x,z) \qquad (5.2)$$

As long as the Kerr nonlinearity is off-resonant and thus weak (typical for polymers) it may be considered as a perturbation. Then, the ansatz

$$E(x,z) = a_s(z) E_s(x)\, e^{ik\beta_s z} + a_a(z) E_a(x)\, e^{ik\beta_a z} \qquad (5.3)$$

transforms equ.(5.1) in the framework of a slowly-varying envelope approximation to a system of coupled-mode equations for the slowly-varying amplitudes $a_i(z)$

$$-i a_s' = C_s |a_s|^2 a_s + 2C_{sa}|a_a|^2 a_s + C_{sa} a_s^* a_a^2 e^{2ik\Delta\beta z} \qquad (5.4)$$

$$-i a_a' = C_a |a_a|^2 a_a + 2C_{as}|a_s|^2 a_a + C_{as} a_a^* a_s^2 e^{-2ik\Delta\beta z} \qquad (5.5)$$

In the nonlinear case the linear mode beating is accompanied by the nonlinear mode coupling. The first two terms on the right-hand side of (5.4) and (5.5) describe the effects of self-and cross-phase modulation, respectively, where the last term yields the energy exchange between the modes. One arrives at (5.4) and (5.5) only if the supermodes are orthogonal. Leaky waves fail to fulfill this condition. In order to guarantee the orthogonality the refractive indices of the outermost regions need to be reduced. This reduction has no significant influence on the real part of the effective indices, because the field at these interfaces is almost vanishing. Because the ARROW is highly antiresonant the imaginary part of β is very small ($\beta'' < 10^{-6}$) and the losses can be neglected in order to bring oput the principal action of the nonlinear ARROW-switch.

In (5.4) and (5.5) the following abbreviations where introduced

$$C_{a,s} = \frac{\omega\varepsilon_0\alpha}{4N_{a,s}} \int\limits_{-\infty}^{\infty} E_{a,s}^4 \, dx \, , \qquad\qquad C_{as} = \frac{\omega\varepsilon_0\alpha}{4N_a} \int\limits_{-\infty}^{\infty} E_a^2 E_s^2 \, dx$$

$$C_{sa} = \frac{N_a}{N_s} \, C_{as} \, , \quad \Delta\beta = \beta_s - \beta_a \tag{5.6}$$

where $N_{a,s}$ is the power guided by the antisymmetric and symmetric mode, respectively, provided that $a(z) \equiv 1$. The system (5.4), (5.5) has two constants of integration

$$\Gamma_1 = N_a|a_a|^2 + N_s|a_s|^2 \tag{5.7}$$

which is the power conservation and

$$\Gamma_2 = \frac{1}{2}(|a_a|^4 + |a_s|^4)(C_{as}C_a - C_{sa}C_s) - |a_a|^2|a_s|^2 \, x$$

$$x\left\{(C_{as}C_s - C_{sa}C_a) - (C_{as}^2 - C_{sa}^2)\left(2 + \frac{1}{2}\cos\psi\right)\right\}$$

$$- k\Delta\beta(C_{as} + C_{sa})(|a_a|^2 + |a_s|^2) \tag{5.8}$$

with $\quad \psi = k\Delta\beta z + \phi_s(z) - \phi_a(z) = k\Delta\beta z + \Delta\phi(z)$

where $a_{a,s} = |a_{a,s}|e^{i\phi_{a,s}}$ was used. Exploiting the constants of integration the system (5.4) and (5.5) can be solved analytically in terms of Jacobian elliptic functions (see for details [SIL 87]). Furthermore, they can be used to calculate the critical power needed for all-optical switching [TRU 91]

$$P_{cr} = \frac{(N_a + N_s)2k\Delta\beta(C_{as} - C_{sa})C_{sa}}{N} \tag{5.9}$$

324

with

$$N = (C_{as} + C_{sa})(C_{as}C_a - C_{sa}C_s) + 2C_{sa}^2 \{ (C_{as} - C_{sa})(\cos^2\Delta\phi - \sin^2\Delta\phi + 2) - 3(C_s - C_a) \}$$

This critical power is plotted in Fig.5.1 together with the corresponding half-beat length as a function of the spacing distance d_0. It can clearly be recognized that there is a trade-off between minimum critical power and minimum coupling length. In Fig.5.2 the switching behaviour is illustrated. The propagation of the field input in channel "A" is sketched well below and above the critical power.

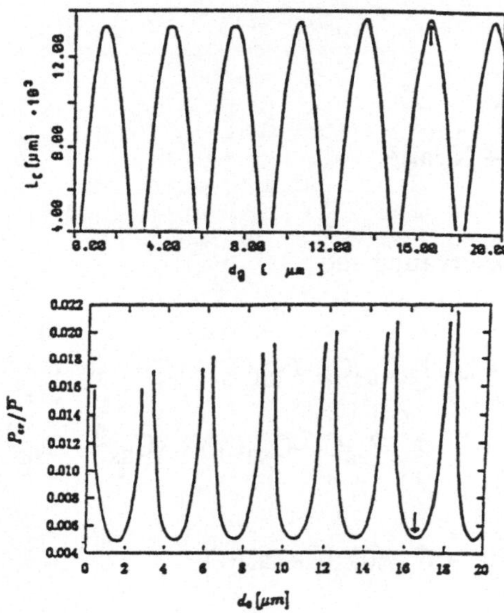

Fig.5.1 Critical power and half-beat length as functions of the spacing distance d_0,

parameters: $\alpha = 6.25 \times 10^{-17}$ m^2/V^2, $\bar{P} = (2\omega\mu_0\alpha)^{-1} = 3.58 \times 10^6$ W/m

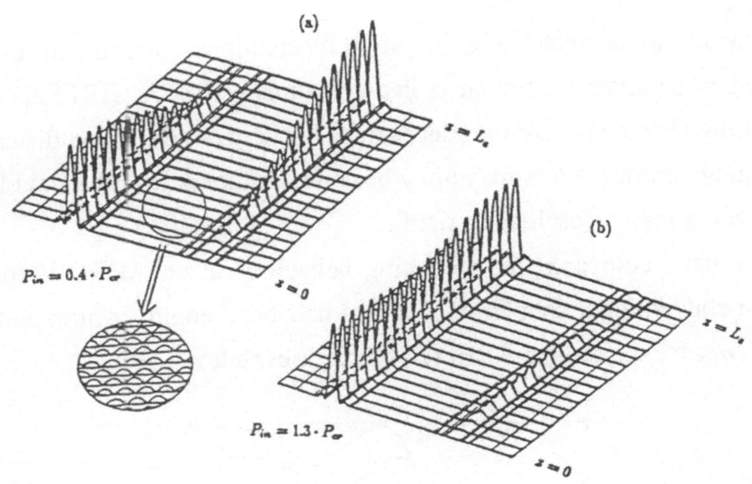

Fig.5.2 Evolution of the field below and above the critical power

The straight-through transmission is shown in Fig.5.3 as a function of the input

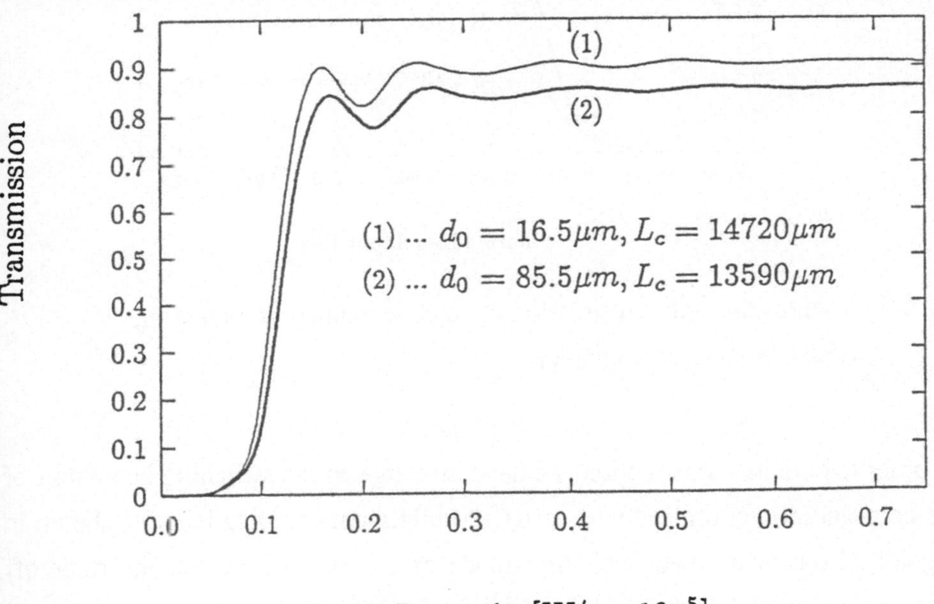

Fig.5.3 Straight-through transmission at the half-beat length as a function of the
normalized input power $P_{in} = (P_a(0) + P_b(0))/\bar{P}$ for two different d_0

326

Evaluating Fig.5.3 it turns out that

a) remote switching is achievable in ARROW-nonlinear directional coupler where the response curves are similar as in conventional NLDC's [JEN 82].

b) The switching characteristics reproduces even for very large spacing distances.

c) The switching contrast is less than unity because a minor part of the field is left within the spacing region (oscillating field).

In Fig.5.4 we have compared the switching behaviour of an ARROW-coupler with that of a conventional coupler of identical half-beat length. It turns out that the critical power for the remote ARROW-switch is even less.

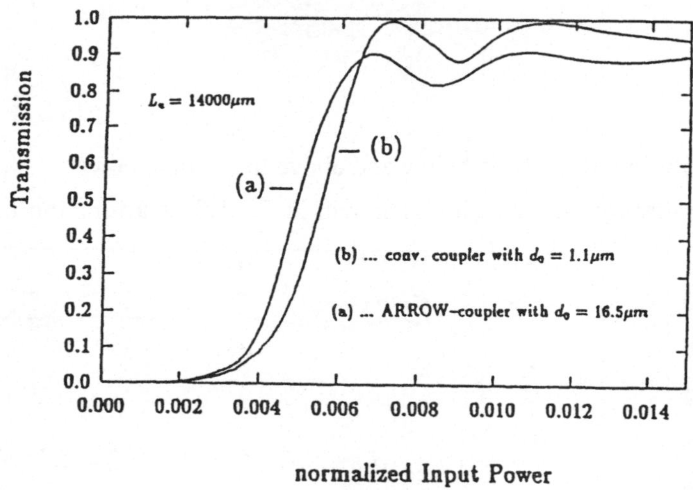

Fig.5.4   Straight-through transmission of an conventional as well as an
ARROW-nonlinear coupler

In order to complete this section, we have investigated the switching behaviour of the configuration optimized with respect to minimum coupling length (shown in Fig.4.7). Both in the two- and three-mode case it turned out that the trade-off between coupling length and critical power has been maintained. As a general rule, the quantity $L_c P_c$ seems to be a fundamental constant in all nonlinear directional couplers. The underlying physics is that the coupling length decreases with increasing mode mismatch $\Delta\beta$ between the supermodes, whereas, on the contrary, the nonlinear phase shift needed to overcome this mismatch has consequentely to increase, thus requiring more power.

In summary, in the framework of a coupled-supermode approach we have shown that the all-optical routing of an input signal is achievable over large spacing distances using a remote ARROW-coupler, but that there is always a trade-off between coupling length and critical power. There is no opportunity to reduce considerably the switching power compared to a conventional nonlinear directional coupler.

6 THE NONLINEAR ARROW-CUT-OFF MODULATOR

One of the motivations to deal with ARROW's was the anticipation that the resonant guiding mechanism can easily be detuned if a nonlinearity comes into the play. Because a weak, off-resonant Kerr nonlinearity is most desirable for all-optical guided wave applications one needs "smart" waveguides which react sensitively on nonlinear changes of the refractive index. The idea is that a slight detuning of the resonant condition changes the very guiding mechanism. We have shown previously [MAN2 91] that an ARROW can be driven from a well-guiding state to an extremly leaking one in dependence on the input power. We have investigated the guiding properties of the configuration depicted in Fig.6.1.

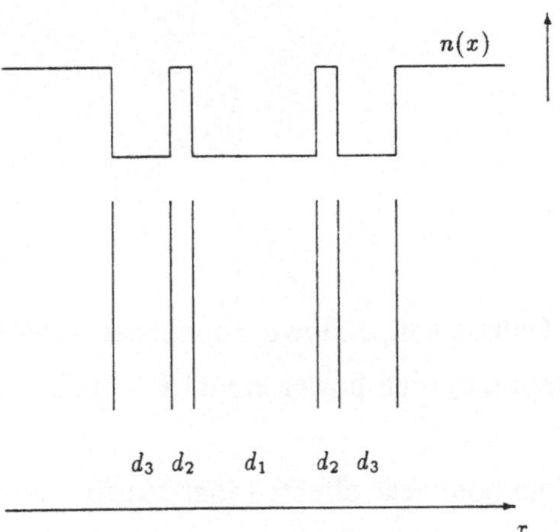

Fig.6.1 ARROW waveguide under investigation, parameters as in Fig.4.1

In a numerical experiment where we have launched a Gaussian input to the core we could show that the input power needed for cut-off modulation is at least one order of magnitude less than for a conventional nonlinear waveguide. A typical result is shown in Fig.6.2.

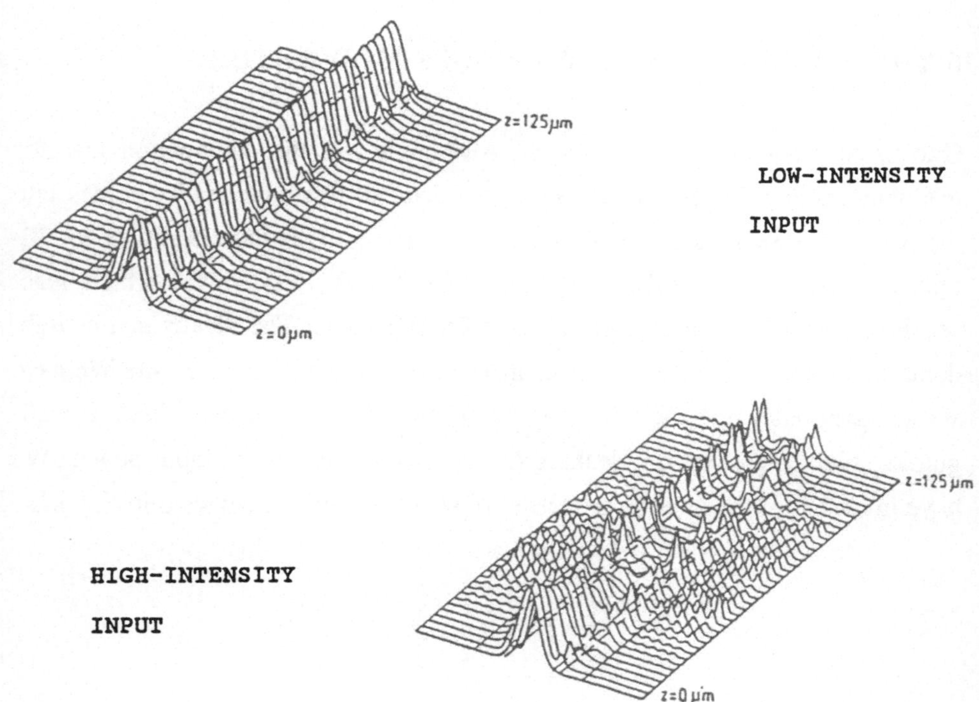

Fig.6.2　Evolution of a Gaussian input down a nonlinear ARROW
a) low-power input, b) high-power input ($P = 16.25 \times 10^6$ W/m)

The investigation of large nonlinear effects (nonlinearily induced change of the very guiding mechanism) in ARROW's is far from being complete. Until now, there is no theoretical proof what film (high-index reflector, low-index reflector, core) should be endowed with a nonlinearity in order to get the most sensitive response. Obviously, the BPM calculations have to be completed by the solution of the exact dispersion relation for nonlinear guided waves in ARROW's. In doing this serious complications might emerge. Due to the leaky nature of the waves the power-dependent effective indices are local ones because the radiation losses in the course of propagation tend to reduce the guided power. This in turn evokes a change in the effective index. Serious efforts have to be focused onto this subject in order to identify the behaviour of nonlinear waves in resonant waveguides.

CONCLUSIONS

We have shown that ARROW's represent a potential alternative to conventional waveguides. They exhibit some unique features that can be exploited in linear as well as in nonlinear schemes.

In the linear regime both remote control of the transmission of an ARROW and remote routing of an input signal in an ARROW-coupler can be achieved. Potential applications in optical sensing and in communication networks may be anticipated.

Furthermore, it turns out that the resonant guiding mechanism responds sensitively on changes of the refractive index caused by weak, off-resonant nonlinearities. The action of a remote all-optical switch as well as a cut-off modulator were investigated in detail.

The challenges consist in a better understanding of the nonlinear effects in ARROW's and in further experimental studies.

An extension of the ARROW-concept towards rib-ARROW's might be of great practical importance and is now under way in our group.

REFERENCES

[AIT 91] Aitchison,J.S.,Kean,A.H.,Ironside,C.N., Villeneuve,A., Stegeman,G.I.:
Nonlinear directional coupler in $Al_{0.18}Ga_{0.82}As$ near half the band
gap. Proc.Top.Meeting on Nonlinear Guided Wave Phenomena,
Cambridge 1991,paper Pd-2

[BAB 88] Baba,T., Kokubun,Y.,Sakaki,T.,Iga,K.: Loss reduction of an AR-
ROW waveguide in shorter wavelength and its stack configuration.
IEEE J.Lightw.Techn.LT-6(1988)1440-45

[BAR 91] Bartuch,U.,Bräuer,A.,Dannberg,P., Hörhold,H.H.,Raabe,D.:
Measurement of high, nonresonant third-order nonlinearity in MP-PPV
waveguides. Int.J.Opoelectronics 5(1991)3

[BOA 91] Boardman,A.D.,Egan,P.,Lederer,F.,Langbein,U.Mihalache,D.:
Third-order nonlinear electromagnetic TE and TM guided waves.
in:Nonlinear surface electromagnetic phenomena,chapter 2
eds.:H.E.Ponath and G.I.Stegeman,Amsterdam 1991, North-Holland

[BUB 91] Bubeck,C.,Kaltbeitzel,A.,Grund,A.,LeClerc,M.: Resonant degenerate
four wave mixing and scaling laws for saturable absorption in thin
films of conjugated polymers and Rhodamine 6G.
Chem.Phys.154(1991)343-350

[CAN91] Cantin,M.,Carignan,C., Côté,R., Duguay,M.A., Larose,R., LeBel,P.,
Ouellette,F.,: Remotely switched hollow-core antiresonant reflecting
optical waveguide. Opt.Lett.16(1991)1738-1740

[DUG 86] Duguay,M.A.,Kokubun,Y.,Koch,T.L.: Antiresonant reflecting optical
waveguides in SiO_2-Si multilayer structures.
Appl.Phys.Lett.49(1986)13-15

[HEA 88] Heatley,D.R.,Wright,E.M.,Stegeman,G.I.: Soliton coupler.
Appl.Phys.Lett.53(1988)172-174

[JEN 82] Jensen,S.M.: The nonlinear coherent coupler.
IEEE J.Quantum Electron. QE-18(1982)1580-83

[LED 91] Lederer,F.,Trutschel,U.,Wächter,Ch.: Prismless excitation of guided
waves. J.Opt.Soc.Am.A8(1991)1536-1540

[MA1 91] Mann,M.,Trutschel,U.,Wächter,Ch., Leine,L., Lederer,F.:
Directional coupler based on an antiresonant optical waveguide.
Opt.Lett.16(1991)805-807

[MA2 91] Mann,M., Trutschel,U., Lederer,F., Wächter,Ch., Leine,L.:
Nonlinear leaky waveguide modulator.
J.Opt.Soc.Am.B8 (1991)1612-1617

[SIL 90] Silberberg,Y.,Stegeman,G.I.: Nonlinear coupling of waveguide modes.
Appl.Phys.Lett.50(1987)801-803

[STE 90] Stegeman,G.I.,Wright,E.M.: All-optical waveguide switching.
Opt.and Quantum Electron.22(1990)95-122

[SHE 91] Sheik-Bahae,M.,Hutchings,D.C.,Hagan,D.J.,VanStryland,E.W.:
Dispersion of bound electronic nonlinear refraction in solids.
IEEE J.Quantum Electron.QE-27(1991)1296-1309

[TRU91] Trutschel,U., Mann,M., Lederer,F., Wächter,Ch., Boardman,A.:
Nonlinear switching in coupled antiresonant optical waveguides.
Appl.Phys.Lett.59(1991)1940-1942

Evolution of Guided Wave Fields in Rib ARROW-Structures

L.Leine [*], M.Mann [**], K.Singh [+], Th.Peschel [*],
U.Trutschel [++], C.Wächter [*],
A.D.Boardman [+], and F.Lederer[*]

[*] FSU Jena, Phys-Astr.Fak., Inst.f.Festkörpertheorie u.Theor.Optik,
Max -Wien-Platz1, O-6900 Jena
[**] Univ. Osnabrück, FB 4 Physik, W-4500 Osnabrück
[+] Univ.of Salford, Department of Pure and Appl.Physics,
Salford M5 4WT, UK
[++] Tufts University, Electro-Optics Tecnology Center,
Medford Masachusetts 02155, USA

1 Introduction

Novel waveguiding schemes basing on resonance effects like the Bragg-reflection waveguide and the Anti-Resonant-Reflecting Optical Waveguide (ARROW) have attracted an increasing interest within the last few years. This is due mainly to the fact, that some drawbacks are inherent in the conventional waveguiding concept. This concerns for instance the diameter mismatch of fiber-waveguide interconnects, the restriction to evanescent coupling, and the requirements for field-induced nonlinear refractive index changes to evoke nonlinear device characteristics. The first items require considerable technological efforts, whereas the latter one restricts the class of materials that can be used remarkably.

ARROW's have inherent peculiarities to be promising candidates to realise effective fiber-waveguide interconnects, remote couplers and sensitivity with respect to field-induced refractive index changes, see [LED 92] and the references therein.

In this paper the physics of a slab-ARROW and the fundamentals of the propagation of optical fields in guiding structures are concisely reviewed in section 2. Because potential application will eventually require rib structures our main concern is to show that the ARROW concept can be extended to these guiding structures as well (section 3). Different models to calculate the propagation constants of rib-ARROW modes are discussed and results of the numerical modelling of guided field propagation in the rib-ARROW are presented, showing the same basic features as the resonant guiding in a slab structure. So, the rib-ARROW design presented here will provide all necessary information to realise such a configuration experimentally.

In what follows stationary optical fields are propagating into z-direction, i.e. $E(x,z,t) \sim \exp(i(k_0 \beta z - \omega t))$, where $\beta = \beta' + i\beta''$ is the complex valued propagation constant of the modal fields, k_0 is the free space wavenumber and a wavelength of $\lambda = 1.06 \mu m$ was used for all calculations. In section 2 the investigations are restricted to the TE-polarisation, whereas in section 3 the hybrid character of the fields in the two-dimensional refractive index profile will be discussed in detail.

2 The ARROW-configuration and adequate field propagation method

The ARROW guiding scheme is based on Fabry-Perot reflections instead of total internal reflection in the case of conventional waveguiding. This implies that the central film (core) may exhibit the minimum refractive index. So, ARROW-modes are leaking and have complex-valued propagation constants with real parts less than the refractive index of the central layer. The basis of the ARROW-design is the use of simple Fabry-Perot resonance conditions in order to get an estimate of appropriate film thicknesses with the subsequent calculation of the propagation constants using the transfer matrix approach [WÄC 87], see [LED 92] for details.

If β' is close to the core refractive index the imaginary part β'' which counts for the radiation loss of the basically leaky ARROW-mode is much less than unity and the roughly designed β' will be changed only slightly. Minimum losses can be realised by either increasing the thickness of the central layer as well as by increasing the number of reflector cells. So, the minimum number of layers forming an ARROW-guide is equal to five, see Fig. 2.1 with $n_0 = n_1$.

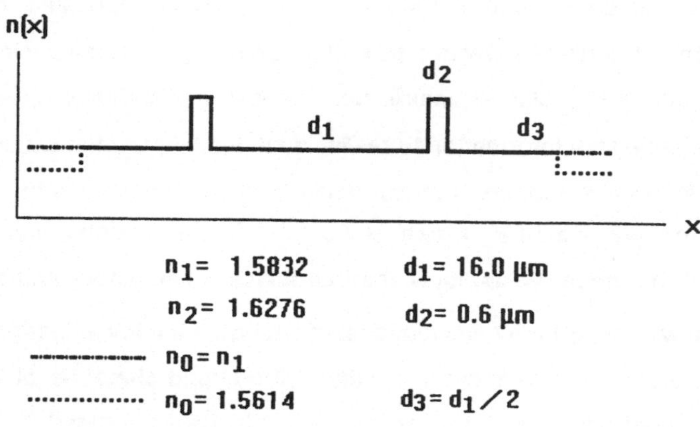

Fig.2.1 ARROW-configuration

Eventually, to remove all losses, the refractive indices of the outermost regions can be reduced below n_1, so that the oscillating nature of the field in the guide remains unchanged, but gets evanescent in the surrounding media. This increase of the number of layers by two, see Fig. 2.1 with $n_0 = 1.5614$, is introduced into the configuration under consideration to be able to compare the results for the slab-ARROW configuration with those for the rib-ARROW presented in the following section. Obviously, the modes of the (planar) configuration, see Figs. 2.2 a-d, will propagate without any change of their field-profile along the ARROW-structure. With an input of a properly tailored Gaussian field profile the mode of Fig. 2.2b as well as a superposition of the modes of Figs 2.2 c and d can be excited, see [LED 92] for applications.

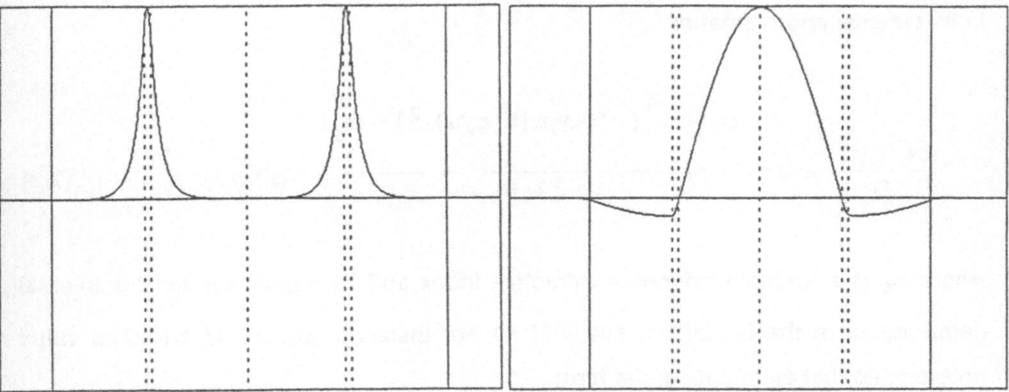

$\beta = 1.596144 \quad \beta = 1.596144$

Fig. 2.2a

$\beta = 1.5828763$

Fig. 2.2b

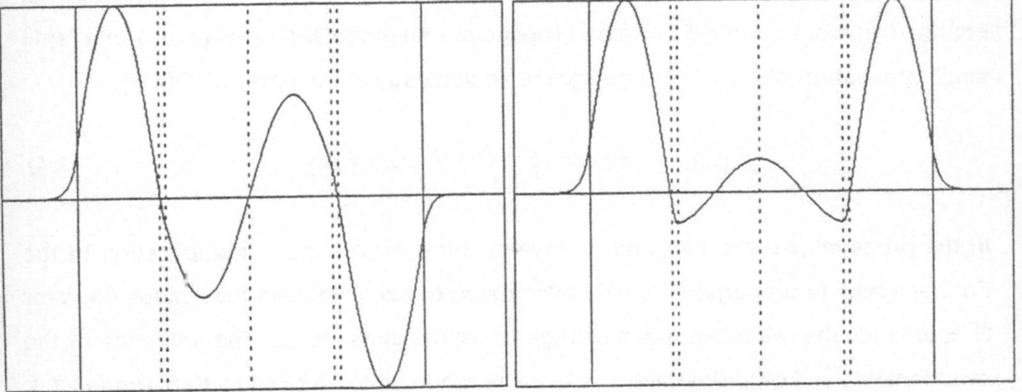

$\beta = 1.5820807$

Fig. 2.2c

$\beta = 1.5820149$

Fig. 2.2d

A common procedure to simulate the propagation of (guided) optical fields in nonlinear (as well as in linear) media is to solve the wave equation

$$\frac{\partial^2 E(x,y,z)}{\partial z^2} = -\left[\frac{\partial^2}{\partial x^2} + \frac{\partial^2}{\partial y^2} + k_0^2 n^2 (x,y,z,|E(x,y,z)|^2) \right] E(x,y,z) \qquad (2.1)$$

in its paraxial approximation

$$\frac{\partial E(x,y,z)}{\partial z} = -i \; \frac{\Delta_t + k_0^2 \, (\, n^2(x,y,z,|E(x,y,z)|^2) - n_e^2 \,)}{2 \, k_0 \, n_e} \; E(x,y,z) \qquad (2.2)$$

where n_e is a suitable reference refractive index and Δ_t stands for the transversal components of the Laplacian, see [FEI 80] for instance. Eq. 2.2 is an initial value problem, so that an ansatz of the form

$$E(x,y,z_0 + \Delta z) = e^{\pm \hat{A} \Delta z} \; E(x,y,z_0) \qquad (2.3)$$

with an operator $\hat{A}$ corresponding to eq. (2.2) is well suited . The classical Fast Fourier Transformation (FFT) Beam Propagation Method (BPM) makes use of a 'split step' formulation of eq. (2.3) to guarantee an accuracy of the order of $O(\Delta z^3)$

$$E(x,y,z_0 + \Delta z) = P \cdot Q \cdot P \; E(x,y,z_0) \qquad (2.4)$$

In the propagator P the FFT and its inverse allow for a simple multiplication in the Fourier space to propagate the field over a step of $\Delta z/2$, whereas the phase operator Q counts for the refractive index changes over the distance Δz. The influence of the electric field itself onto the refractive index is assumed to be positive Kerr-like ($a > 0$), i.e.

$$n^2(x,y,z,|E(x,y,z)|^2) = n_{lin}^2(x,y,z) + a(x,y,z) \; |E(x,y,z)|^2 \quad . \qquad (2.5)$$

This assumption is based upon recent results [SHE 91] which show that semiconductor materials exhibit an off-resonant, ultrafast Kerr nonlinearity (no two photon absorption) provided the wavelength is near half the band-gap. Very recently it have been outlined that just this kind of nonlinearity is most advantegeous for photonic applications in waveguides [STE 90].

The basic limitations of the FFT-BPM are the restriction to small index changes Δn even in the case of propagation steps $\Delta z < \lambda$ (typical values are: $0.01 < \Delta n < 0.1 \rightarrow 0.1\mu m < \Delta z < 0.01\mu m$) as well as the constraints to have vanishing field amplitudes at the edges of the computational window to minimise reflections. So, all field components radiating towards the boundaries of the window have to be cancelled out by the use of an (unphysical) non-reflecting absorber.

Within the last years serious attempts have been made to remedy those shortcomings of the FFT-BPM. Very recently, a powerfull technique, based on a series expansion of the term $\exp(\hat{A}\Delta z)$ was proposed [SPL 91].

The benefit of this method is, at least, twofold. Firstly, the restrictions with respect to Δn and Δz are reduced remarkably ($|| \hat{A} || \cdot \Delta z < 1$ is the requirement left for the numerics), so that configurations with large refractive index changes can be investigated. Secondly, transparent boundary conditions [HAD 92] can be introduced as well to allow reflectionless radiation out of the computational window, which is essential for leaky modes.

3 Evolution of guided fields in rib ARROW - structures

The confinement of guided optical fields not only in one (slab structures) but even in two directions (rib-structures) will provide advantages in the integration of optical components in miniaturised systems, in the reduction of the dimension of integrated optical devices and in the decrease of total guided power, for instance. Therefore, much progress in the experimental as well as in the theoretical work concerning waveguiding in three-dimensional structures was made within the last years. An adequate description of the guiding characteristics of those structures requires the determination of the propagation constants of guided modes in two-dimensional refractive index structures. To this end, a number of different methods does exist. The simplest one is the effective index method (EIM), where the refractive index

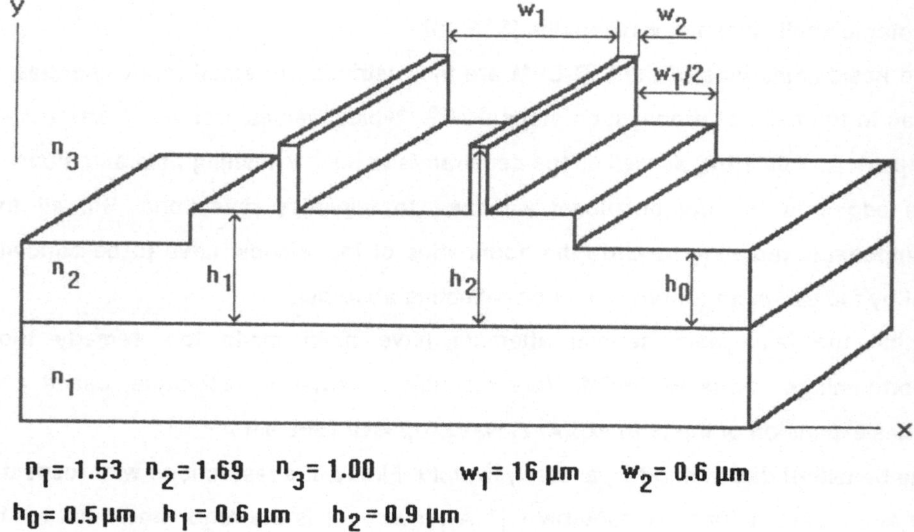

$n_1 = 1.53$ $n_2 = 1.69$ $n_3 = 1.00$ $w_1 = 16\ \mu m$ $w_2 = 0.6\ \mu m$

$h_0 = 0.5\ \mu m$ $h_1 = 0.6\ \mu m$ $h_2 = 0.9\ \mu m$

Fig. 3.1 Rib-ARROW-structure

profile n(x,y) is 'transformed' into an one-dimensional profil $n_{eff}(x)$ with effective refractive indices equal to the propagation constants of the guided modes in all of the slab waveguides with refractive index profiles n(x=const.,y). So, the configuration of Fig. 3.1 will be transformed into a sequence of 7 slabs with 3 different n_{eff}-values corresponding to the values of h_0, h_1 and h_2. The resulting propagation constant is provided by the solution of the dispersion relation in the $n_{eff}(x)$-profile.

From Maxwell's equations the fields of guided modes in a rib-structure in the EIM-picture can be either E_xH_y or E_yH_x. Thus, different polarisation directions for the dispersion relations in the n(x=const.,y) and $n_{eff}(x)$ profiles have to be used. The resulting $n_{eff}(x)$-profile for Fig. 3.1 is just the profile of Fig. 2.1, when the TM-polarisation for the dispersion relation in the n(x=const.,y)-profile was chosen. Consequently, the TE-dispersion relation in the $n_{eff}(x)$-profile yields the EIM-values for the E_yH_x- modes of the rib-ARROW with just the values known from Figs. 2.2 b-d. The shortcomings of the EIM ar due mainly to the fact, that uncoupled one-dimensional scalar wave equations are considered in y- and x-direction instead of a two-dimensional vectorial one.

In order to get more accurate results, one has to solve the vectorial problem by the use of adequate techniques. The benchmarks concerning the β-values of guided modes and their fields in rib-structures are provided by the Finite Element Method (FEM), see [MCl 90]. Fig. 3.2 shows the H_x-field component of the basic ARROW-mode of configuration 3.1. computed with a FEM-program.

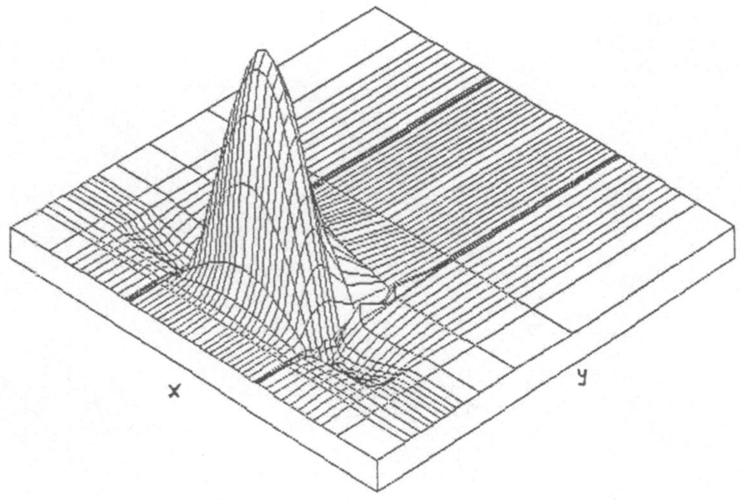

Fig. 3.2 Basic ARROW mode in the rib-structure

The limit of usual FEM-codes is the restriction to real propagation constants. Low losses can be involved into the FEM-scheme by the use of sophisticated pertur-bation-methods, where one needs to use high-order complex-valued systems of equations when the losses are increasing. Therefore the configuration (Fig. 3.1) was chosen in order to have real valued propagation constants only. The results of the FEM and the EIM calculations are:

FEM:	EIM:
$\beta = 1.58062$	$\beta = 1.5828763$
$\beta = 1.57987$	$\beta = 1.5820807$
$\beta = 1.57983$	$\beta = 1.5820149$

Obviously, the differences between both methods are in the order of $2 \cdot 10^{-3}$ only. In the case of the basic ARROW-mode this is accompanied by the disappearance of the hybrid character of the field (the H_y-field component is about 3% of the H_x one in the average). This gives the justification to use a less complicated two-dimensional scalar series expansion BPM for the propagation calculations presented below.

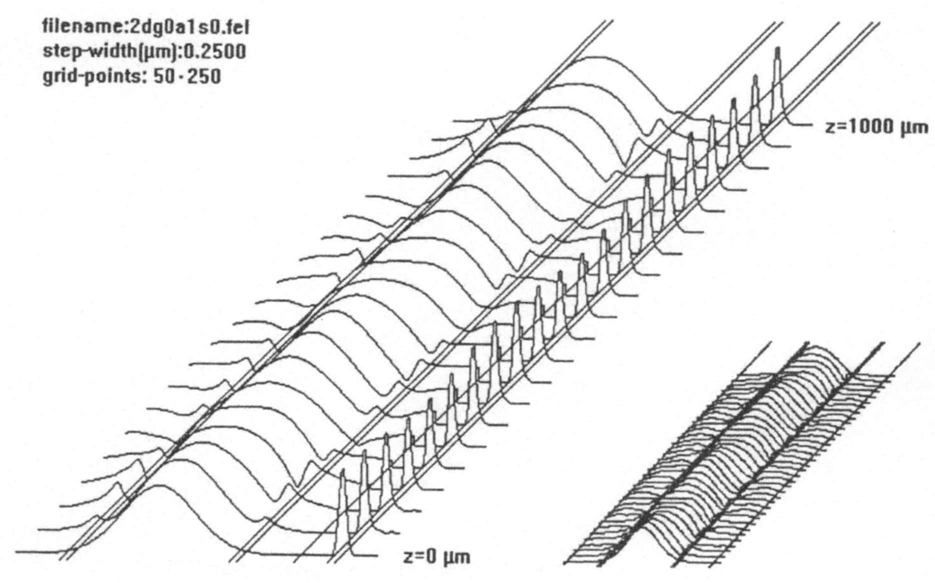

Fig.3.3 Evolution of Gaussian input field profile

Fig. 3.3 shows the evolution of a Gaussian field profile with half-width' $hw_x = 5.0\mu m$ and $hw_y = 0.5\mu m$ as input. Within the wide area, bounded by two double lines, the field is plotted versus the x-coordinate, i.e. versus the direction of resonant confinement. The region shown presents the central part of the structure. The narrow peak is the plot versus the y-coordinate, i.e. versus the direction of conventional waveguiding. The inset in the lower right corner depicts the evolution in the corresponding slab structure. The excitation of the basic ARROW-mode can be recognised easily. Some minor changes in the field profile are due to the process of the formation of the guided mode, where radiation losses play a minor role, shown by the conservation of the guided power of about 99%. Fig. 3.4 shows the results of the same calculation but in an iso-line plot (10% decrease of $|E|$ from line to line) at selected values of z.

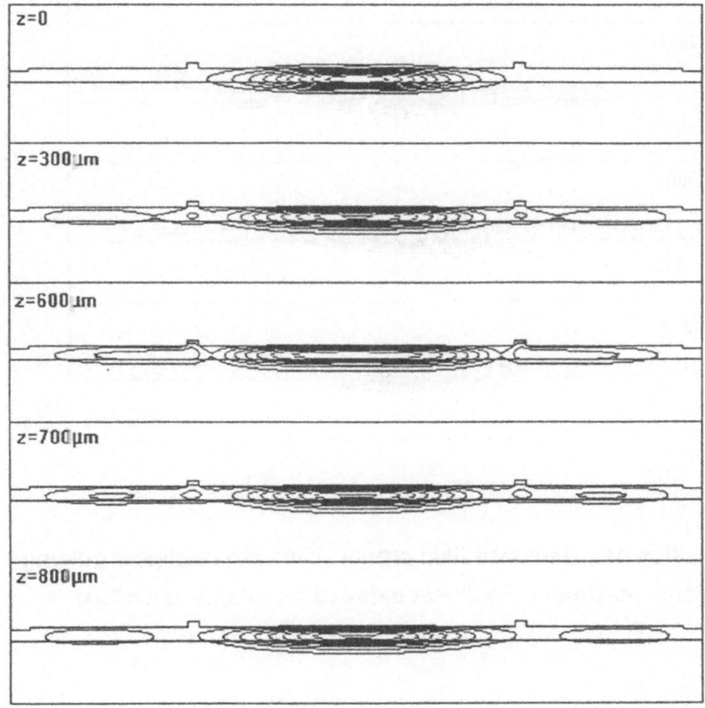

Fig. 3.4 Excitation of the basic ARROW-mode with symmetry adapted input, iso-plot

In Fig. 3.5 the same excitation is presented, but with a Kerr-like nonlinearity taken into account. The competition between the self-focusing and the underlying linear guiding mechanism is evident, comparing the field concentration in the rib-regions at $z=300µm$ as well as in the central region at $z=800µm$ to the linear case.

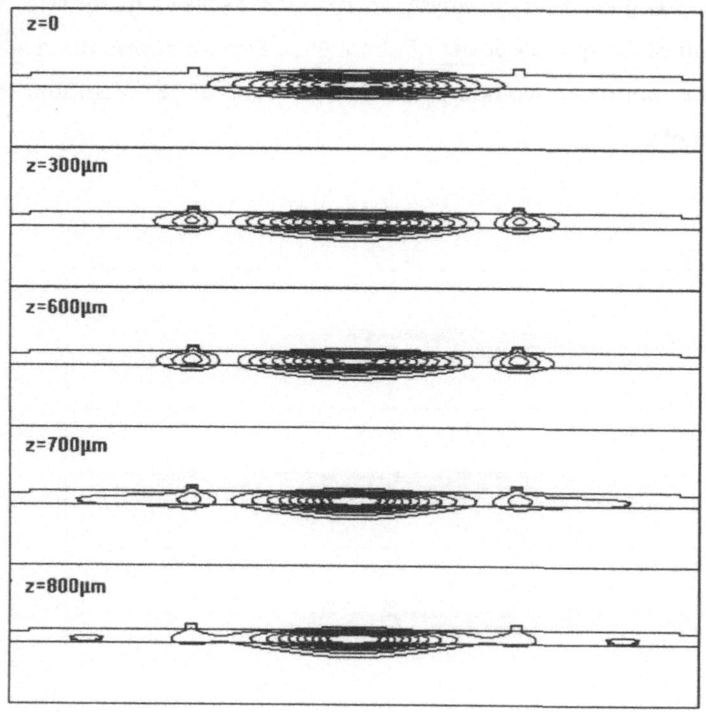

Fig.3.5 Evolution of a Gaussian field profile , Kerr-like nonlinear guiding region,   maximum nonlinear induced $\Delta n = 0.005$ at $z = 0\mu m$

4 Conclusions

We have shown that the ARROW-concept which holds for dielectric slab-structures can be extended to rib-structures as well. The design of the rib-structure, where the guiding effect in (at least) one of the transversal directions is due to Fabry-Perot resonances can be managed within the EIM-picture as long as the other transversal direction provides guiding, too. The propagation of optical fields in rib-ARROW's can be simulated with well-adapted BPM-methods. It has been demonstrated numerically, that Gaussian input field profiles which correspond to an end-firing of a laser-beam onto a waveguide cross-section, will excite ARROW-modes. When optically nonlinear media are involved, nonlinear induced refractive index changes much less than the refractive index changes within the linear limit of the structure cause appreciable effects. Hence, rib-ARROW structures are supposed to have the same field of applications as their dimension-reduced counterparts. To show this, the next investigations should cover the setup of a rib-ARROW coupler in the linear as well as in the nonlinear case.

Concerning experiments with rib-ARROW's polymeric materials are promising candidates because of their inherent peculiarities as adjustable refractive indices, high third-order nonlinear coefficients and the possibilities to fabricate three-dimensional waveguides with different technologies, etching as well as the use of dies.

REFERENCES

[LEN 90]   Lenz,G.;Salzman,J.: Bragg Reflection Waveguide Composite structures. IEEE JQE $\underline{26}$ (1990) , 519-531

[DUG 86]   Duguay,M.A.;Kokubun,Y.;Koch,T.L.: Antiresonant reflecting optical waveguides in SiO_2-Si multilayer structures. Appl.Phys.Lett.$\underline{49}$(1986) , 13-15

[BAB 88]   Baba,T.;Kokubun,Y.;Sakaki,T.;Iga,K.: Loss reduction of an ARROW waveguide in shorter wavelength and its stack configuration. IEEE J.Lightw.Techn.$\underline{LT-6}$(1988) , 1440-45

[LED 92]   Lederer,F.;et al.: Linear mode beating and nonlinear mode coupling in resonant optical waveguides; see this volume

[WÄC 87]   Wächter,C.;Hehl,K.: General Treatment of Slab Waveguides Including Lossy Materials and Arbitrary Refractive Index Profiles. phys.stat.sol(a) $\underline{102}$ (1987) , 835-842

[FEI 80]   Feit,M.D.;Fleck,J.A.jr.: Mode properties and dispersion for two optical fiber index profiles by the propagating beam method. Appl.Optics $\underline{19}$ (1980) , 3140-3150

[SHE 91]   Sheik-Bahae,M. et al. : Dispersion of Bound Electronic Nonlinear

Refraction in Solids. IEEE JQE $\underline{27}$ (1991) , 1296-1309

[STE 90]   Stegeman,G.I.;Wright,E.M.: All-optical waveguide switching
Opt.and Quantum Electronics $\underline{22}$ (1990) , 95-122

[SPL 91]   Splett,A.;Majd,M.;Petermann,K.: A novel Beam Propagation Method
for Large Refractive Index Steps and Large Propagation
Distances. IEEE Phot.Techn.Lett. $\underline{3}$ (1991)

[MCI 90]   McIlroy,P.W.A.;Stern,M.S.;Kendall,P.C.: Spectral Index Method for
Polarized Modes in Semiconductor Rib Waveguides.
Journ.Lightw.Techn. $\underline{8}$ (1990), 113-117